The Application of Programmable DSPs in Mobile Communications

The Application of Programmable DSPs in Mobile Communications

Edited by

Alan Gatherer and Edgar Auslander
Both of
Texas Instruments Inc., USA

JOHN WILEY & SONS, LTD

Other Wiley Editorial Offices

John Wiley & Sons, Inc., 605 Third Avenue,
New York, NY 10158-0012, USA

WILEY-VCH Verlag GmbH
Pappelallee 3, D-69469 Weinheim, Germany

John Wiley & Sons Australia Ltd, 33 Park Road, Milton,
Queensland 4064, Australia

John Wiley & Sons (Canada) Ltd, 22 Worcester Road
Rexdale, Ontario, M9W 1L1, Canada

John Wiley & Sons (Asia) Pte Ltd, 2 Clementi Loop #02-01,
Jin Xing Distripark, Singapore 129809

British Library Cataloguing in Publication Data

A catalogue record for this book is available from the British Library

ISBN 0471 48643 4

Typeset in Times by Deerpark Publishing Services Ltd, Shannon, Ireland.
Printed and bound in Great Britain by T. J. International Ltd, Padstow, Cornwall.

This book is printed on acid-free paper responsibly manufactured from sustainable forestry, in which at least two trees are planted for each one used for paper production.

Contents

Biographies   xiii

List of Contributors   xv

1 Introduction   1
   Edgar Auslander and Alan Gatherer
1.1 It's a Personal Matter   2
1.2 The Super Phone?   3
1.3 New Services   6
1.4 The Curse and Opportunity of Moore's Law   8
1.5 The Book   9

2 The History of DSP Based Architectures in Second Generation Cellular Handsets   11
   Alan Gatherer, Trudy Stetzler and Edgar Auslander
2.1 Introduction   11
2.2 A History of Cellular Standards and Wireless Handset Architectures   11
   2.2.1 1G and 2G Standards   11
   2.2.2 2.5G and 3G Standards   12
   2.2.3 Architecture Evolution   14
2.3 Trends in Low Power DSPs   17
   2.3.1 Process Improvement   17
   2.3.2 Instruction Set Enhancement   19
   2.3.3 Power Management   21
References   21

3 The Role of Programmable DSPs in Dual Mode (2G and 3G) Handsets   23
   Chaitali Sengupta, Nicolas Veau, Sundararajan Sriram, Zhenguo Gu and Paul Folacci
3.1 Introduction   23
3.2 The Wireless Standards   24
3.3 A Generic FDD DS Digital Baseband (DBB) Functional View   25
3.4 Functional Description of a Dual-Mode System   28
3.5 Complexity Analysis and HW/SW Partitioning   29
   3.5.1 2G/3G Digital Baseband Processing Optimized Partitioning   31
3.6 Hardware Design Approaches   32
   3.6.1 Design Considerations: Centralized vs. Distributed Architectures   32
   3.6.2 The Coprocessor Approach   33
   3.6.3 Role of DSP in 2G and Dual-Mode   37
3.7 Software Processing and Interface with Higher Layers   38
3.8 Summary   39
3.9 Abbreviations   39
References   40

4 Programmable DSPs for 3G Base Station Modems   **41**
Dale Hocevar, Pierre Bertrand, Eric Biscondi, Alan Gatherer, Frank Honore, Armelle Laine,
Simon Morris, Sriram Sundararajan and Tod Wolf
4.1 Introduction   41
4.2 Overview of 3G Base Stations: Requirements   42
    4.2.1 Introduction   42
    4.2.2 General Requirements   42
    4.2.3 Fundamental CDMA Base Station Base Band Processing   43
    4.2.4 Symbol-Rate (SR) Processing   44
    4.2.5 Chip-Rate (CR) Processing   44
4.3 System Analysis   46
    4.3.1 SR Processing Analysis   46
    4.3.2 CR Processing Analysis   46
4.4 Flexible Coprocessor Solutions   48
    4.4.1 Viterbi Convolutional Decoder Coprocessor   48
    4.4.2 Turbo Decoder Coprocessor   50
    4.4.3 Correlator Coprocessor   52
4.5 Summary and Conclusions   54

5 The Use of Programmable DSPs in Antenna Array Processing   **57**
Matthew Bromberg and Donald R. Brown
5.1 Introduction   57
5.2 Antenna Array Signal Model   58
5.3 Linear Beamforming Techniques   62
    5.3.1 Maximum Likelihood Derivation   62
    5.3.2 Least Mean Square Adaptation   66
    5.3.3 Least Squares Processing   67
    5.3.4 Blind Signal Adaptation   71
    5.3.5 Subspace Constraints   73
    5.3.6 Exploiting Cyclostationarity   75
    5.3.7 Transmit Beamformer Techniques   77
5.4 Multiple Input Multiple Output (MIMO) Signal Extraction   83
    5.4.1 MIMO Linear System Model   83
    5.4.2 Capacity of MIMO Communication Channels   86
    5.4.3 Linear Estimation of Desired Signals in MIMO Communication Systems 87
    5.4.4 Non-linear Estimation of Desired Signals in MIMO Communication Systems 90
    5.4.5 Conclusions   93
References   93

6 The Challenges of Software-Defined Radio   **97**
Carl Panasik and Chaitali Sengupta
6.1 Cellular Communications Standards   98
6.2 What is SDR?   98
6.3 Digitizing Today's Analog Operations   101
6.4 Implementation Challenges   103
6.5 Analog and ADC Issues   103
6.6 Channel Filter   104
6.7 Delta-Sigma ADC   104
6.8 Conclusion   105
References   105

7 Enabling Multimedia Applications in 2.5G and 3G Wireless Terminals: Challenges and Solutions   **107**
Edgar Auslander, Madhukar Budagavi, Jamil Chaoui, Ken Cyr, Jean-Pierre Giacalone, Sebastien de Gregorio, Yves Masse, Yeshwant Muthusamy, Tiemen Spits and Jennifer Webb
7.1 Introduction   107
   7.1.1 "DSPs take the RISC"   107
7.2 OMAP H/W Architecture   111
   7.2.1 Architecture Description   111
   7.2.2 Advantages of a Combined RISC/DSP Architecture   113
   7.2.3 TMS320C55x and Multimedia Extensions   113
7.3 OMAP S/W Architecture   114
7.4 OMAP Multimedia Applications   116
   7.4.1 Video   116
   7.4.2 Speech Applications   116
7.5 Conclusion   117
Further Reading   117

8 A Flexible Distributed Java Environment for Wireless PDA Architectures Based on DSP Technology   **119**
Gilbert Cabillic, Jean-Philippe Lesot, Frédéric Parain, Michel Banâtre, Valérie Issarny, Teresa Higuera, Gérard Chauvel, Serge Lasserre and Dominique D'Inverno
8.1 Introduction   119
8.2 Java and Energy: Analyzing the Challenge   120
   8.2.1 Analysis of Java Opcodes   120
   8.2.2 Analyzing Application Behavior   121
   8.2.3 Analysis   125
8.3 A Modular Java Virtual Machine   127
   8.3.1 Java Implantation Possibilities   127
   8.3.2 Approach: a Modular Java Environment   129
   8.3.3 Comparison with Existing Java Environments   131
8.4 Ongoing Work on Scratchy   132
   8.4.1 Multi-Application Management   133
   8.4.2 Managing the Processor's Heterogeneity and Architecture   133
   8.4.3 Distribution of Tasks and Management of Soft Real-Time Constraints   133
   8.4.4 Energy Management   133
8.5 Conclusion   133
References   134

9 Speech Coding Standards in Mobile Communications   **137**
Erdal Paksoy, Vishu Viswanathan and Alan McCree
9.1 Introduction   137
9.2 Speech Coder Attributes   138
9.3 Speech Coding Basics   139
   9.3.1 Waveform Coders   141
   9.3.2 Parametric Coders   141
   9.3.3 Linear Predictive Analysis-by-Synthesis   143
   9.3.4 Postfiltering   146
   9.3.5 VAD/DTX   146
   9.3.6 Channel Coding   146
9.4 Speech Coding Standards   147
   9.4.1 ITU-T Standards   147
   9.4.2 Digital Cellular Standards   148
   9.4.3 Wideband Standards   152

9.5 Speech Coder Implementation   153
  9.5.1 Specification and Conformance Testing   153
  9.5.2 ETSI/ITU Fixed-Point C   154
  9.5.3 DSP Implementation   155
9.6 Conclusion   155
Acknowledgements   156
References   156

10 Speech Recognition Solutions for Wireless Devices   **160**
  Yeshwant Muthusamy, Yu-Hung Kao and Yifan Gong
10.1 Introduction   160
10.2 DSP Based Speech Recognition Technology   160
  10.2.1 Problem: Handling Dynamic Vocabulary   161
  10.2.2 Solution: DSP-GPP Split   161
10.3 Overview of Texas Instruments DSP Based Speech Recognizers   161
  10.3.1 Speech Recognition Algorithms Supported   161
  10.3.2 Speech Databases Used   161
  10.3.3 Speech Recognition Portfolio   162
10.4 TIESR Details   165
  10.4.1 Distinctive Features   165
  10.4.2 Grammar Parsing and Model Creation   166
  10.4.3 Fixed-Point Implementation Issues   167
  10.4.4 Software Design Issues   168
10.5 Speech-Enabled Wireless Application Prototypes   168
  10.5.1 Hierarchical Organization of APIs   169
  10.5.2 InfoPhone   171
  10.5.3 Voice E-mail   172
  10.5.4 Voice Navigation   173
  10.5.5 Voice-Enabled Web Browsing   174
10.6 Summary and Conclusions   175
References   176

11 Video and Audio Coding for Mobile Applications   **179**
  Jennifer Webb and Chuck Lueck
11.1 Introduction   179
11.2 Video   181
  11.2.1 Video Coding Overview   182
  11.2.2 Video Compression Standards   186
  11.2.3 Video Coding on DSPs   187
  11.2.4 Considerations for Mobile Applications   188
11.3 Audio   190
  11.3.1 Audio Coding Overview   191
  11.3.2 Audio Compression Standards   193
  11.3.3 Audio Coding on DSPs   195
  11.3.4 Considerations for Mobile Applications   196
11.4 Audio and Video Decode on a DSP   198
References   200

12 Security Paradigm for Mobile Terminals   **201**
  Edgar Auslander, Jerome Azema, Alain Chateau and Loic Hamon
12.1 Mobile Commerce General Environment   202
12.2 Secure Platform Definition   203
  12.2.1 Security Paradigm Alternatives   204
  12.2.2 Secure Platform Software Component   204
  12.2.3 Secure Platform Hardware Component   205

12.3 Software Based Security Component   205
    12.3.1 Java and Security   205
    12.3.2 Definition   205
    12.3.3 Features for Security   206
    12.3.4 Dependency on OS   207
12.4 Hardware Based Security Component: Distributed Security   207
    12.4.1 Secure Mode Description   208
    12.4.2 Key Management   210
    12.4.3 Data Encryption and Hashing   211
    12.4.4 Distributed Security Architecture   212
    12.4.5 Tampering Protection   213
12.5 Secure Platform in Digital Base Band Controller/MODEM   214
12.6 Secure Platform in Application Platform   215
12.7 Conclusion   215

13 Biometric Systems Applied To Mobile Communications   **217**
    Dale R. Setlak and Lorin Netsch
13.1 Introduction   217
13.2 The Speaker Verification Task   219
    13.2.1 Speaker Verification Processing Overview   219
    13.2.2 DSP-Based Embedded Speaker Verification   224
13.3 Live Fingerprint Recognition Systems   225
    13.3.1 Overview   225
    13.3.2 Mobile Application Characterization   226
    13.3.3 Concept of Operations   226
    13.3.4 Critical Performance Metrics   228
    13.3.5 Basic Elements of the Fingerprint System   233
    13.3.6 Prototype Implementation   247
    13.3.7 Prototype System Processing   248
13.4 Conclusions   251
References   251

14 The Role of Programmable DSPs in Digital Radio   **253**
    Trudy Stetzler and Gavin Ferris
14.1 Introduction   253
14.2 Digital Transmission Methods   254
14.3 Eureka-147 System   255
    14.3.1 System Description   255
    14.3.2 Transmission Signal Generation   262
    14.3.3 Receiver Description   265
14.4 IBOC   279
14.5 Satellite Systems   284
14.6 Conclusion   285
References   286

15 Benchmarking DSP Architectures for Low Power Applications   **287**
    David Hwang, Cimarron Mittelsteadt and Ingrid Verbauwhede
15.1 Introduction   287
15.2 LPC Speech Codec Algorithm   288
    15.2.1 Segmentation   288
    15.2.2 Silence Detection   288
    15.2.3 Pitch Detection Algorithm   289
    15.2.4 LPC Analysis – Vocal Tract Modeling   289
    15.2.5 Bookkeeping   290

15.3 Design Methodology   290
  15.3.1 Floating-Point to Fixed-Point Conversion   290
  15.3.2 Division Algorithm   292
  15.3.3 Hardware Allocation   293
15.4 Platforms   293
  15.4.1 Texas Instruments TI C54x   293
  15.4.2 Texas Instruments TI C55x   294
  15.4.3 Texas Instruments TI C6x   294
  15.4.4 Ocapi   294
  15.4.5 A|RT Designer   294
15.5 Final Results   294
  15.5.1 Area Estimate   295
  15.5.2 Power Estimate   295
15.6 Conclusions   297
Acknowledgements   298
References   298

16 Low Power Sensor Networks   **299**
  Alice Wang, Rex Min, Masayuki Miyazaki, Amit Sinha and Anantha Chandrakasan
16.1 Introduction   299
16.2 Power-Aware Node Architecture   300
16.3 Hardware Design Issues   302
  16.3.1 Processor Energy Model   303
  16.3.2 DVS   304
  16.3.3 Leakage Considerations   306
16.4 Signal Processing in the Network   311
  16.4.1 Optimizing Protocols   312
  16.4.2 Energy-Efficient System Partitioning   313
16.5 Signal Processing Algorithms   317
  16.5.1 Energy–Agile Filtering   318
  16.5.2 Energy–Agile Data Aggregation   319
16.6 Signal Processing Architectures   320
  16.6.1 Variable-Length Filtering   321
  16.6.2 Variable Precision Architecture   322
16.7 Conclusions   324
References   324

17 The Pleiades Architecture   **327**
  Arthur Abnous, Hui Zhang, Marlene Wan, George Varghese, Vandana Prabhu, Jan Rabaey
17.1 Goals and General Approach   327
17.2 The Pleiades Platform – The Architecture Template   329
17.3 The Control Processor   331
17.4 Satellite Processors   332
17.5 Communication Network   334
17.6 Reconfiguration   338
17.7 Distributed Data-Driven Control   339
  17.7.1 Control Mechanism for Handling Data Structures   342
  17.7.2 Summary   345
17.8 The Pleiades Design Methodology   345
17.9 The P1 Prototype   348
  17.9.1 P1 Benchmark Study   350
17.10 The Maia Processor   352
  17.10.1 Control Processor   353
  17.10.2 Address Generator Processor   353

  17.10.3 Memory Units   354
  17.10.4 Multiply-Accumulate Unit   354
  17.10.5 Arithmetic/Logic Unit   354
  17.10.6 Embedded FPGA   354
  17.10.7 Maia Results   355
 17.11 Summary   357
 References   358

18 Application Specific Instruction Set Architecture Extensions for DSPs **361**
 Jean-Pierre Giacalone
 18.1 The Need for Instruction Set Extensibility in a Signal Processor   361
 18.2 ISA Extension Capability of the TMS320C55x Processor   362
  18.2.1 Control Modes   364
  18.2.2 Dataflow Modes   366
  18.2.3 Typical C55x Extension Datapath Architecture   367
  18.2.4 Integration in Software Development Tools   370
 18.3 Domains of Applications and Practical Examples   372
 18.4 ISA Extensions Design Flow   376
 References   377

19 The Pointing Wireless Device for Delivery of Location Based Applications **379**
 Pamela Kerwin, John Ellenby and Jeffrey Jay
 19.1 Next Generation Wireless Devices   379
 19.2 The Platform   379
 19.3 New Multimedia Applications   379
 19.4 Location Based Information   380
 19.5 Using Devices to Summon Information   380
 19.6 Pointing to the Real World   380
 19.7 Pointing Greatly Simplifies the User Interface   381
 19.8 Uses of Pointing   382
 19.9 Software Architecture   382
  19.9.1 Introduction   382
  19.9.2 Assumptions   382
  19.9.3 Overview   383
  19.9.4 Alternatives   383
 19.10 Use of the DSP in the Pointing System   383
 19.11 Pointing Enhanced Location Applications   384
  19.11.1 Pedestrian Guidance   385
  19.11.2 Pull Advertising   386
  19.11.3 Entertainment   386
 19.12 Benefits of Pointing   387
  19.12.1 Wireless Yellow Pages   387
  19.12.2 Internationalization   387
  19.12.3 GIS Applications   387
  19.12.4 Entertainment and Gaming   388
  19.12.5 Visual Aiding and Digital Albums   388
 19.13 Recommended Data Standardization   388
  19.13.1 Consideration of Current Standards Efforts   388
  19.13.2 Device Data Types and Tiered Services   388
  19.13.3 Data Specifications   389
  19.13.4 Data Format   391
  19.13.5 Is it Sufficient?   393
 19.14 Conclusion   393

Index   **395**

Biographies

Alan Gatherer received his BEng degree in Electronic and Microprocessor Engineering from Strathclyde University (Scotland) in 1988. He then moved to Stanford University and obtained MS and PhD degrees, both in Electrical Engineering in 1989 and 1993. He joined Texas Instruments in 1993. Since 1993, Dr. Gatherer has worked on digital communications research and development in the areas of digital subscriber line, cable modem and wireless. Presently he is a distinguished member of technical staff and manager of systems development within Wireless Infrastructure at Texas Instruments where he leads a team involved in the development of technology for third generation cellular telephony. He presently holds 11 patents in the field of digital communications.

Edgar Auslander is the Worldwide Strategic Marketing Director at Texas Instruments Wireless Communications Business Unit, Dallas, Texas, working for the general manager of the business unit and Texas Instruments' senior vice-president. Some of the strategic plans he has written have resulted in Texas Instruments acquiring or investing in companies. He first worked at Texas Instruments in Nice, France, as European product marketing manager for the TMS320 digital signal processors product line in 1990. Mr. Auslander became Texas Instruments' Worldwide GSM marketing manager in 1993 prior to his current position. He obtained his MEEng and MBA degrees from Cornell University and Columbia Business School in 1988 and 1990, respectively. While at Cornell, he held several teaching assistant positions in signal processing and digital communications courses. While at Columbia Business School, he was a teaching assistant in statistics and worked at the Office of Science and Technology Development, selling Columbia Telecommunications Research Center's patents rights.

List of Contributors

Arthur Abnous
Care of Jan Rabaey
EECS Dept.
511 Cory Hall
Berkeley, CA 94720
USA
abnous@broadcom.com

Edgar Auslander
Texas Instruments Inc.
12500 TI Boulevard, MS 8723
Dallas, TX 75243
USA
ea282@columbia.edu

Jerome Azema
Texas Instruments France
821 avenue Jack Kilby - BP 5
06271 Villeneuve-Loubet Cedex
France
j-azema@ti.com

Michel Banâtre
INRIA-Rennes
Campus Universitaire de Beaulieu
35042 Rennes Cedex
France
michel.banatre@irisa.fr

Pierre Bertrand
Texas Instruments France
821 Jack Kilby Ave - BP 5
06271 Villeneuve-Loubet Cedex
France
p-bertrand@ti.com

Eric Biscondi
Texas Instruments France
821 Jack Kilby Ave - BP 5
06271 Villeneuve-Loubet Cedex
France
e-biscondi@ti.com

Matthew Bromberg
Dept. of Electrical and Computer Engineering
Worcester Polytechnic Institute
100 Institute Road
Worcester MA 01609-2280
USA
mattbro@bigfoot.com

Donald R. Brown
Dept. of Electrical and Computer Engineering
Worcester Polytechnic Institute
100 Institute Road
Worcester MA 01609-2280
USA
drb@ece.wpi.edu

Madhukar Budagavi
Texas Instruments Inc.
12500 TI Boulevard, MS 8649
Dallas, TX 75243
USA
madhukar@ti.com

Gilbert Cabillic
INRIA-Rennes
Campus Universitaire de Beaulieu
35042 Rennes Cedex
France
gilbert.cabillic@irisa.fr

Anantha Chandrakasan
Massachusetts Institute of Technology
50 Vassar St. Rm 38-107
Cambridge, MA 02139
USA
anantha@mtl.mit.edu

Jamil Chaoui
Texas Instruments France
821 Jack Kilby Ave - BP 5
06271 Villeneuve-Loubet Cedex
France
j-chaoui1@ti.com

Alain Chateau
Texas Instruments France
821 avenue Jack Kilby - BP 5
06271 Villeneuve-Loubet Cedex
France
a-chateau@ti.com

Gerard Chauvel
Texas Instruments France
821 avenue Jack Kilby - BP 5
06271 Villeneuve-Loubet Cedex
France
g-chauvel@ti.com

Ken Cyr
Texas Instruments Inc.
12500 TI Boulevard, MS 8650
Dallas, TX 75243
USA
kencyr@ti.com

Sebastien de Gregorio
Texas Instruments France
821 Jack Kilby Ave - BP 5
06271 Villeneuve-Loubet Cedex
France
s-de-gregorio@ti.com

Dominique D'Inverno
Texas Instruments France
821 avenue Jack Kilby - BP 5
06271 Villeneuve-Loubet Cedex
France
d-dinverno@ti.com

John Ellenby
GeoVector Corporation
601 Minnesota Street
#212 San Francisco
CA 94107
USA
john@geovector.com

Gavin Ferris
RadioScape Ltd.
2 Albany Terrace
Regents Park
London NW1 4DS
UK
gavin.ferris@radioscape.com

Paul Folacci
Texas Instruments France
821, avenue Jack Kilby - BP 5
06271 Villeneuve-Loubet Cedex
France
p-folacci@ti.com

Alan Gatherer
Texas Instruments Inc.
12500 TI Boulevard, MS 8723
Dallas, TX 75243
USA
gatherer@ti.com

Varghese George
Care of Jan Rabaey
EECS Dept.
511 Cory Hall
Berkeley, CA 94720
USA
varghese.george@st.com

Jean-Pierre Giacalone
Texas Instruments France
821 Jack Kilby Ave – BP 5
06271 Villeneuve-Loubet Cedex
France
j-giacalone@ti.com

Yifan Gong
Texas Instruments Inc.
12500 TI Boulevard, MS 8723
Dallas, TX 75243
USA

Zhenguo Gu
Texas Instruments Inc.
12500 TI Boulevard, MS 8723
Dallas, TX 75243
USA
zgu@ti.com

Loic Hamon
Texas Instruments France
821 avenue Jack Kilby - BP 5
06271 Villeneuve-Loubet Cedex
France
l-hamon@ti.com

Teresa Higuera
INRIA-Rennes
Campus Universitaire de Beaulieu
35042 Rennes Cedex
France
teresa.higuera@inria.fr

Dale Hocevar
DSPS R&D Center
Texas Instruments Inc.
MS 8632 South Campus
12500 TI Boulevard
Dallas TX 75243
USA
hocevar@ti.com

Frank Honore
MIT
50 Vassar St., Rm 38-107
Cambridge, MA 02139
USA
honore@mtl.mit.edu

David Hwang
427 Veteran Ave.
Los Angeles, CA 90024
USA
dhwang@icsl.ucla.edu

Valérie Issarny
INRIA-Rennes
Campus Universitaire de Beaulieu
35042 Rennes Cedex
France
valerie.issarny@inria.fr

Jeffrey Jay
GeoVector Corporation
601 Minnesota Street
#212 San Francisco
CA 94107,
USA
jj@geovector.com

Yu-Hung Kao
5505 Morehouse Drive
San Diego, CA 92121
USA
yhkao@ti.com

Pamela Kerwin
GeoVector Corporation
601 Minnesota Street
#212 San Francisco,
CA 94107,
USA
pam@geovector.com

Armelle Laine
Texas Instruments France
821 Jack Kilby Ave - BP 5
06271 Villeneuve-Loubet Cedex
France
ar-laine@ti.com

Serge Lasserre
Dominique D'Inverno
Texas Instruments France
821 avenue Jack Kilby - BP 5
06271 Villeneuve-Loubet Cedex
France
slasserre@ti.com

Jean-Phillipe Lesot
INRIA-Rennes
Campus Universitaire de Beaulieu
35042 Rennes Cedex
France
jean-philippe.lesot@irisa.fr

Chuck Lueck
6041 Village Bend Dr. #806
Dallas, TX 75206
USA
lueck@ti.com

Yves Masse
Texas Instruments France
821 Jack Kilby Ave - BP 5
06271 Villeneuve-Loubet Cedex
France
y-masse@ti.com

Alan McCree
Texas Instruments Inc.
12500 TI Boulevard, MS 8649
Dallas, TX 75243
USA
a-mccree@ti.com

Rex Min
Massachusetts Institute of Technology
50 Vassar St. Rm 38-107
Cambridge, MA 02139
USA
rmin@mtl.mit.edu

Cimarron Mittelsteadt
27150 Silver Oak Lane, Apt. #2012
Santa Clarita, CA 91351
USA
cimarron@icsl.ucla.edu

Masayuki Miyazaki
Hitachi Ltd., Central Research Laboratory
System LSI Research Dept.
1-280 Higashi-Koigakubo
Kokubunji-shi
Tokyo, 185-8601
Japan
mmiya@crl.hitachi.co.jp

Simon Morris
Lumic Electronics Inc.
18 Antares Dr Suite 200
Ottawa, Ontario
K2E 1A9
Canada
smorris@lumictech.com

Yeshwant Muthusamy
Texas Instruments Inc.
12500 TI Boulevard, MS 8649
Dallas, TX 75243
USA
yeshwant@ti.com

Lorin Netsch
Texas Instruments Inc.
12500 TI Boulevard, MS 8649
Dallas, Texas 75243
USA
netsch@ti.com

Erdal Paksoy
Texas Instruments Inc.
12500 TI Boulevard, MS 8649
Dallas, TX 75243
USA
paksoy@ti.com

Carl Panasik
Texas Instruments Inc.
12500 TI Boulevard, MS 8723
Dallas, TX 75243
USA
panasik@ti.com

Frédéric Parain
INRIA-Rennes
Campus Universitaire de Beaulieu
35042 Rennes Cedex
France
frederic.parain@irisa.fr

Vandana Prabhu
Care of Jan Rabaey
EECS Dept.
511 Cory Hall
Berkeley, CA 94720
USA
vandana@tensilica.com

Jan Rabaey
EECS Dept.
511 Cory Hall
Berkeley, CA 94720
USA
jan@eecs.berkeley.edu

Chaitali Sengupta
Texas Instruments Inc.
12500 TI Boulevard, MS 8723
Dallas, TX 75243
USA
chaitali@ti.com

Dale R. Setlak
709 S. Harbor City Blvd, 4th floor
Melbourne, FL 32902-2719
USA
dsetlak@authentec.com

Amit Sinha
Massachusetts Institute of Technology
50 Vassar St. Rm 38-107
Cambridge, MA 02139
USA
sinha@mtl.mit.edu

Tiemen Spits
Texas Instruments Inc.
12500 TI Boulevard, MS 8723
Dallas, TX 75243
USA
tts@ti.com

Sundararajan Sriram
DSPS R&D Center
Texas Instruments Inc.
MS 8632 South Campus
12500 TI Boulevard
Dallas TX 75243
USA
sriram@ti.com

Trudy Stetzler
Texas Instruments
12203 Southwest Freeway
MS 701
Stafford, TX 77477
USA
t-stetzler1@ti.com

Nicolas Veau
Texas Instruments France
821, avenue Jack Kilby - BP 5
06271 Villeneuve-Loubet Cedex
France
n-veau@ti.com

Ingrid Verbauwhede
University of California, Los Angeles
Electrical Engineering Department
7440B Boelter Hall
420 Westwood Plaza
Los Angeles, CA 90095-1594
USA
ingrid@ee.ucla.edu

Vishu Viswanathan
Texas Instruments Inc.
12500 TI Boulevard, MS 8649
Dallas, TX 75243
USA
v-viswanathan@ti.com

Marlene Wan
Morphics Technology Inc.
675 Campbell Technology Parkway
Suite 100
Campbell, CA 95008-5059
USA
mwan@morphics.com

Alice Wang
Massachusetts Institute of Technology
50 Vassar St. Rm 38-107
Cambridge, MA 02139
USA
aliwang@mtl.mit.edu

Jennifer Webb
7730 Woodstone Ln.
Dallas, TX 75248
USA
webb@ti.com

Tod Wolf
Texas Instruments Inc.
12500 TI Boulevard, MS 8632
Dallas, TX 75243
USA
todw@ti.com

Hui Zhang
Care of Jan Rabaey
EECS Dept.
511 Cory Hall
Berkeley, CA 94720
USA
tomz@zsmc.com

1

Introduction

Edgar Auslander and Alan Gatherer

This book is about two technologies that have had, and will increasingly have, a significant impact on the way we all live, learn and play: personal wireless communications and signal processing. When it comes to both markets, history has shown that reality has often surprised the most optimistic forecasters.

We draw on the experience of experts from MIT, Berkeley, UCLA, Worcester Polytechnic Institute, INRIA, Authentec, Radioscape, Geovector and Texas Instruments, to give a description of some of the important building blocks and implementation choices that combine both technologies, in the past and in the future. We highlight different perspectives, especially regarding implementation issues, in the processing of speech, audio, video, future multimedia and location-based services as well as mobile commerce and security aspects.

The book is roughly divided into three sections:

- Chapters describing applications and their implementations on what might be described as "today's" technology. By this, we mean the use of programmable Digital Signal Processors (DSPs) and ASICs in the manner in which they are being used for today's designs. In these chapters, we highlight the applications and the role of programmable DSPs in the implementation.
- Chapters that present challenges to the current design flow, describing new ways of achieving the desired degree of flexibility in a design by means other than programmable DSPs. Whether these new approaches will unseat the programmable DSP from its perch remains to be seen, as the commercial value of these approaches is less certain. But they give a detailed overview of the directions researchers are taking to leap beyond the performance curve of the programmable DSP approach.
- We conclude with a practical yet innovative application example, a possible flavor of the exciting new personal communications services enabled by digital signal processing.

In this introduction, we overview the aspects of mobile communications that make it a unique technology. We describe how the applications associated with mobile communications have evolved from the simple phone call into a slew of personal technologies. These technologies, and their implementation, are described in more detail in the subsequent chapters.

1.1 It's a Personal Matter

The social impacts and benefits of personal wireless communications are already visible. When phones were not portable and used to only sit on a desk at home or at work, people would call places: work or home; but when phones became portable and accessible anywhere, people began to call *people* rather than *places*: today, when we call people we even often start by asking "Hello, where are you?". The mobile phone has become a safety tool: "I will bring the phone with me in case I need to call for an emergency, if anxious family members want to reach me, or if I am lost". The mobile phone has become a social tool, enabling more flexible personal life planning: "I do not know where I will be at 2 p.m. and where you will be, but I will call you on your mobile and we will sync". A recent survey has shown that when people forget their mobile phone at home, a vast majority is willing to go back home to get it, even when it implies a 30-minute drive. The mobile phone has become a personal item you carry with you like your wallet, your drivers' license, your keys, or even wear, like a watch, a pen, or glasses: it made it to the list of the few items that you carry with you. If you are a teenager, a gaming device or an MP3 player also made their room in your pocket, and if you are a busy executive a personal organizer is maybe more likely to have this privilege. Figure 1.1 illustrates the integration of new features trend; conversely, the wireless communication technology will be pervasive in different end-equipments and create new markets for wireless modules embedded in cars for example.

To some, the use of a mobile phone in public places is an annoyance. Peer pressure "dictates" you have a mobile phone to be reachable "anywhere any time"; not having a mobile phone becomes anti-social in Scandinavian countries for example, where penetration is higher than 70% of the whole population. Like for every disruptive technology widely used, a new etiquette has to be understood and agreed upon, e.g. phones have to be turned off or put

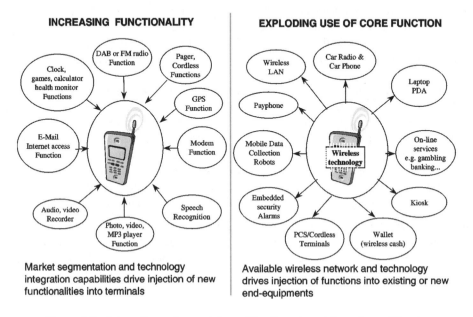

Figure 1.1   Integration and exportation of functions to and from the mobile phone

in silent mode at concerts or in restaurants. Phones are now programmed with different ringing profiles that are "environment friendly" (e.g. meeting mode rings only once and makes the phone vibrate). In the future, we might see phones that are environment aware, with sensors that detect if the phone is in a bag and needs to ring louder for example. In the past, Matra-AEG, now Nokia Mobile Phones, introduced a GSM phone that had an infra-red sensor that served as a proximity detector so as to put the phone automatically on or off hands-free mode. Ringing profiles have also other nice applications: paired with CallerID, they enable users to have different ringing tones for different callers (friends, family, business partners, unknown...).

1.2 The Super Phone?

To the vast majority, the mobile phone is the ultimate telecommunication tool, via voice or short messages, soon to become multimedia messages or multimedia communications.

For some, it is a foregone conclusion that wireless terminals will continue their mutation from fairly simple, voice-oriented devices to smarter and smarter systems capable of increasingly more complex voice and data applications. The argument goes that wireless phones will take on the capabilities of Personal Digital Assistants (PDAs) and PDAs will subsume many of the voice communications capabilities of mobile phones. This line of reasoning proclaims that the handsets of the future eventually will become some sort of super-phone/handheld computer/PDA. But in the end, the marketplace is never nearly as neat and tidy as one might imagine. Rather than an inexorable quest for a one-size-fits-all super-phone, the fractious forces of the market, based as they are on completely illogical human emotions, no doubt will lead handset manufacturers down a number of avenues in support of 2.5G and 3G applications (2.5 and 3G refer to coming phone standard genera-tions to be described later in this book). Many mobile handsets will be capable of converged voice/data applications, but many will not. Instead, they will fulfill a perceived consumer need or perform a certain specialized function very well. Rather than a homogenous market of converged super-phones, the terminal devices for next generation applications will be as diverse as they are today, if not more so. And they will be as diverse as the applications that will make up the 2.5G and 3G marketplace. Mobile device OEMs must be prepared to meet the challenge of a diverse and segmented market. Figure 1.2 illustrates how wireless phone service started to be affordable to a few privileged business professionals and how it diversified in time to become a consumer item. The high-end phone of today is the classic phone of tomorrow as fashion and technology evolve and as people become used to inno-vations brought to them.

We believe that the increasing need for function diversification will drive the program-mable DSP into an even more integrated role within the mobile devices of tomorrow. Non-programmable DSP architectures will have to take on many traits of the programmable DSP in order to compete with it. The later chapters of this book highlight that the future of programmable DSPs in mobile applications hinges on their ability to bring the right level of flexibility, along with low power performance.

Over the last several years, the market for terminals first became polarized and then stratified. The market first polarized at the high and low ends of the spectrum. As more features and functions could be added to handsets, they were and this made up the high end. But to attract new subscribers, wireless carriers still wanted low-end, low-cost yet robust

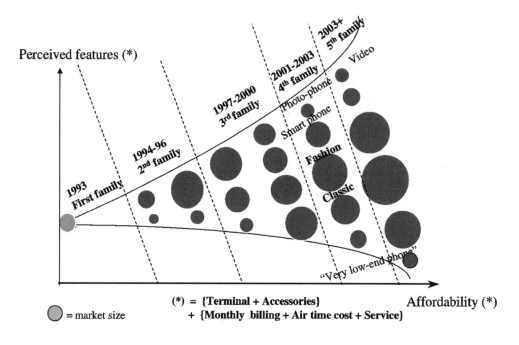

Figure 1.2 Digital cellular phones segments dynamics

mobile phones. In fact, for the service provider offering free handsets to each new subscriber, the lower the cost of the handset, the better off the service provider would be.

In the last few years though, the market has shown that it will splinter and stratify with several different layers or market segments between the poles. Some of the distinct segments that are emerging can be defined as:

- Data-centric devices: evolving from the PDA, these advanced palmtop computers will be integrated with cellular voice and retain or even expand upon their computing capabilities. Data-centric devices can also be modem cards (no keyboard, no display!) that can be plugged into laptops.
- Smart-phones: migrating from the cellular telephone segment of today's market, smart-phones will perform their voice communications functions quite effectively, but they also will be equipped with larger display screens so they can begin to perform new applications like e-mail access, Internet browsing and others.
- Fashion phones: these devices will use fashion techniques to appeal to several segments of consumers. The businessperson, for example, will be attracted to a certain look and feel to make a fashion statement. Younger consumers will have quite different tastes. Although they will cross several demographic market segments, these types of phones will appeal to buyers who are fashion-conscious and who will use fashion to make a statement about their lifestyles.
- Classic mobile phones: for users who are looking for a workhorse mobile phone, the classic handset will be small and easy-to-handle, and it will perform effectively the most frequently used communications features.

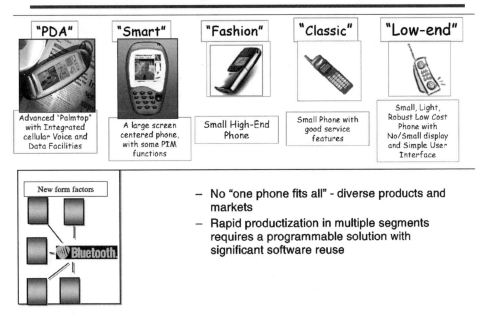

Figure 1.3   New form factors

- Low-end phones: service providers will continue to offer free phones with service contracts. These small, light and robust phones will remain a mainstay in the market because they perform a very valuable function. They often come with a pre-paid calling plan bundle. They attract first-time users. In the future, we might see such phones without a keyboard or a display (to save cost): phone calls would be made via an operator sitting in a call center or a voice dialing/recognition system, most likely in the network.
- Bluetooth-enabled phones: Bluetooth is a short range, low-cost, low power wireless technology operating in the 2.4 GHz unlicensed band. Bluetooth-enabled phones can be any of the above categories, but the form factors may change dramatically as the phone will now be distributed around your body.

The types of handsets that can be identified are illustrated in Figure 1.3 (concept phones courtesy of Nokia). What is not known is what tomorrow may hold and the effects new applications will have on the size, shape and function of future terminal devices.

One thing is for certain: new technologies will be developed that will alter the form factors in use today. For example, a Bluetooth-enabled phone maybe a belt-attached controller/ gateway device linked to an ear piece that communicates audio information. A display unit of some sort could be connected to the user's eye glasses for communicating visual data. And beyond these fairly new applications, medical sensors could be deployed to monitor the person's heartbeat or other vital functions.

A small box, comparable to a flat pager in size, will incorporate cellular and Bluetooth (or another technology such as IEEE802.11B or IEEE802.15) functionalities combined, to communicate with a collection of fashionable accessories; the accessories, of the size and

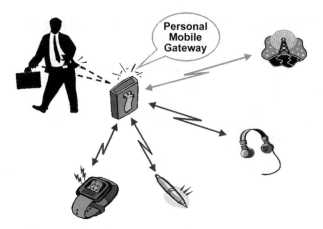

Figure 1.4   Personal Mobile Gateway™ (IXI Mobile Inc.)

weight of a pen, or a flat screen for example, will form a personal area network of thin clients communicating via Bluetooth with the small box, the Personal Mobile Gateway (Figure 1.4, courtesy of IXI Mobile Inc.). That way the "all-in-one" terminal, often too big to be a phone and too small to be a PDA, will become a collection of smart yet thin, fashionable and low cost devices. The concept would appeal to both mobile professionals and teenagers, the primary target for the ever increasing replacement market.

1.3 New Services

We have discussed wireless devices, but what users really care about are the services those devices will bring to them, and industry players care about how money will be made. Before describing the new services that are likely to be offered thanks to personal mobile terminals, a little history lesson will be useful and remind us to be humble, especially when it comes to predicting the future! When the telephone was invented, it was originally to improve the telegraph system. The fundamental idea of the electrical transmitting of sound was published by Charles Bourseul first in 1854 in the magazine *L'Illustration de Paris*. Alexander Graham Bell patented his telephone on the 14 February 1876, just 3 hours before Elisha Gray. Nobody was interested in his invention first. When he asked the Western Telegraph Company in 1877 to buy his patent for $100,000, the response was "What shall we do with a toy like that?". There was some doubt as to the use to which telephones might actually be put in practice. Demonstrations often included speech, song and music, and it was not uncommon for the musical demonstrations to be technically the most successful. "The musical telephone" was a major attraction at the International Electrical Exhibition in Paris in 1881, where the French engineer Clément Ader demonstrated stereophonic transmission by telephone direct from the stages of the Paris Opera House and the Comédie Française. It was believed to be the major application of telephony. In 1890, a commercial company, Compagnie du Theatro-phone (Figure 1.5), was established in Paris, distributing music by telephone from various theatres to special coin-operated telephones installed in hotels, cafés, etc. and to domestic subscribers. The service continued until 1932, when it was made obsolete by radio broad-

The "Theatrophone": One of the first uses of the Phone...

Figure 1.5   The Theatrophone

casting. The phone has come a long way since then, and the first mass market application is simply... talking with other people.

With the advent of the Internet and wireless data services, a new realm of possibilities are already offered, that go far beyond "just talking with other people", as witnessed by the recent success of NTT DoCoMo's I-mode service in Japan. Service categories of the near future will encompass personalized information delivery for news, location-dependant services, travel, banking and personal hobbies; it will also include productivity-related services such as Virtual Private Network (VPN) with the office or the family, personal assistant, agendas, and address books; extended communication, including e-mail, postcard transmission, and of course entertainment. Nokia has already introduced phones with games such as "the snake", but the future will bring much more exciting games (on-line as well as off-line, puzzles, gambling) and new forms of entertainment: music (ringtones, clips and songs), TV (schedules, clips), chat groups, astrology, dating services and what is sometimes called "adult entertainment". Figure 1.6 shows some of the service categories.

The successful deployment of the services will depend on ease of use, convenience, pertinence, and clear affordable billing. The pertinence of the service will require personalization; profiling technology can be used to match content to the needs of the users. Location-based services will enable or facilitate such profiling. Of course localization will have to be volunteered and "legally-correct" information. Most mobile location-based services today use positioning based on Cell of Origin (COO), but the precision is often mediocre, linked to cell size; in some cases, this is acceptable enough. Another method, known as Enhanced Observed Time of Difference (EOTD) is used in some GSM networks. Time of arrival signals from base stations are measured by the phone and what is called a Location-Measurement Unit (LMU). In future UMTS systems, a similar technique will be used that is known as

Service/Feature Clustering

Transactional	**Personal**	**Travel-related**
Tele-banking/shopping	Personalized News	Roaming
Hotel/restaurant/show	Budgeting	Compass, City guide
booking	Health monitor	GPS/CPS
Buy/sell, auctions	Pollution/Bio Analysis	Weather
Money transfers	Electronic key	Local info
Micro-payments	Passport / ID	On-line directory search
	Communication	Language translation

Security	**Mobile Office**	**Fun**
Monitoring	Modem/Fax	Gambling
Alarm	Memo/e-mail	Off-line games
GPS / CPS	Organizer/ Agenda	On-line games
Emergency calls	Dictionary/call mgt	Quizzes
---------------------	Productivity tools	Photo/Video
Contact with	Personal assistant	Radio/TV/Music
anxious family	Alarm/clock	Chat/Dating

Figure 1.6   Service categories

Observed Time Difference of Arrival (OTDOA). The location methods we just talked about only use the network and LMUs as a means to get location information; the use of Global Positioning System (GPS) gives better results, but the cost of a GPS receiver has to be added to the phone. An illustration of an innovative way to exploit and present location-based services is given in the last chapter of the book.

1.4 The Curse and Opportunity of Moore's Law

Moore's law predicted the rapid increase in transistor density on silicon chips. Along with this increase in transistor density, came an increase in clock speed, chip size, and component density on boards. All this has given the system designer an exponentially increasing amount of processing power to play with in his or her quest for more and more sophisticated systems. The design community has reacted to this explosion by making less and less efficient use of the transistors offered to it. This has been true since we first moved from hand laid out transistors to logic gates. The latter is less efficient in terms of silicon area and speed optimization, but is much more efficient in terms of a more precious resource: human intellect. From logic to RTL to microprocessors, the designer has moved to an increasingly high level of abstraction in order to design more and more complex devices in reasonable timeframes. Despite this, designers continue to lag behind process engineers in their ability to consume the transistors being made available to them. This can be clearly seen in Figure 1.7 which plots the ability of a designer to use transistors against the availability of transistors that can be used. This trend makes the use of programmable devices within mobile communications systems inevitable for the foreseeable future. The only question is, what will these

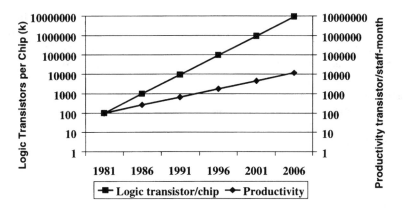

Figure 1.7   The widening productivity gap

programmable devices look like? Programmable DSPs are programmable devices that include features that enable efficient implementation of systems within the special class of signal processing problems. By focusing on signal processing DSP designers have put programmable DSPs at the heart of many consumer devices, including mobile communication systems. Recently DSPs have been specialized to perform specifically in the domain of signal processing for mobile communications (more details are given in Chapter 2). The balance between specialization and flexibility is important for any DSP to succeed.

As DSPs are programmable, they are not "just pieces of silicon," they come with a development environment. In the early 1980s, DSP was considered black magic, used by gurus who wrote all applications in assembly language. Now, powerful development tools including application boards, emulators, simulators, debuggers, optimizing High Level Language (HLL) compilers, assemblers, linkers, block diagram environments, code generators, real-time operating systems (enabling easier multitasking, preemptive scheduling, high-speed context switching, low interrupt latency, and fast, flexible intertask communication) as well as many DSP-related books and application notes and innovative visual tools have made DSP technology a tool for rapid design of increasingly complex systems.

In competition to DSPs, "silicon compilers" have arisen. These compilers promise to take high level descriptions of a system, and output a design ready for synthesis, usually with a certain amount of user feedback along the way. Though such tools have shown some success and are no doubt a useful tool in a designers arsenal, they do not provide a way to modify a system once it has been fabricated. This is becoming an increasingly important requirement because systems evolve quickly and are increasingly difficult to specify at design time. For instance, a mobile handset may not be fully tested until it has been used in the field. The increasing cost of mask sets for the fabrication of chips means any change that cannot be done by reprogramming may cost millions of dollars and months of time. This is unacceptable in today's marketplace.

1.5 The Book

In this book we attempt to cover some of the important facets of mobile communications design. We start of with five chapters covering various aspects of the design of the communications engine itself for 2G, 2.5G, and 3G phones. We then move onto the applications that will exist on top of the communications engine, covering a wide range of applications from video through biometric identification to security, for the next seven chapters. Then, after a chapter on digital radio broadcast, we move onto the architecture section of the book, with four chapters covering competitors, extensions and comparisons to programmable DSPs. The final chapter gives a taste of the completely new applications that are waiting to be discovered in the unique environment created when mobility meets signal processing.

We would like to thank all the contributing authors to this book for all the hard work that went into producing the excellent chapters within. They are a great example of the expertise and intelligence that is setting alight the field of mobile computing today.

2

The History of DSP Based Architectures in Second Generation Cellular Handsets

Alan Gatherer, Trudy Stetzler and Edgar Auslander

2.1 Introduction

Programmable Digital Signal Processors (DSPs) are pervasive in the second generation (2G) wireless handset market for digital cellular telephony. This did not come about because everyone agreed up front to use DSPs in handset architectures. Rather, it was a result of a battle between competing designs in the market place. Indeed, the full extent of the use of programmable DSPs today was probably not appreciated, even by those who were proposing DSP use, when the 2G market began to take off.

In this chapter we present the argument from a pro-DSP perspective by looking at the history of DSP use in digital telephony, examining the DSP based solution options for today's standards and looking at future trends in low power DSPs. We show that some very compelling arguments in favor of the unsuitability of DSPs for 2G digital telephony turned out to be spectacularly wrong and that, if history is to teach us anything, it is that DSP use increases as a wireless communications standard matures. As power is the greatest potential roadblock to increased DSP use, we summarize trends in power consumption and MIPS.

Of course, history is useless unless it tells us something about our future. Moreover, as the DSP debate starts to rage for third generation (3G) mobile communication devices we would like to postulate that the lessons of 2G will apply to this market also.

2.2 A History of Cellular Standards and Wireless Handset Architectures

2.2.1 1G and 2G Standards

The first commercial mobile telephone service in the US was established in 1946 in St. Louis, Missouri. This pre-cellular system used a wide-area architecture with one transmitter covering 50 miles around a base station. The system was beset with severe capacity problems. In

1976, Bell Mobile offered 12 channels for the entire metropolitan area of New York, serving 543 customers, with 3700 on a waiting list.

Although the concept of cellular telephony was developed by Bell Labs in 1947, it was not until August 1981 that the first cellular mobile system began its operations in Sweden, using a standard called Nordic Mobile Telephone system (NMT). NMT spread to Scandinavia, Spain and Benelux. It was followed by Total Access Communication System (TACS) in Austria (1984), Italy and the UK (1985), by C-450 in Germany (1985) and by Radiocom2000 in France (1985). These European systems were incompatible with each other, while trans-border roaming agreements existed between countries using the same standard (e.g. Denmark, Finland, Norway and Sweden with NMT-450 or NMT-900 systems, and Belgium, Luxembourg, and the Netherlands with NMT-450).

The US began cellular service in 1983 in Chicago with a single system called Advanced Mobile Phone System (AMPS). The market situation for the US was more favorable than Europe as a single standard provided economies of scale without incompatibility problems. The European model became a disadvantage, pushing Europe to unify on a single digital pan-European standard in the early 1980s and deployed in 1992. Later, this spread far beyond Europe: Global System for Mobile telecommunications (GSM). According to the GSM Association, more than a half billion GSM wireless phones are in use worldwide as of 11 May 2001; the standard accounts for more than 70% of all the digital wireless phones in use worldwide and about 60% of the world's GSM users are in Europe, but the single largest group of GSM users is in China, which has more than 82 million users.

Ironically, while Europe went from a fragmented, multiple-standard situation to a unified standard in the 1990s with seamless roaming structures in place (use of SIM cards), the US went from a single standard to multiple incompatible standards (IS54/136, IS95, GSM1900) with some inconvenient roaming schemes (use of credit cards). The IS136 operators have recently announced (March 2001) that they will overlay their network with GSM.

All the standards that were deployed in the 1980s were analog Frequency Division Multiple Access (FDMA) based, aimed at voice communication. As such, they belong to the first generation (1G). The standards deployed in the 1990s were digital Time Division Multiple Access (TDMA), FDMA, Frequency Division Duplex (FDD) or Code Division Multiple Access (CDMA). These standards enabled data capabilities from 9.6 to 14.4 kb/s, and were called 2G.

2.2.2 2.5G and 3G Standards

As demand for capabilities requiring higher data rates percolated in the mid-1990s, We experienced the evolution of standards to 2.5G with higher data rates, enabled by multi-slot data. High Speed Circuit Switched Data (HSCSD) is the first multi-slot data deployed. HSCSD is circuit switched based and combines 2–8 time slots of one channel on the air interface for each direction. The problem with circuit switched data is that circuits are dedicated to a communication, thus "reserved" to two customers for all the time of the communication: this results in costly communication for the users and sub-optimal use of capacity for the operators as users book circuits even if they do not use them. Another drawback of the technology, is that a RAS connection is needed before each data connection, and a bad communication can result in dropping the data communication all together, forcing the user to redial the RAS connection and paying for all the wasted time for the poor connection. Packet data enables these problems to be overcome, as packets of data belonging to different users can be distributed during what would be idle times in a circuit switched model; this enables billing to be based on data transferred rather than time, allowing better user experience and an always-on-always-connected model; a little bit like the difference between a RAS connection to Internet with a 14 kb/s modem and an always on connection with DSL or cable. The first real successful deployment of wireless packet data has been demonstrated with NTT DoCoMo's I-mode service, which relies on PDC-P (PDC-Packet data, where PDC stands for personal digital communications, the major Japanese digital cellular 2G standard).

GSM packet data standard is known as General Packet Radio Service (GPRS). GPRS was anticipated to be deployed in 2000 but will in practice be really used commercially in 4Q2001. In theory, data rates could be as high as 115 kb/s, but in practice, we will rather experience up to 50 kb/s. Enhanced Data rate for Global Evolution (EDGE) can be implemented over GPRS for even higher data rates, up to 384 kb/s, as a result of a change in the modulation scheme used. Next, 3G, driven by data applications, supports multi-mode and multi-band for Universal Mobile Telecommunication System (UMTS)/GSM as well as CDMA2000/IS95. 3G was supposed to be a single "converged standard" under the FPLMTS initiative, soon re-named IMT2000 and the 3GPP initiative; but then came 3GPP2 as the world could not agree on a single standard... after all, even though Esperanto was a good concept, historical, political and economical reasons are such that very few people do speak that language! The world of cellular will remain multi-mode, multi-band and complex. Figure 2.1 illustrates the path from 1G to 3G systems.

The 3G wireless systems will be deployed first in Japan in mid-2001 for capacity reasons and later in the rest of the world mainly for wireless multimedia, and will deliver a speed up to 2 Mb/s for stationary or 384 kb/s for mobile applications. Many questions remain as far as profitability and business models are concerned, so actual deployment might take longer than anticipated.

Evolution of Cellular Standards

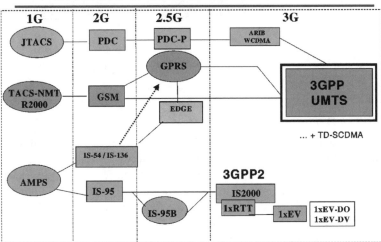

Figure 2.1   From 1G to 3G

The applications anticipated for 2.5G and 3G will require terminals to move from a closed architecture to an open programmable platform (for details, read Chapter 7).

2.2.3 Architecture Evolution

As we mentioned in the introduction, there is a continuing debate over the role of DSPs in wireless communications. To provide a historical basis for our arguments, in this section we examine the case of GSM evolution. The assumption is, of course, that 3G products will evolve in a similar manner to GSM, which is in itself debatable, but we believe that history does have some good points to make with respect to 3G.

A common functional block diagram of a GSM system is given in Figure 2.2. We recognize a classical digital communication model with signal compression, error correction, encryption, modulation, and equalization [11]. In the early days of GSM it was assumed that the low power requirement would mean that most of the phone would be implemented in ASIC. In what follows we show that the power difference between DSP and ASIC was not significant enough compared to other factors that were driving GSM phone evolution.

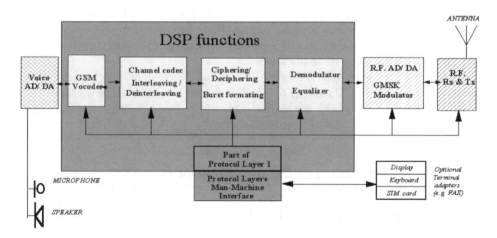

Figure 2.2   Functional block diagram of a GSM phone

2.2.3.1 Mission Creep

The early GSM phones were mostly ASIC designs. However, attempts to design vocoders with standard ASIC design techniques were not very successful and the voice coder was the part of the architecture that most engineers agreed should be done on a DSP. Hence, in early designs the DSP was included mainly to do the vocoding. The coder used in GSM phase 1 compressed the speech signal at 13 kb/s using the Regular Pulse Excited Linear Predictive Coding with Long Term Prediction (RPE-LTP) technique as per GSM 06-10 specification. So the DSP migrated from the vocoder engine to the central role as seen in Figure 2.2 over a period of a few years. Why did this happen?

One reason is that once a programmable device gets its "foot in the door" of an architecture a certain amount of "mission creep" starts to occur. The DSP takes on more functionality that was previously done in ASIC. Why this happens is a debatable subject, but the authors believe that several factors can be identified:

- *DSPs harness process improvement more rapidly than ASIC.* This is because the DSP tends to be hand designed by a much larger team than one would normally find on one ASIC block. This is a side effect of the amortization of the cost of DSP development over several markets.
- *DSP scale better with process improvement.* This is because a programmable device, when migrating to a higher clock rate, is capable of increased functionality. Many ASIC designs on the other hand do not gain functionality with increased clock speed. An example might be a hardware equalizer that is a straightforward ASIC filter implementation. If this device is run faster, it is just an equalizer that runs too fast. Even if you wish to perform another equalization task with the same device, you will probably have to redesign and add a considerable amount of control logic to allow the device to time share between two equalization operations. Indeed, in order to achieve future proof flexibility, ASIC designers tend towards development of devices with a degree of programmability. This increases the design effort considerably. Recently there has been a flurry of reconfigurable architecture proposals (for instance, Chapter 17) that are trying to bridge the gap between the efficiency of ASIC and the programmability of DSP, without the associated design cost.
- *DSPs are multitasking devices.* A DSP is a general purpose device. As process technology improves, two different functions that were performed on two DSPs, can now be performed on a single DSP by merging the code. This is not possible with ASIC design. The development of operating systems (OS) and real time OS (RTOS) for DSPs also have reduced the development costs of multitasking considerably. After 1994, a single DSP was powerful enough to do all the DSP baseband functions, making the argument for a DSP only solution for the baseband even more compelling.
- *DSPs are a lower risk solution.* Programmable devices can react to changes in algorithms and bug fixes much more rapidly, and with much lower development costs. DSPs also tend to be used to develop platforms that support several handset designs, so that changes can be applied to all handset designs at once. Testing of DSP solutions is also easier than ASIC solutions.

2.2.3.2 The Need for Flexibility

Flexibility was also important in the evolving standard. GSM phase 2 saw the introduction of Half Rate (HR) and Enhanced Full Rate (EFR). HR was supposed to achieve further compression at a rate of 5.6 kb/s for the same subjective quality, but at the expense of an increased complexity and EFR had to provide better audio qualities and better tandeming performance, also at the expense of higher complexity, using an enhanced Vector-Sum Excited Linear Prediction (VSELP) algorithm. Along with these changes came changes in the implementation of the physical layer as better performance, cost, and power savings combinations were found. As a result, each generation of phone had a slightly different physical layer from the previous, and upgrades to ASIC based solutions became costly and difficult.

A good example of this is the evolution of the adaptive equalizer in the GSM receiver, from a simple Least Mean Squares (LMS) based linear equalizer through Recursive Least Squares (RLS) adaptation to maximum likelihood sequence estimators. Indeed the performance of adaptive equalizers and channel estimators is difficult to predict without field trials, as the models used for the channel are only approximate. Implementation of equalization varies from company to company and has changed over time within companies. This comment also applies to other adaptive algorithms within the physical layer, such as timing recovery and frequency estimation. None of these algorithms appear within the standards as they do not affect the transmitted signal. Each company therefore developed their own techniques based on what was available in the literature.

Because the DSPs were now being designed with low power wireless applications in mind, the power savings to be had from ASIC implementation of the DSP functions were not significant enough that system designers were willing to live with the lack of flexibility. To improve system power consumption and board space, several DSPs such as the Motorola 56652 [1] and the Texas Instruments Digital Baseband Platform [2] integrate a RISC micro-controller to handle the protocol and man–machine interface tasks to free the DSP for communication algorithm tasks. The presently most popular partitioning of GSM is shown in Figure 2.3. Apart from algorithmic changes, the DSP was seen as an attractive component for a handset architecture for the following reasons:

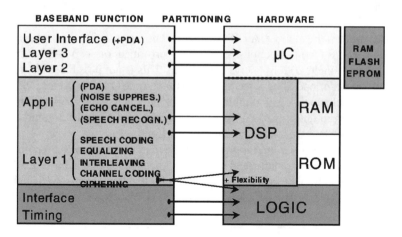

Figure 2.3   GSM function partitioning

- As GSM phones have evolved they have gradually moved beyond the simple phone function and this has lead to an increase in the fraction of the DSP MIPs used by something other than physical layer 1. This evolution is shown in Figure 2.4. With the advent of wireless data applications and the increased bandwidth of 3G we expect this trend to accelerate.
- Flexibility is also required when the product life cycle decreases. It becomes more and more difficult to manage the development of new and more complex devices in shorter and shorter time periods, even if the cost of development is not an issue. In GSM the product life cycle shortened from 2.5 years to 1 year thanks to the phone becoming a personal fashion statement.

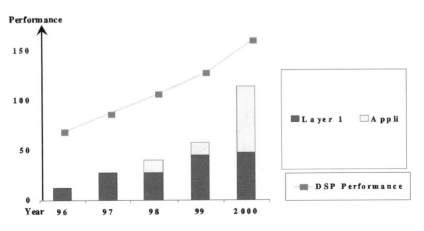

Figure 2.4   Layer 1 and application MIPS with time

- Different worldwide standards related to GSM and the need for product families addressing different market segments called for a platform based architecture so that OEMs could spin different products quickly. Development of a platform based system implies that the platform is also flexible in order to implement several standards. This is hard to achieve without some level of programmability.
- A DSP based baseband approach can cope better with different RF and mixed-signal offerings which occur due to technology improvements and market changes (e.g. AGC and AFC will change with different front ends).
- Spare DSP MIPS come for free and enable product differentiation (echo cancellation, speech recognition, noise cancellation, better equalizers).

2.3 Trends in Low Power DSPs

DSPs continue to evolve and compete with each other for the lucrative wireless market. Performance improvement can be achieved in several ways. Process improvement, instruction set enhancement and development of effective peripherals (such as DMA and serial ports) are three important ways to improve the performance of the device. Of course development of better software tools for development, debugging and simulation of DSP code cannot be underestimated as an incentive to pick one DSP over another.

2.3.1 Process Improvement

The digital baseband section is critical to the success of wireless handsets and, as we saw in Section 2.2, programmable DSPs are essential to providing a cost-effective, flexible upgrade path for the variety of evolving standards. Architecture, design, and process enhancements are producing new generations of processors that provide high performance while maintaining the low power dissipation necessary for battery powered applications. Many communications algorithms are Multiply-Accumulate (MuAcc) intensive. Therefore, we evaluate DSP power dissipation using mW/MMuAcc, where a MuAcc consists of fetching two operands

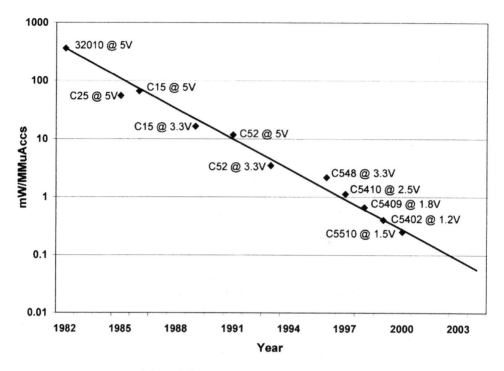

Figure 2.5   Power dissipation trends in DSP

from memory, performing a MuAcc, and storing the result back in memory. A MMuAcc is 1 million MuAccs. As shown in Figure 2.5, DSP power dissipation is following a trend of halving the power every 18 months [3]. As the industry shifts from 2G to 3G wireless we are seeing the percentage of the physical layer MIPs that reside in the DSP going from essentially 100% in today's technology for GSM to about 10% for WCDMA. However, the trend shown in Figure 2.5 along with more efficient architectures and enhanced instructions sets implies that the DSP of 3 years from now will be able to implement a full WCDMA physical layer with about the same power consumption as today's GSM phones.

Since these DSPs use static logic, the main power consumption is charging and discharging load capacitors on the device when the device is clocked. This dynamic (or switching) power dissipation is given by:

$$\text{Power} = \alpha C \times V_{\text{swing}} V_{\text{supply}} \times f$$

where α is the number of times an internal node cycles each clock cycle, and V_{swing} is usually equal to V_{supply}. The dynamic power for the whole chip is the sum of this power over all the nodes in the circuit. Since this power is proportional to the voltage squared, decreasing the supply voltage has the most significant impact on power. For example, lowering the voltage from 3.3 to 1.8 V decreases the power dissipation by a factor of 3.4. However, if the technology is constant, then lowering the supply voltage also decreases performance. Therefore, technology scaling (which decreases capacitance) and power supply scaling are combined to improve performance while decreasing the total power consumption of the DSP. In addition, parallelism can be used to increase the number of

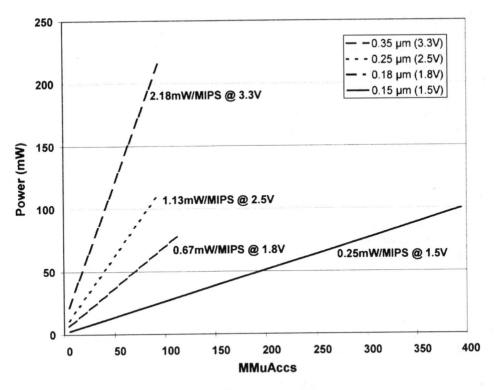

Figure 2.6   C5000 power vs. MMuAccs

MuAcc operations that can be performed in a single cycle, further improving processor efficiency as shown in Figure 2.6. This combination of techniques is used to enable the current TMS320C55x to achieve 400 MMuAccs at 1.5 V and 0.25 mW/MMuAcc in 0.15 μm CMOS technology.

2.3.2 Instruction Set Enhancement

In what follows we use the TI TMS320C55x [4,5] as an example of an evolving DSP that is optimized for wireless applications. However, the reader should note that because of the growing importance of the wireless market (more than 400 million units projected for 2000 [6]), there are now several DSPs on the market that have been designed with wireless applications in mind, for instance the Agere Systems (formally Lucent) 16000 series [7] and the ADI21xx series. IBM has also announced a TMS320C54x clone. This level of effort by several companies is a sign that the collective wisdom of the marketplace has chosen to bet on a programmable DSP future for wireless technology. We should also note that though designed for wireless applications, these DSPs are finding major markets in other low power applications such as telephony modems, digital still camera, and solid-state audio players.

As was mentioned in Section 2.2, the power difference between DSP and ASIC solutions was significantly reduced by designing the DSP for low power wireless applications. Several power saving features are built into the TMS320C55x architecture and instruction set to reduce the code size and processor cycles required. The core uses a modified Harvard architecture that incorporates five data memory buses (three read, two write), one program memory bus, and six address buses. This architecture leads to high memory bandwidth and enables multiple operand operations, resulting in fewer cycles to complete the same function. The TMS320C55x also contains two MuAcc units, each capable of a 17-bit × 17-bit multiplication in a single cycle. The central 40-bit Arithmetic/Logic Unit (ALU) can be split to perform dual 16-bit operations, and it is supplemented with an additional 16-bit ALU. Use of the ALU instructions is under instruction set control, providing the ability to optimize parallel activity and power management.

Another strategy used by DSP designers is to add instructions that, though fairly generic in themselves, allow efficient implementation of algorithms important to wireless applications. For instance in the TMS320C55x, one of the ALU inputs can be taken from a 40-bit barrel shifter, allowing the processor to perform numerical scaling, bit extraction, extended arithmetic, and overflow prevention. The shifter and exponent detector enable single-cycle normalization of values and exponential encoding to support floating-point arithmetic for voice coding. A compare-select-store unit contains an accelerator that, for channel decoding, reduces the Viterbi "butterfly update" to three cycles. This unit generally provides acceleration for any convolutional code based on a single shift register, which accounts for all the codes commonly in use in wireless applications today. Using this hardware accelerator, it is possible to decode one frame of a GSM voice channel (189 values) with coding rate 1/2 and constraint length 5 in approximately 6800 cycles, including traceback. The TMS320C55x also contains core level multimedia-specific extensions, which facilitate the demands of the multimedia market for real-time, low-power processing of streaming video and audio. There are also three hardware accelerators for motion estimation, Discrete Cosine Transform (DCT), Inverse Discrete Cosine Transform (IDCT) and 1/2-pixel interpolation to improve the efficiency of video applications. In addition, it contains four additional data registers that can be used with the 16-bit ALU for simple arithmetic and logical operations typical of control code, avoiding the use of higher power units.

The TMS320C55x instruction set also contains several dedicated instructions including single and block repeat, block memory move, conditional instructions, Euclidean distance calculation, Finite Impulse Response (FIR) and LMS filtering operations. The trend towards more specialized instructions will continue increasing as the cost of supporting these instructions goes down. Other instruction enhancements for bit manipulation, which is traditionally done much more efficiently in ASIC, will occur in the near future.

Another trend in DSP evolution is towards VLIW processors to support a compiler based, programmer friendly environment. Examples of this include TI's TMS320C6x [8], ADI's TigerSHARC [9] and Agere Systems and Motorola's Star*Core [10]. These VLIW processors use Explicitly Parallel Instruction Computing (EPIC) with predication and speculation to aid the compilers. The processors are also statically scheduled, multiple-issue implementations to exploit the instruction level parallelism inherent in many DSP applications. Though the application of this to physical layer processing in the handset is not apparent so far, these devices allow very efficient compilation of higher level code so reducing the need for DSP specific assembly level coding of algorithms. As explained in Chapter 7, the trend of wireless towards an open, applications driven system will make this kind of DSP much more compelling as a multimedia processor in the handset.

2.3.3 Power Management

Power management is very important in a low power DSP and several new advanced power management methods are implemented in the TMS320C55x. First, the TMS320C55x monitors all the peripherals, memory arrays, and individual CPU units and automatically powers down any units not in use. Memory accesses are reduced through the use of a 32-bit program bus and instruction cache with burst fill to minimize off-chip accesses. In addition, the user can configure the TMS320C55x processor for 64 combinations enabling or disabling six key functional domains: CPU, instruction cache, peripherals, DMA, clock generator, and External Memory Interface (EMIF). This enables customization of the power consumption for a specific application. The TMS320C55x also supports variable length instructions, from 8 bits to 48 bits, to allow optimization of code density and power consumption. The instruction buffer automatically unpacks the instructions to make the most efficient use of each clock cycle. The reduction in DSP core memory bus activity decreases the power consumption while longer instructions can carry out more functions per clock cycle. A flexible digital PLL based clock generator and multiplier allows the user to optimize the frequency and power for their application. In general these techniques allow a DSP that is not designed for a specific function to optimize its power usage for that function bringing its power level closer to that of a dedicated ASIC design.

References

[1] http://www.mot.com/SPS/WIRELESS/products/DSP56652.html
[2] http://www.ti.com/sc/docs/wireless/97/digbase.htm
[3] Gelabert, P. and Stetzler, T., *Industry's Lowest Power General Purpose DSP*, Embedded Processor Forum, 3–4 May 1999.
[4] TMS320C55x Technical Overview, Texas Instruments, Literature Number SPRU393, February 2000.
[5] TMS320C55x Functional Overview, Texas Instruments, Literature Number SPRU312, June 2000.
[6] Dataquest, *Mobile Communications Semiconductor Applications Markets, 1997–2002*, 12 April 1999, WSAM-WW-MT-9901.
[7] http://www.lucent.com/micro/wireless.html
[8] http://dspvillage.ti.com/docs/dspproducthome.jhtml
[9] http://www.analog.com/industry/dsp/
[10] http://www.starcore-dsp.com/
[11] Auslander, E. and Couvrat, M., Take the LEAD in GSM, in 'Applications of Digital Signal Processing', *Proceedings of DSP94 UK*, 1994, and in *Technologist Proceedings*, Herzlya, Israel, November 1995.

3

The Role of Programmable DSPs in Dual Mode (2G and 3G) Handsets

Chaitali Sengupta, Nicolas Veau, Sundararajan Sriram, Zhenguo Gu and Paul Folacci

3.1 Introduction

Third generation (3G) mobile radio standards are the result of a massive worldwide effort involving many companies since the mid-1990s. These systems will support a wide range of services, with voice and low rate data to high data rate services up to 144 Kbps in vehicular outdoor environments, 384 Kbps in pedestrian outdoor environments, and 2 Mbps in indoor environments. Both circuit and packet switched services with variable quality of service requirements will be supported.

The key challenges in designing 3G modems arise from the signal processing dictated by the underlying CDMA-based air interface with a chip rate of 3.84 Mcps (for the FDD DS mode explained later), the high data rate requirements, and the multiple and variable rate services that need to be supported simultaneously. Due to the various service scenarios – low-end voice to high-end high data rate – flexibility of the design is imperative.

In telecommunications, a "multi-mode" mobile is one that can support many different telecommunication standards with different radio access technologies. For example, the dual-band mobiles GSM + DCS are not considered as multi-mode mobiles because it uses the same radio access technology and the difference is only on the frequencies. By looking at the origin of the dual-mode system, we find two main drivers.

Operator driven: when ETSI developed the GSM specifications, it wasn't expected that the second generation (2G) mobile would be backward compatible with their analog 1G counter-parts. This was acceptable because the number of 1G users was negligible compared to the forecasted 2G users. On the other hand, in the 1980s it was quite easy for the small number of European members to agree on a single radio access technology because nobody then had an existing digital cellular network, so no compatibility was required. But when the success of GSM expanded outside Europe, the constraints changed and some operators decided to couple other standards with GSM. The main examples are GSM + DECT, GSM + AMPS, and GSM + ICO. However, such dual subsystems were not well adapted to allow a good integration for lowering the cost and reducing the size, and the two standards weren't allowed seamless handover.

Standardization committee driven: for the 3G Partnership Project (3GPP), the objective was to build an international standard with the ambition that a mobile could be used anywhere on the earth. The best solution was to agree on a single radio access technology for all the countries in the world. This was unfortunately impossible because it was too difficult to find a single radio access technology which could be backward compliant with all the different 2G radio access technologies already used by billions of customers all around the world. The best solution found by 3GPP to be backward compatible with 2G and allow a global roaming was to select a few radio access technologies (five) and to specify the mechanisms to allow intersystem handover. This solution is technically very difficult and needs to overcome many problems. But this solution compared to the operator driven one has more chance of leading us towards a viable solution.

From an operator point of view, the multi-mode mobile has many advantages. When an operator buys a UMTS license it gets the authorization to use the five possible air interfaces in its band. Depending on its strategy, the multi-mode could exploit many configurations. If the operator already has a 2G network (most cases), it could protect its 2G network investment (and its 2G mobile users) by using a dual-mode mobile. It also permits a smooth transition from 2G to 3G. The last interest is to increase its capacity and its coverage.

In this chapter we focus on the 3G FDD DS option as defined by 3GPP. This option is most likely to be the first deployed 3G mode. We present the salient features of the 3GPP FDD DS (popularly called WCDMA) mode followed by an overview of the requirements for the 3G-handset architecture and the role of a programmable DSP to meet those requirements as well as that of a GSM/WCDMA dual mode handset.

3.2 The Wireless Standards

Since the 3G standardization activities began [1–3], three main parallel development efforts have progressed in Europe (ETSI), Japan (ARIB) and the US. However, through the harmonization efforts of several groups, there are now three (harmonized) modes of the 3G standard (Table 3.1).

The FDD-DS mode is widely accepted as the mode that will be deployed first starting in Japan in 2001. In the rest of the chapter, we base our discussions about design of a 3G handset, on this mode. Table 3.2 lists the salient features of this mode. Table 3.3 lists the salient features of GSM.

Table 3.1 The three CDMA based modes of 3G

Parameter	Mode 1: FDD direct sequence	Mode 2: FDD multi-carrier	Mode 3: TDD
Chip rate (Mcps)	3.84	3×1.2288	3.84
Channel structure	Direct spread	Multi-carrier	Direct spread
Spectrum allocation	Paired bands	Paired bands	Unpaired band

Table 3.2 Parameters defining the FDD-DS (WCDMA) 3G standard

Parameter	Description/value
Carrier spacing (MHz)	5
Physical frame length (ms)	10
Spreading factor	2^k, $k = 2$–8: uplink, 2^k, $k = 2$–9: downlink
Channel coding	Convolutional and Turbo
Multirate	Variable spreading and multicode
Diversity techniques	Multiple transmit antennas, multipath
Maximum data rates	384 Kbps outdoor, 2 Mbps indoor

Table 3.3 Parameters defining the GSM (2G) standard

Parameter	Description/value (GSM)
Multiple access	TDMA/FDMA
Channel spacing (kHz)	200
Physical frame length (ms)	4.615
Channel coding	Convolutional
Multirate	None
Diversity techniques	Frequency hopping
Maximum data rates	9.6/14.4 Kbps (2.5G/GPRS: 171.2 Kbps)

The key features of the 2.5G and 3G standards illustrate the major differences between the two. Later we will highlight the commonalities between the two and the operation of inter-system measurements and handover.

3.3 A generic FDD DS Digital Baseband (DBB) – Functional View

The radio interface is layered into three protocol layers:

- Physical layer (Layer 1), responsible for data transfer over the air interface.
- Data link layer (Layer 2), responsible for determining the characteristics of the data being transferred, such as, handling data flow and quality of service requirements. The MAC is the Layer 2 entity that passes data to and from Layer 1.
- Network layer (Layer 3), responsible for control exchange between the handset and the UTRAN, and allocating radio resources. RRC is the Layer 3 entity that controls and allocates radio resources in Layer 1.

In this chapter, we will concentrate on the physical layer receiver processing, the most demanding layer in terms of hardware–software resources, and real-time constraints. Also we will not talk about the RF and analog portions that convert the radio signal at the antenna to a suitable stream of bits for DBB processing.

Figure 3.1 presents an overview of the various functional components of the physical layer processing in digital baseband. The rest of this section describes the main processing modules

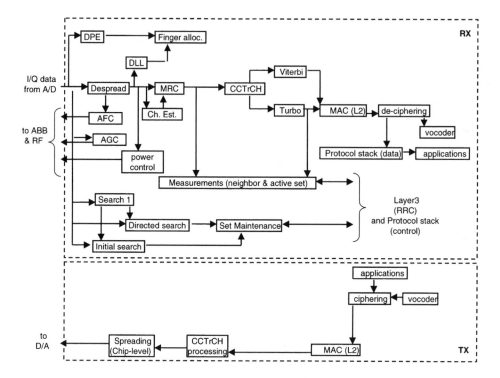

Figure 3.1   Functional overview of physical layer processing in DBB

in the receiver section, which is the more demanding part of the modem in terms of resource requirements.

Despreading: the despreading process consists of correlating the complex input data with the channelization code (Walsh code) and scrambling code, and dumping the result every SF chips, where SF is the spreading factor. Every significant received path of every downlink physical channel must be despread. Whether a path is significant depends upon the strength of the path compared to the strongest path.

Maximal ratio combination: one of the properties of CDMA signals is their pseudo-noise behavior due to the spreading process. As a result, signal paths that are separated by more than one chip interval appear uncorrelated. Maximal Ratio Combining (MRC) is the process of combining such paths to exploit time diversity against fading and increase the effective SNR. The contribution from each path to the final decision statistic is proportional to its SNR. The MRC step also needs to take into account any forms of antenna diversity in use.

Multipath search or Delay Profile Estimation (DPE): once the cell search unit has provided the strongest path that the mobile receives from a base station, the mobile must be able to find the next strongest paths in the vicinity of the main path, in order to perform maximal ratio combining. To facilitate soft hand-off, multipath search must be performed simultaneously for several base stations.

CCTrCH processing: in the downlink transmitter at the base station, data arrives from the MAC (Layer 2 entity) to the coding/multiplexing unit in the form of transport block sets once every transmission time interval {10 ms, 20 ms, 40 ms, and 80 ms}. In the handset receiver, the following steps must be performed to reverse each of the corresponding steps in the transmitter:

- De-multiplexing of transport channels
- De-interleaving (inter-frame and intra-frame)
- Rate detection (explicit and implicit) and de-rate matching
- CRC checking

Channel decoding: this step actually occurs in between the CCTrCH processing steps of rate detection and CRC checking. Channels may be either Turbo or convolution coded at the transmitter, thus necessitating both Turbo and Viterbi decoders. The former is usually used for the higher data rates and channels requiring a higher degree of protection.

Cell search: during cell search, the mobile station determines the downlink scrambling code and frame synchronization of a cell. The cell search is typically carried out in three steps: slot synchronization, frame synchronization, and cell specific scrambling code identification (popularly referred to as Search 1, 2, 3).

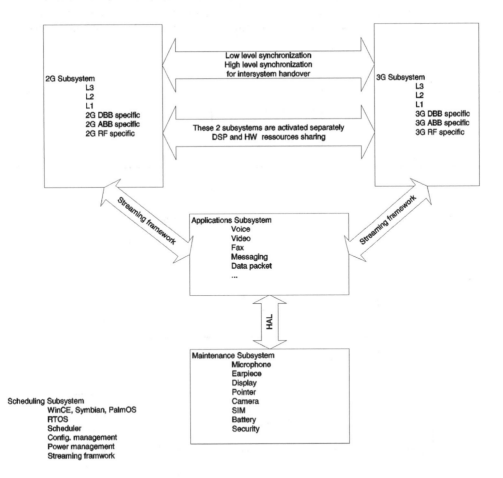

Figure 3.2   The dual-mode concept

3.4 Functional Description of a Dual-Mode System

The following description shows a system level view of a dual-mode handset (i.e. no algorithm, processors, partitioning are discussed at this level, Figure 3.2).

A dual-mode system is the combination of a GSM mobile [6] and a UMTS mobile. From a UE centric point of view, all these subsystems must share the maximum of hardware devices to reduce the die size and the BOM. Therefore the scheduling becomes a key part of a dual-mode system because it has to deal with very different time scale domains. On the other hand it must provide an efficient way to use a complex multiprocessor architecture, with multiple memories and data paths.

Compressed mode is the mechanism specified by 3GPP to allow intersystem handover preparation when the mobile is in WCDMA dedicated mode (Figure 3.3). This is a very tricky process of handover preparation and has not yet been proved in implementation. As such, it is one of those areas that will require much fine-tuning and evolution in the field.

A Type 2 dual-mode UE is defined by 3GPP, as a handset that can receive data from a cell in one mode (e.g. WCDMA) while at the same time it can monitor neighbor cells in another mode (e.g. GSM). Such UEs have one single subscription, which is common for all modes of operation. The different modes are related to different radio access technologies on the same

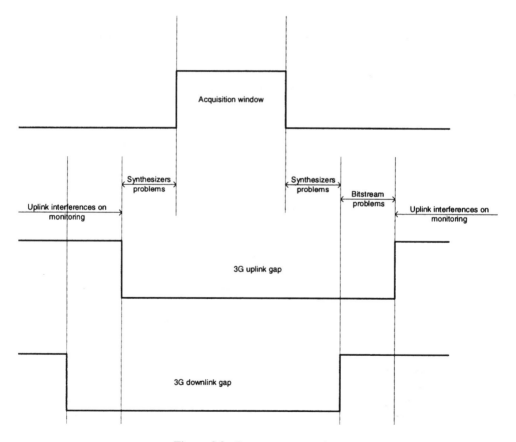

Figure 3.3   Intersystem operation

type of core network (UTRA/FDD and GSM radio on a Mobile Application Part (MAP) based core network).

Multi-mode operation is based on the separation of the Public Land Mobile Network (PLMN) selection from the mode/cell selection. Once the PLMN is selected, the choice of the mode has to be decided among the ones offered by the selected PLMN (controlled by operator through parameter settings). The user can choose a PLMN and request certain types of services. However, the user cannot choose the serving cell or the radio access technology and its mode.

3.5 Complexity Analysis and HW/SW Partitioning

3G terminals must be able to handle a wide range of service scenarios from low-end voice only to high data rate multimedia. In this section, we identify three representative scenarios in *steady state* and present a comparison of the processing requirements of the receiver functional blocks described in the previous section.

Scenario A: this scenario addresses a voice only terminal with only one 8 Kbps circuit switched voice service. This data rate was chosen to illustrate the requirements of a low-end handset.

Scenario B: this scenario supports 12.2 Kbps voice and 384 Kbps packet switched video. This is a high end but realistic case with multiple service bearers with different quality of service requirements.

Scenario C: this scenario supports a 2 Mbps service – the ultimate challenge that the 3G standards set for designers.

In addition to the dedicated services in each scenario, the handset is assumed to be receiving the required control information from the UTRAN.

The processing requirements of some of the most demanding modules, shown in Figure 3.4, depend not only upon the data rate, but also other factors such as number of services, number of strong cells in the vicinity, characteristics of the wireless channel, e.g. number of multipaths, etc. The despread unit includes despreading of all channels including the common pilot for channel estimation, time tracking, etc.

The HW/SW partition of the required processing – i.e. modules mapped to dedicated ASIC gates and modules mapped to SW, typically a programmable DSP are influenced by various factors. It must be chosen for a particular product meant for a specific service scenario. The key factor for handsets is processing requirements vs. target power budget. Additional factors include flexibility requirements, data I/O requirements, memory requirements, processing latency requirements, possibility of the function evolving in future, etc.

The basic trade-off involves that between target power and flexibility. For handsets, power is of course of primary concern. In general, lowest power is achieved by mapping functions to dedicated HW specifically designed to perform that function and nothing else. However, such dedicated HW also has lower flexibility to change (either due to feedback from the field or due to evolution of standards) when compared to a low power programmable DSP (e.g. Texas Instruments TMS320C54x and TMS320C55x series of processors, specifically designed to achieve low power for handsets, but high enough performance in terms of MHz to meet the challenge of 2G/3G).

The above requirements suggest some hardware–software partitioning options for a WCDMA receiver, as indicated in Figure 3.5. The figure shows modules that are:

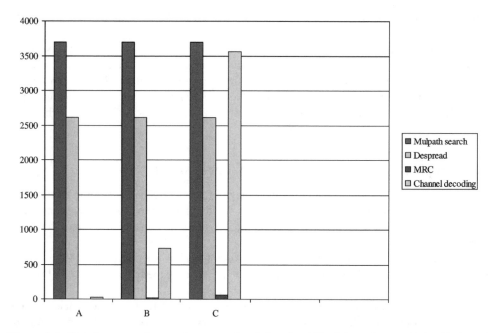

Figure 3.4   Relative processing requirements of each functional block in various scenarios (A, B, and C). The processing is shown in operations (millions per second)

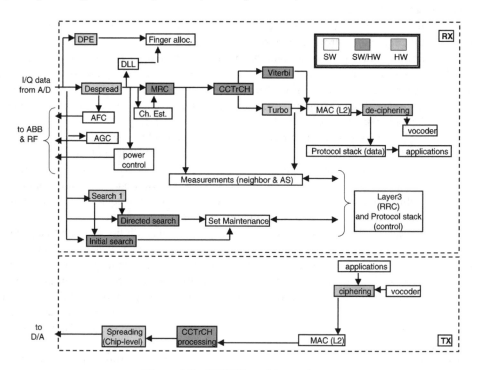

Figure 3.5   HW/SW partition options

- Definitely in HW in the near term, based on factors such as very high MIPS or data bandwidth requirements that a general purpose device such as a DSP is unable to meet;
- Definitely in SW, based on reasonable processing requirements, and more importantly a need for flexibility that requires a programmable device;
- In HW or SW based on total power targets and service scenarios for a specific implementation.

It must be remembered that 3G standards are new and yet to be deployed. Historically, it has been seen, as the DSP performance improves, functionality is moved from the ASIC to the DSP. However, 3G designers still have to face the problem of designing systems that will meet high processing requirements as well as have the flexibility required to meet a evolving standard, growing and new markets, and new service scenarios. This issue will be addressed in a later section.

3.5.1 2G/3G Digital Baseband Processing Optimized Partitioning

The upper part of Figure 3.6 shows a block diagram of the W-CDMA signal processing chain and the lower part shows a block diagram of the GSM signal processing chain. The shaded blocks represent functions, which could favorably be parameterized to be used by both the modem subsystems. The configuring of these parameters could be advantageously performed in the DSP while the main stream is performed in parameterized hardware attached to the DSP. This approach has the following advantages:

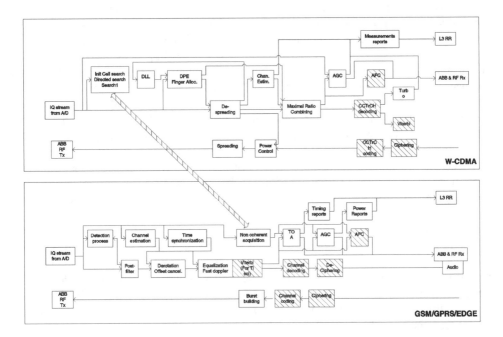

Figure 3.6 Common operations between modes

- The GSM sub-system reuses embedded W-CDMA accelerators in order to reduce power consumption and release DSP MIPS for applications.
- Software parameterization could help to patch the signal processing functions in case of specification change, algorithm improvement, and bugs.

Again, the GSM standard is quite mature compared to 3G and DSP technology has evolved to the point where a GSM modem can be very much SW based (example: extensive use of the TMS320C54x in GSM handsets). However, in dual mode, with the existence of GSM and WCDMA on the same platform, the partition for GSM needs to be reconsidered and re-mapped to the most appropriate architecture with the least cost.

3.6 Hardware Design Approaches

3.6.1 Design Considerations: Centralized vs. Distributed Architectures

By nature, CDMA systems are parallel. For a communication link between the base station and handset, there exists multi-code channels, and each channel is received via multiple propagation paths. The design challenge is the sharing or distribution of system resources between these parallel functional streams. In the handset the problem must be solved with the additional constraints imposed by the requirements of low power consumption and small silicon area.

This problem can be solved using two different hardware approaches: centralized or distributed architectures. In the centralized approach, a piece of hardware can be programmed for more than one CDMA modem function, say the searcher and fingers, so that the resources can be shared for different functions (if they have a common core function unit, for example, the correlation operator). On the other hand, a distributed architecture involves less resource sharing so that each functional module is relatively independent and autonomous.

Both approaches have their advantages and disadvantages. In general, a more centralized architecture will require less silicon area but more complex control in both software and hardware. Power consumption is proportional to both area and frequency. Therefore, to have the same amount of processing power, a centralized (more general purpose) architecture may have less area than a more functionally distributed architecture but will consume more power than a distributed system. This is because in addition to added control complexity, a general purpose architecture has to consider accommodating all supported functions while dedicated modules can be designed most efficiently for their own functions only. Also, it is easier to turn off sections of a distributed architecture, when not in use. The operating frequency of the hardware would also affect the differences of power consumption between the two architectures. A distributed architecture would need a lower clock rate than a centralized architecture.

Another factor that must be considered is the stand-by or sleep mode of a mobile handset, in which only a small number of channels need to be processed for a short period of time, between longer periods of inactivity. The system architecture should also consider how to efficiently partition the functional modules so that no hardware module with redundant functionality is activated in sleep mode, to maximize the total length of standby time. Meanwhile, these modules should be able to support heavy channel traffic when in normal mode.

Timing and latency of required response may also be considered in system architecture design. Under the condition of meeting system throughput requirements, trade-offs should be made between a centralized architecture but with higher frequency and a distributed one with lower clock rate. Generally, higher clock rate may cause more design difficulty and overhead so that sufficient manpower should be allocated.

No specific system architecture can claim to be a purely centralized or distributed system, there is a difference of the degree of centralized vs. distributed architecture. Trade-offs must be made for CDMA system architecture design based on the various system level constraints.

3.6.2 The Coprocessor Approach

In this section we discuss how coprocessors can complement the function of programmable DSPs in the implementation of a flexible 3G platform. For a WCDMA voice rate terminal, if we make a rough count of the "operations" required, only about 10% are suitable for implementation on a current DSP. But a fixed function solution would be a high-risk option due to a lack of flexibility, especially in a new standard. Therefore the system designer is faced with the problem of balancing the power and flexibility requirements. If we assume a long-term trend to increased use of more powerful DSPs then the designer also requires a roadmap for his design to migrate towards these devices.

One appealing solution to this problem is a coprocessor based architecture with a single programmable device at its core. The coprocessors enhance the computational capabilities of the architecture. At the same time they provide the desired amount of software program-mability, flexibility, and scalability required to meet standard evolution, provide product or service differentiation, and ease the process of prototyping, final integration, and validation.

We divide the world of coprocessors into "loosely coupled" and "tightly coupled" [4], which are defined relative to the average time to complete an instruction on the DSP and the type of interface it has with its host processor. With a Tightly Coupled Coprocessor (TCC) the DSP will initiate a task on the coprocessor that completes in the order of a few instruction cycles. A task initiated on a Loosely Coupled Coprocessor (LCC) will run for many instruction cycles before it requires more interaction with the DSP.

TCCs can be viewed as an extension of the host DSP instruction set by which macro-instructions, such as butterfly decoding or complex 16 bits multiply-accumulate operations, run on a specific hardware closely tied to the DSP through a standardized interface. Therefore TCCs benefit from the DSP addressing capability, DSP address/data bus bandwidth, internal registers and common DSP memory space. Additionally the DSP development toolset is re-used for developing and testing purposes. As each task in TCC only takes a few cycles it will naturally only involve a small amount of data. Also, parallel scheduling of tasks on the DSP and TCC will be difficult, as the DSP will interrupt its task after a few cycles to service the TCC. Therefore the DSP will generally freeze during the operation of the TCC. The TCC is therefore a user definable instruction set enhancement that provides power and speed improvements for small tasks where there is no data bottleneck through the DSP. A TCC also may have a very specific task and be relatively small compared to the DSP. With time, the function of the TCC may be absorbed into the DSP by either replacing it with code in a faster, lower power DSP, or by absorbing the function of the TCC into the core of the DSP and giving it a specific instruction. An example of this sort of function would be a Galois arithmetic unit for coding purposes or a bit manipulation coprocessor providing data to symbol mappings that are not presently efficiently implemented in the DSP instruction set.

TCC to main processor communication typically occurs through register reads and writes, and control is transferred back to the main processor upon completion of the TCC task.

There are processors now commercially available that allow the native instruction set to be enhanced through specially added hardware TCC units by means of a "Coprocessor Port". Examples of these are the ARM processor (the ARM7TDMI), and the TMS320C55x processor. The coprocessor port provides access to the processor register set, internal busses, and possibly even the data cache memories. In the ARM7TDMI, the coprocessor is attached to the memory interface of the ARM core. The coprocessor intercepts instructions being read by the ARM core and executes instructions meant for it. The TCC also has access to the ARM registers through the memory interface.

In the C55x processor on the other hand, the TCC connects to the main core via a dedicated port, through which it has access to the processor memory and register file (Figure 3.7). The main instruction decode pipeline of the processor sends control information to the TCC when it encounters a coprocessor instruction during program execution. A TCC may consume multiple clock cycles to execute its function, during which the main processor pipeline is idled. Examples of C55x coprocessors are accelerators for Discrete Cosine Transform (DCT), Variable Length Decoding (VLD), and Motion Estimation. These image processing TCCs result in between four- and seven-fold performance improvement as compared to the native C55x instruction set.

Loosely coupled coprocessors are more analogous to a subroutine call than an instruction. As they perform many operations without further DSP intervention, they will generally operate on large data sets. Unlike the TCC, the LCC will have to run in parallel with the DSP if it is to achieve its full benefit. This means the programmer will have to be more careful with the scheduling of LCC instructions. But, as the LCC has minimal contact with the DSP this should not be a problem. The main advantage of the LCC is that it solves the serious problem of bus bandwidth that can occur when either the raw input data rate to the system is very high or else the number of times data is reused in calculations is very high. In either case the bus bandwidth becomes the bottleneck to performing the computation because the data is stored at the other end of the bus from the computational units. An LCC removes this

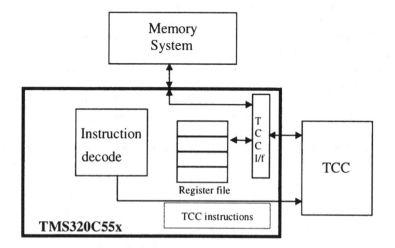

Figure 3.7   Tightly coupled coprocessor example

bottleneck having the computational units local to the data and arranged specifically for the data access required for a class of computations. In time the DSP will evolve to a point where its bus bandwidth and computational power is sufficient for the LCC's task and the pseudo subroutine implemented by the LCC will become a real subroutine.

The LCC design tends to be closely tied to the external bus interface and Direct Memory Access (DMA) capability of the native processor. Modern DSPs such as the TMS320C6x include highly sophisticated DMA engines that can perform multi-dimensional data transfers, and have the ability to perform a chain of transfers autonomously. Such DMA engines are ideal for transferring data in and out of LCC units with minimal DSP intervention. This reduces or even eliminates DSP overhead in performing data movement, and reduces the interrupt rates seen by the DSP.

The LCC concept applies easily at the chip rate to the symbol rate boundary of a CDMA system. In the WCDMA physical layer the DSP would still perform much of the symbol rate processing tasks such as the timing recovery, frequency and channel estimation, finger allocation, etc. The chip rate processing tasks such as despreading, path delay estimation, acquisition, etc. would be farmed out to a coprocessor that is designed to perform such tasks efficiently. For chip rate processing, TI has proposed a Correlator Coprocessor (CCP), which performs the common despreading tasks for fingers and path delay estimation operations in a CDMA receiver (both for the handset as well as the base station). The coprocessor can also perform some simple but high MIPS tasks that occur directly at the chip–symbol boundary. Examples of these are coherent and non-coherent averaging for channel estimation. However, the DSP still chooses the type of averaging that should occur and how to post-process the data to produce the final channel estimate. In effect the system is fully programmable within the domain of CDMA chip rate processing. The DSP also has control of how the correlation-MIPS provided by the CCP are allocated. For instance in a base station context, the DSP may choose to allocate a portion of the MIPS to one user with six multipaths. Alternatively it may reallocate these same MIPS to several users with fewer multipaths to despread. Similarly, multicode de-spreading for high data rate reception can be flexibly handled by the CCP. In the handset context, the correlation MIPS may be flexibly allocated between search tasks and RAKE finger despreading tasks, thus providing the flexibility to handle various channel conditions and data rates. Apart from allowing different WCDMA chip-set manufacturers to differentiate and improve their WCDMA solutions completely in software, such a flexible coprocessor allows the same system to be reprogrammed to perform WCDMA, CDMA2000, IS95, GPS and other CDMA based demodulation systems. It also provides a common platform for both low cost voice-only terminals and high-end multimedia terminals, and the same basic CCP architecture is applicable to both handsets as well as base stations.

A simplified block diagram of the CCP along with its system environment is shown in Figure 3.8. Note that the coprocessor is connected directly to the analog front end to remove the chip rate data completely from the bus. The incoming chips are stored in an input buffer, and the CCP processes a vector of N chips in each clock cycle, where N can be 16 or 32 for example. Another important feature is that the instruction and output buffers are memory mapped to allow flexible access to the coprocessor by the DSP. The DSP writes tasks (for setting up RAKE fingers, or search functions) into the task buffer. The CCP controller reads the tasks from the task buffer and performs the corresponding operation on the set of N chips stored in the input buffer. All the tasks in the task buffer are processed before the CCP moves

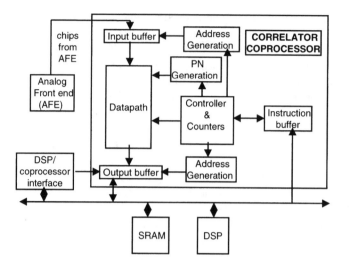

Figure 3.8   Loosely coupled (correlator) coprocessor based system

on to the next set of N chips. A task written into the task buffer is therefore executed "forever" until either it is overwritten with another task by the DSP, or it is explicitly disabled. The despread symbols are stored in memory mapped output buffers; the DSP can allocate this output memory flexibly among tasks running on the CCP. This flexibility becomes very useful for handling the rich variety of data rates supported by the 3G CDMA standards.

Comparisons to fixed function designs show that, with careful design of the coprocessor logic, there is no significant power penalty to be paid for the flexibility. This is essentially because the data flow dominates the power budget and this is independent of the flexibility of the design. The size of the coprocessor is somewhat larger than a dedicated design, but not significantly so within the complete system budget. The controller in a centralized design such as the CCP is somewhat more complicated compared to a hardwired "distributed" implementation approach, but the increase in complexity is more than made up for by the resulting increase in flexibility.

Decoding is another area which can benefit from the application of LCCs. Voice rate Viterbi decoding is easily performed on today's DSPs but the higher data rate requirements in 3G make decoding hard to do programmably. Nevertheless, it is possible to find a DSP/ coprocessor partition that maintains the flexibility required along with a reasonable MIPS level on the DSP. As an example, for Viterbi decoding in the base station, the DSP could perform all the data processing up to the branch metric generation and a coprocessor could perform the remaining high MIPS tasks of state metric update and trace back. This allows the DSP to define a decoder for any code based on a single shift register, including puncturing to other rates. Such a Viterbi coprocessor has already been implemented as part of the TMS320C6416 base station DSP.

3.6.3 Role of DSP in 2G and Dual-Mode

When GSM phones were first being designed, the ETSI specifications were stable enough that building a GSM mobile was realistic but there were no guarantees regarding perfect functionality. It was expected that the standard would evolve and get refined over time. To cope with this uncertainty the best way was to use a flexible signal-processing platform. The processing power required for GSM signal was fortunately compatible with the available DSP technology.

This technical model allowed the manufacturers to rapidly set up working handsets and fix the specifications and implementation problems on the field. This approach is more cost effective than spending a long time in simulations, or going through several ASIC prototyping cycles.

Moreover, the DSP presented another big advantage by allowing dissociation of the hardware platform problems from the GSM application problems. It is a definite advantage because the platform can evolve independently, gathering many improvements from its large fields of wireless applications, related to architecture and power saving features and gaining in reliability because of its large test coverage. In reality, a modification of a modem algorithm doesn't require full hardware test coverage to be rerun and on the other hand, a hardware technology improvement doesn't require full software testing. In the software centric model for a GSM modem, most of the terminal problems are related with the software design or specification interpretations, which are less critical than a hardware problem.

For 3G, the DSP role has changed somewhat because the available technology doesn't allow complete signal processing on a programmable DSP device. As explained earlier, many hardware coprocessors have been designed to compensate for the lack of processing power. They offer a good trade-off between performance and flexibility and will therefore fill the gap before a full software solution on DSP will be possible.

To build a dual-mode (2G and 3G) terminal, one can consider the "Velcro" solution consisting of assembling two single mode terminals in the same case, with minimal hooks needed to allow inter-system monitoring. This simplifies the software and hardware integration, but this solution is not cost-effective.

A better way would be to integrate all the DSP routines in the same DSP core. We call this solution an "integrated" solution. For the dual-mode terminal, the "integrated" DSP centric solution has several advantages:

Efficient memory usage: a multi-mode mobile is composed of a software subsystem per each supported RAT. Each subsystem has two main modes: The active one for all the usual single mode activities and an inter-RAT monitoring dedicated for measurement under gaps constraints. Depending on what subsystems and modes are used, the requirement for available memory changes dynamically. If the buffers are all in the DSP internal memory, it is easier to dynamically manage it and limit the maximum memory requirement. The DSP MMU will prevent inter-subsystem corruption.

Efficient power management: to reduce power consumption we need to take benefit of and predict periods of device inactivity. In a multi-mode system where most of the scheduling is centralized and DSP driven, the power management layer can have accurate information to switch unused devices off.

Bit stream management: in a multimedia system, a key requirement is the transfer of a large amount of data. State-of-the-art DSPs and DSP-Mega-cell, are sensitive to this requirement. The DSP is optimized for data transfer due to its embedded DMA capabilities and provides a lot a flexibility in using these channels. Such capabilities can be fully utilized only by an integrated dual-mode solution.

Resynchronization mechanism: in a dual-mode system, an active subsystem can help the other subsystems in inter-RAT monitoring mode by providing them with information about the cells to monitor. This requires a time exchange mechanism, which is easier to implement if all the signal-processing routines are running on the same core.

Common functions: some signal processing routines need to be reworked from an algorithm or from an interface standpoint to be usable by the other subsystems, instead of rewriting entire functions.

Future evolution: the applications to be run on a 3G or dual mode terminal are still uncertain. An integrated solution will allow more efficient management of system resources to accommodate yet unknown "killer apps" on the same platform.

At the same time, a DSP centric dual-mode solution has certain drawbacks. The constraints on the scheduler increase with the number of tasks. So, by merging tasks from many subsystems it is more difficult to guarantee correct concurrent code execution and can cause resource contentions that are hard to predict.

3.7 Software Processing and Interface with Higher Layers

The coprocessor based approach described earlier, or any programmable ASIC implementing any modem function, must meet the needs of an evolving 3G standard, with multiple modes, and for various service scenarios. In order to respond to these varying and changing needs quickly, it is necessary to have efficient software APIs to interface with these hardware modules. These APIs will allow easy reconfiguration of the hardware from software running on the DSP to meet system demand. On the other side, these APIs interface with the rest of the modem control structure (control-plane) as well as the signal processing algorithms operating on the data (data plane).

One commonly used approach for implementing the modem processing, due to its combination of signal processing algorithms, and a complicated control structure, is the use of a DSP and micro-controller combination [5]. A good example is the Texas Instruments OMAP^{TM} architecture consisting of an ARM9 and a C55x processor. In this approach, the DSP is responsible for the heavy-duty signal processing part it is best suited for, whereas the control plane is divided between the DSP and the micro-controller. The part of the control plane in the DSP typically deals with low latency hard real time functions. On the other hand, the control plane in the micro-controller provides a centralized control of all physical layer resources (hardware and software) on one side and provides an interface to the higher layers in the protocol stack (Layer 2 or MAC, and the Radio Resource Controller in Layer 3). The real time content of the system decreases as one goes up the protocol stack, which is typically implemented on the micro-controller.

Another point to note is that 2G has been primarily voice centric, whereas 3G is expected to be more data centric. However, it is still to be determined what the killer application for 3G will be. Several applications are good candidates: MP3, MPEG4, still-camera photos, video, etc. There has been considerable debate about the ideal platform for modem functions as well as applications. One approach is to have two different platforms for each – thus providing a lot of resources for applications, but at a higher cost. The other approach is to have a common platform that will be lower in cost but more difficult to achieve. The difficulty lies in protecting the real time nature of the modem being interfered with by the applications. In reality, there will possibly be both types of approaches, the former reserved for high end phones, and the second for low end primarily voice with suitably less demanding applications.

3.8 Summary

The dual-mode 2G/3G handset is very demanding in terms of processing requirements that will be hard to meet solely using programmable DSPs today. However, due to the lack of maturity of the 3G standards, flexibility of the implementation is imperative. Hence the most prudent approach will be to carefully map the functions consisting of very high operations per second (e.g. de-spreading) to hardware that is dedicated but parameterized (TCC, LCC) and attached to a programmable DSP. The rest of the signal processing functions that require a lot of flexibility (e.g. cell search processing) and will fit into the DSP within the target DBB power budget will be mapped to DSP-SW. As the standard matures and DSP technology improves, this picture will change with the DSP taking on more of the signal processing functions and providing the necessary flexibility required by a standard with a large deployment covering a multitude of service scenarios.

3.9 Abbreviations

AFC	Automatic Frequency Control
AGC	Automatic Gain Control
API	Application Programming Interface
ASIC	Application Specific Integrated Circuits
BOM	Bill of Materials
CCTrCH	Coded Composite Transport Channel
CDMA	Code Division Multiple Access
DBB	Digital Base Band
DLL	Delay Locked Loop
DSP	Digital Signal Processor
ETSI	European Telecommunications Standards Institute
FDD	Frequency Division Duplex
GPR	General Packet Radio Service
GSM	Global System for Mobile Communication
LCC	Loosely Coupled Coprocessor
MAC	Medium Access Layer (Layer 2 Component)
MAP	Mobile Application Part: GSM-MAP Network
MIPS	Million Instructions Per Second
PLMN	Public Land Mobile Network
RF	Radio Frequency
RRC	Radio Resource Controller (Layer 3 Component)
SNR	Signal to Noise Ratio

TCC   Tightly Coupled Coprocessor
TDD   Time Division Duplex
UTRAN   UMTS Terrestrial Radio Access Network

References

[1] Dahlman, E. Gudmundson, B., Nilsson, M. and Skold, A., 'UMTS/IMT-2000 based on Wideband CDMA',
    IEEE Communications Magazine, September 1998, Vol. 36, No. 9, pp. 70–80.
[2] Ojanpera, T. and Prasad, R., 'An overview of air interface multiple access for IMT-2000/UMTS', *IEEE
    Communications Magazine*, September 1998, Vol. 36, No. 9, pp. 82–86.
[3] Knisely, D.N., Kumar, S., Laha, S. and Nanda, S., 'Evolution of wireless data services: IS-95 to CDMA2000',
    IEEE Communications Magazine, October 1998, Vol. 36, No. 10, pp. 140–149.
[4] Gatherer, A., Stetzler, T., McMahan, M. and Auslander, E., 'DSP based architectures for mobile communica-
    tions: past, present, and future', *IEEE Communications Magazine*, January 2000.
[5] Baines, R., 'The DSP bottleneck', *IEEE Communications Magazine*, May 1995.
[6] Mouly, M. and Pautet, M.-B., *The GSM System for Mobile Communication*, Telecom Publishing, Palaiseau,
    France, 1992.

4

Programmable DSPs for 3G Base Station Modems

Dale Hocevar, Pierre Bertrand, Eric Biscondi, Alan Gatherer, Frank Honore, Armelle Laine, Simon Morris, Sriram Sundararajan and Tod Wolf

4.1 Introduction

Third generation (3G) cellular systems will be based on Code Division Multiple Access (CDMA) approaches and will provide significant data services as well as increased capacity for voice channels. This results in considerable computational requirements for 3G base stations. This chapter discusses an architecture that provides the needed computation together with significant flexibility. At the same time, this approach is one of the most cost effective known. Based upon a Texas Instruments TMS320C64x™ as the core DSP, the architecture utilizes three Flexible Coprocessors (FCPs): a Correlation Coprocessor for the CDMA portion, a Turbo Decoder Coprocessor for the data services, and a Viterbi Decoder Coprocessor for the voice services. The solution can be used for the two main flavors of 3G cellular as well as for second generation systems.

The explosive growth in wireless cellular systems is expected to continue. There will be 1 billion mobile users perhaps as early as 2003. 3G wireless systems will play a key role in this growth and roll-out of 3G should begin within 1 year. The key feature of 3G systems is the integration of significant amounts of data communication with voice communication, all at higher user capacities than previous systems. More recently, IP networking has become a key interest and such capabilities will become 3G services as well. These new 3G standards come under the coordination of the International Telecommunication Union (ITU) under the name of IMT-2000. Wideband CDMA techniques form the core of the higher capacity portions of these new standards and are the primary focus of this chapter.

3G base stations are more difficult to build compared to 2G due to their increased computational requirements. The increased computation is due to more complex algorithms and higher data rates, and the desire for more channels per hardware module. This chapter presents our approach for providing a very cost-effective solution for the physical layer (radio access) portion of the base station. It is based upon a partitioning of the workload between a TMS320C64x™ and three FCPs. The concept is to utilize a coprocessor when there are regularized functions that can be realized with very high silicon efficiencies relative to the DSP. Another feature is to incorporate a high degree of flexibility into each coprocessor so that it can be used as a platform for multiple base station solutions developed by multiple OEMs with differing requirements. This allows each DSP to handle a larger number of channels and/or to incorporate advanced algorithmic approaches, e.g. smart antennas and interference cancellation.

First we will provide an overview of the requirements of 3G systems and some system level analysis to give an understanding of the computational needs. Then each flexible coprocessor will be described: Viterbi Decoder, Turbo Decoder and Correlation Coprocessor. We conclude with a summary of advantages of this hybrid approach to 3G base station architectures.

4.2 Overview of 3G Base Stations: Requirements

4.2.1 Introduction

The objective of 3G wireless networks is to provide wideband services (Internet, video, etc.) together with voice services to mobile users. Thus, the downlink (base station (BTS) to mobile) data flow is predominant compared to the uplink (mobile to BTS) and is the primary limiter of 3G cell capacity. However, the BTS computation budget is limited by the uplink because of the much greater algorithmic complexity on the receiver (Rx) side. A key manufacturer careabout is achieving a high channel density, that is, a large number of mobile users processed in a single hardware module (RF interface + DSP + coprocessors). This motivates a highly efficient computational solution.

There are two primary 3G standards under IMT-2000: IS-2000 (CDMA2000), originated by Qualcomm in North America, and 3GPP (UMTS) originated by international standards bodies in Europe and Asia. Both use a wideband Direct Sequence CDMA (DS-CDMA) access system at the physical layer and implement similar base band functions such as despreading, finger allocation, maximal ratio combining, channel coding, interleaving, etc. This motivates a highly flexible DSP-based implementation to support both standards and their future evolutions using the same hardware.

The main issue is that some of these functions (such as, the despreader, convolutional decoder and turbo decoder) are very computationally intensive so that, at current DSP rates, a DSP-only solution cannot achieve sufficient channel density. However, because these functions can be realized with known fixed algorithms, often with regular, repeated operations, they can be implemented in flexible/semi-programmable coprocessors, FCPs, thus alleviating the DSP load and significantly increasing the channel density. This also achieves more optimal and efficient usage of silicon area thus providing a more cost-effective solution. And, it allows the powerful capabilities of the DSP to be used for more advanced algorithms.

4.2.2 General Requirements

In general, the basic 3G base station system requirements are as follows:

- *Performance*: the basic technical requirements are set by ITU's IMT-2000 initiative. The important factors are as follows:

– Evolution from 2G and global roaming capability
– Support of high speed data access up to 2 Mbps
– Support of packet mode services

• *Cost*: the cost per channel goals are every aggressive and must be more competitive compared to 2G channel costs. This means that cheaper voice service must be provided in order to justify the added costs for providing high-rate data service to users;
• *Flexibility*: the requirement for flexibility is being driven by a number of factors such as:

– More than one radio access technology (i.e. multiple CDMA techniques)
– Ease of product improvement, migration
– In-field maintainability
– An evolving standard

• *Time-to-market*: the initial schedule for 3G roll-out was aggressive thus surprising many market analysts who claimed that 2.5G services would delay the deployment. At present however, the roll-out schedule has slowed and 2.5G services are part of the reason. But 3G licenses have been awarded this past year and NTT DoCoMo in Japan is deploying its 3G networks for service roll-out in 2001. In any event, base station manufacturers are working on production ready systems including cost reductions.

4.2.3 Fundamental CDMA Base Station Base Band Processing

Although its partitioning can vary, the basic functionality of a 3G base station CDMA baseband processing is shown in Figure 4.1. The base band processing card(s) are connected to a backplane network bus and to an IF/RF front-end. On the baseband processing card(s) there are usually one or more DSPs which may be interfaced to a control processor that runs the main application code to implement the standard air interface and handles upper layer processing. The DSP generally performs the physical layer of the baseband signal processing. In CDMA there are two categories of digital baseband signal processing to consider:

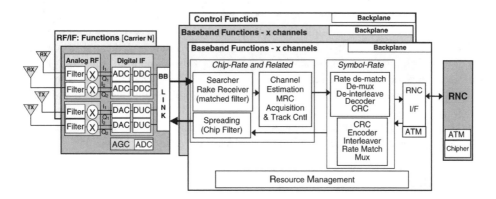

Figure 4.1   Block Diagram for Wideband CDMA Base Station Depicting Major Functions

- Processing at the *chip (spreading) rate*
- Processing at the *symbol rate*

Though much of the symbol rate processing can be done in the DSP, it still requires some key hardware acceleration. Essentially all of the chip rate processing requires hardware acceleration.

4.2.4 Symbol-Rate (SR) Processing

The challenge for 3G base station TDMA and CDMA Symbol-Rate (SR) processing is the requirement to not only process multiple channels, but to process very high data rate channels ($\geq$ 384 kbps). The argument for programmability is even greater for the SR processing as many channels at various data rates must be dynamically formatted, rate matched and multiplexed. DSPs can perform the SR processing for multiple channels in a flexible and cost effective manner for many SR functions. However, one important set of functions, Forward Error Correction (FEC) channel decoding (convolutional and turbo), is presently a challenge for the DSP when the data rates are high or when hundreds of voice channels need to be processed. Thus it is common practice to implement channel decoding in external hardware interfaced to the DSP. If this external hardware is a separate ASIC then this leads to increased board space area, however, this hardware can also be closely coupled to the DSP core and integrated within the DSP itself.

The SR solution is not complete without the correct peripherals to meet the board interface requirements. In particular the following is required:

- A host processor interface, debug interface, and timers
- High-speed, wide bandwidth memory interfaces to the spreader/despreader solution and external memory
- Serial ports for inter-DSP communications and/or downlink/forward link transmit data
- A network interface like the ATM physical interface Utopia II

4.2.5 Chip-Rate (CR) Processing

The Chip-Rate (CR) functions provide despread symbols to the symbol rate functions. At current chip rates (3.6864 Mchips/s for IS-2000 and 3.84 Mchips/s for 3GPP), many DSPs would be required to execute multiple channels of uplink chip rate receiver functions for such CDMA systems. Therefore, it is best to use an optimized solution dedicated to real-time processing of high-rate correlations (i.e. >2 MHz). This correlation function (RAKE despreader and searcher) can be implemented today in an ASIC. The challenge is to implement the CR processing in a cycle efficient, flexible (semi-programmable) and cost-effective manner.

On the receiver side, the main CR functions, which demand hardware acceleration, can be partitioned into the *searcher* functions and the *RAKE despreader* functions.

4.2.5.1 Searcher: Access Searcher&Traffic Searcher

There are two types of searcher functions: access searcher functions and traffic searcher functions. The access searcher has the function of observing and then connecting the users into a base station's set of active users. Providing statistics on the multi-path components for the delay profile management is the job of the traffic searcher.

Access searcher: after having successfully completed the downlink synchronization, a mobile station enters the cell network by sending a request on a common uplink access channel according to certain schemes. There are several types of access channels but they all have the same global structure: a preamble made of a non-modulated pilot, followed by an encapsulated message. The access searcher's function is to detect this new user in the cell by monitoring these access channels. Thus, a relatively large search window, proportional to the cell radius, is used. The access searcher searches for the preamble, whose structure differs from one standard to another. The IS-2000 access channel preamble is a simple non-data-bearing PN-spread pilot, while in 3GPP, a 16-chip Walsh signature randomly selected by the mobile station is superimposed on the PN-spread pilot.

Traffic searcher: after access is obtained, a 3G base station continues search operations for each user in the cell. The goal is to update periodically the delay profile of each user (i.e., identify each multi-path and certain related statistics). The traffic searcher function processes smaller search windows than the access searcher (typically 64 PN chips). In IS-95 (Radio Configurations (RC) 1&2 of IS-2000), the traffic searcher looks for the 64-ary Walsh–Hadamard modulated traffic channel of the user. Otherwise, the traffic searcher searches for the traffic pilot channel of the user, PN-multiplexed with the traffic data channel. The pilot channel is generally time-multiplexed with modulated symbols carrying information such as power control or spreading factor. Consequently, search tasks have to optimally exploit the pilot channel structure, either taking the modulated bit values into account or being scheduled only to search for the non-modulated bits. In IS-2000, the traffic and access searchers can share the same post-despreading hardware (non-coherent accumulation) and search task implementation. This is unlike the traffic searcher in IS-95 (RC 1&2 of IS-2000) and the 3GPP RACH preamble which have specific channel structures that require dedicated post-despreading hardware for Fast Hadamard Transformation (FHT).

4.2.5.2 RAKE Despreader

The RAKE despreads, via chip correlations against the various code sequences, as many replicas delayed in time of one user's signal as identified by the user's delay profile estimated by the traffic searcher. Channel estimation is performed on each of these "fingers" before they are combined by Maximal Ratio Combining (MRC) to provide the resulting (matched filtered) symbols. Channel estimation can be performed by the DSP. MRC can be implemented on the DSP or on a dedicated coprocessor. Channel estimation is performed on the traffic pilot channel of the user which is PN-multiplexed with the traffic data channel. Hence, the pilot channel is despread in parallel for each finger. In addition, each finger despreading requires despreading of the signal at the early/on-time/late positions. The energy or IQ measurements from the early and late despreaders feed a Delay Lock Loop (DLL). Time granularity for the DLL is typically 1/8th of a chip. The DLL function is performed on the DSP.

4.3 System Analysis

The CDMA system analysis is divided into two sections: SR processing and uplink CR.

4.3.1 SR Processing Analysis

The SR signal processing functions are as follows:

- FEC channel encoding and decoding: this can include a CRC, convolutional and turbo encoding/decoding;
- Interleaving/de-interleaving: there can be two levels of interleaving before and after channel multiplexing;
- Rate matching/de-matching;
- Multiplexing/de-multiplexing; and,
- Channel MRC: for the purpose of this study these are treated in the CR processing analysis as they are closely related.

The most DSP intensive functions are the two types of channel decoding: convolutional decoding and turbo decoding. Convolutional encoding is used for low data rate frames such as voice while turbo encoding is used for high data rate frames such as video. Though channel decoding appears to be well suited for implementation on a general purpose DSP it has typically been implemented in external ASICs for cost effectiveness and lower total process delay. Convolutional decoding has been implemented in coprocessors due to the large number of low-rate channels that need to be decoded while turbo decoding has been too computationally intensive for today's DSPs. The analysis below uses two common scenarios to compare the software only solution (DSP for all functions) to a DSP + FCP solution.

1. Support for 64 × 8 kbps voice channels (81 bit, Class A, AMR frames).
2. Support for 4 × 384 kbps data channels.

Table 4.1 shows the analysis results for these scenarios. One can see that for just the SR processing one needs more than 1000 MHz of a four MAC/cycle DSP like the TMS320C64x™. Therefore, a solution was proposed that would augment the DSP resources with flexible coprocessors. The solution using these FCPs takes approximately 118 MHz. This is a reduction by 10 × in the processing load. These channel decoding FCPs could be implemented externally, but cost, power and performance can be further optimized by integrating the flexible coprocessors on the DSP and designing them to take advantage of the DSP's architecture.

4.3.2 CR Processing Analysis

The CR processing contains multiple functions. On the uplink, the RAKE despreader, the access and traffic searcher, the channel estimation and the MRC are the most intensive in terms of operations per second. Other functions (acquisition, finger allocation, DLL) are considered as control functions and do not require much processing power. In the downlink, the most intensive function is the spreader, which is also implemented in hardware. As explained in Section 4.2, the BTS computation budget is dominated by the uplink receiver, so the downlink spreader is not considered in this analysis.

Table 4.1 Symbol-rate analysis for two scenarios comparing the DSP-only approach with the DSP + FCP approach

	64 × 8 kbps		4 × 384 kbps		Memory
	C64x (MHz)	C64x + FCPs (MHz)	C64x (MHz)	C64x + FCPs (MHz)	
Symbol rate encoding[a]	29	29	53	53	5 Mbits (data)
Symbol rate decoding (excluding convolutional and turbo decoders)[b]	17	17	16.5	16.5	20 kbytes (Pgm)
Convolutional decoder	211	~2[c]	N/A	N/A	18 kbytes (data)
Turbo decoder	N/A	N/A	~800+	~5[d]	46 kbytes (data)
Total DSP only	~257		~870		
Total DSP + coprocessors		~48		~75	

[a] Symbol rate encoding comprises: CRC encoder, convolutional or turbo encoder, 1st interleaver, rate matching, 2nd interleaver, muxing (for voice).

[b] Symbol rate decoding comprises: 2nd de-interleaver, de-muxing, rate de-matching, 1st de-interleaver, CRC check. Convolutional and Turbo decoder requirements are shown apart comparing the SW and HW implementations.

[c] For control in the DSP and 20% of a Viterbi coprocessor running at C64x CPU/4.

[d] For control in the DSP and 10% of a flexible coprocessor running at C64x CPU/2.

4.3.2.1 Uplink Receiver Analysis

Basically, the RAKE despreader and the access/traffic searcher use the same basic operation; Pseudo Noise (PN) and Walsh despreading. This operation consists of generating the properly timed pseudo noise and Walsh sequences and performing a correlation between the generated sequences and the incoming chip sequences. These correlations are performed at the CR. The RAKE despreader and the access/traffic searcher also perform energy estimation and non-coherent accumulation, but these functions require less processing power than the correlations.

The channel estimation algorithm determines the phase correction coefficients that have to be applied during the MRC. The channel estimation algorithm is based on a Weighted Multi-Slot Average (WMSA) filter and the complexity of this filter is that of an FIR that operates on a slot basis (considering one phase correction coefficient per slot). Using the previously computed phase correction coefficients for each path, the MRC can recombine all paths together to provide symbols to the SR processing portion of the implementation. The MRC performs a complex multiply per path (complex multiply of the despread signal with the phase correction coefficient) and then sums all corrected symbols together to provide the combined symbols to the remaining SR processing functions. The MRC typically runs at the SR; that is, one complex multiply is performed for each path at the SR. At times this rate may be higher due to changing or unknown spreading factors.

As stated earlier, the chip rate of the 3G standard is 3.6864 Mcps for IS-2000 and 3.84 Mcps for 3GPP. These high chip rates obviously increase the number of operations per second necessary for the CR processing. When considering these chip rates and the required number of users to be supported (as specified by base station manufacturers) the processing power required for the RAKE despreader and the access/traffic searcher is in the ballpark of 10–30 GOPS for 64 users. As stated in an earlier section, it would require many high-performance DSPs to execute multiple channels of uplink CR receiver functions of a CDMA system. Therefore, it appears that a full software based approach for the CR processing cannot be implemented in a cost-effective manner.

4.3.2.2 Using a Coprocessor

To support a large number of users per DSP, a hardware solution is necessary for the CR processing to minimize cost. This solution can take the form of an external ASIC correlation coprocessor to the DSP. Flexibility must be achieved however. To provide a high level of flexibility in the solution, the functions implemented on the coprocessor must remain under the DSP control, must provide a high level of programmability and must be well parameterized.

A Correlation Coprocessor (CCP) can be implemented to assist the DSP in the CR functions for RAKE despreading and access/traffic searching. Flexibility can be maintained in a cost-effective manner by carefully designing flexibility, by various means, into the correlation machine. The DSP can program this CCP using a set of tasks or instructions. This CCP will be discussed in a later section.

Flexibility within the overall solution can be achieved in part by allowing the channel estimation and MRC to be implemented in software on the DSP. Likewise, the DSP performs all control tasks such as finger allocation, timing recovery and correction based on the results obtained from the CCP. This flexibility allows the system designers to implement proprietary algorithms and approaches for improving performance. It also allows for later changes and upgrades. Channel estimation is just one example of a function that could be implemented with improved approaches that would increase performance.

Table 4.2 shows the primary CR computational requirements, assuming 64 users with four fingers for each user. Two situations are given: TMS320C64x™ DSP only and DSP with CCP. The CCP is one of a class of FCPs described in the next section.

4.4 Flexible Coprocessor Solutions

The concept behind FCPs is to couple the idea of hardware acceleration with substantial flexibility of the implemented function, perhaps to the point of semi-programmability. This includes the strategy of developing well conceived and efficient interfaces with the core DSP, both at the physical level and at the upper operational levels. For the 3G base station architecture a very cost effective and synergistic solution has been devised utilizing a TMS320C64x™ DSP with the three FCPs: Viterbi Decoder, Turbo Decoder and CCP. These are described in the following sections. In addition, a new DSP communications processor from Texas Instruments, the TMS320C6416, incorporates this Viterbi decoder and turbo decoder in a closely coupled fashion within the DSP itself.

4.4.1 Viterbi Convolutional Decoder Coprocessor

A Viterbi decoder is typically used to decode the convolutional codes used in these wireless

Table 4.2 CR analysis comparing the DSP-only approach with the DSP + CCP approach for key functions

	C64x (BOPS or MHz)	C64x + CCP (MHz)	Memory
RAKE despreader (CCP)[a]	~10 BOPS	Negligible	3 Mbits
Access/traffic searcher (CCP)[b]	~20 BOPS	Negligible	1 Mbits
MRC	200 MHz	200	5 Mbits
Channel estimation based on WMSA	10 MHz	10	64 Kbits
Control functions (acquisition, finger allocation, tracking, ...)	20 MHz	20	80 Kbits

[a] RAKE despreader estimated to take 250 K gates in CCP at 80 MHz.
[b] Access/traffic searcher estimated to take 275 K gates in CCP at 80 MHz.

applications. This algorithm comprises two steps: (1) computing state or path metrics forward through the code's trellis, and (2) using the stored results from step 1, traversing backwards through this data to construct the most likely codeword transmitted (known as traceback). State metric calculation is much more computationally intensive than traceback and consists mainly of Add, Compare and Select (ACS) operations. An ACS operation determines the next value of each state metric in the trellis and does this by selecting the largest of two candidate metrics, one from each branch entering the state. The candidate metrics come from adding the respective branch metric to the respective previous state metric. The branch metrics are derived from the received data to be decoded. In addition, the ACS operation stores the branch which was chosen for use in the traceback process.

The top level architecture of our flexible Viterbi Coprocessor (VCP) is shown in Figure 4.2 and consists of three major units: state metric unit, traceback unit and DSP interface unit. When operating at 80 MHz (160 MHz for its memory) the state metric unit can perform 320 ×

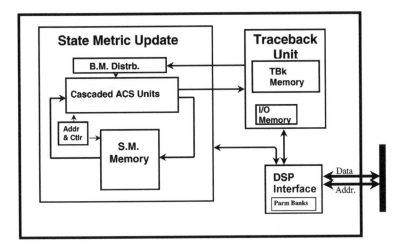

Figure 4.2   Viterbi Decoder CoProcessor Top Level Architecture

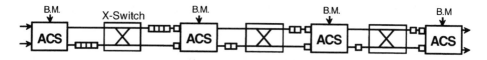

Figure 4.3 Radix 16 Cascade Datapath for State Metric Computation

10^6 ACS butterfly operations per second, and the VCP can decode at a rate of 2.5 Mbps. This is equivalent to well over 200 voice channels for 3G wireless systems.

To accomplish this while reducing the metric memory bandwidth to a reasonable level, the cascade structure in Figure 4.3 was implemented. This structure actually operates on a radix 16 subtrellis (16 states over four stages) and thus skips memory I/O for three of the four trellis stages resulting in a 75% reduction in bandwidth. This datapath also incorporates a unique register exchange over four trellis path bits (called the pretraceback). The 4-bit segments will need no further traceback later, they can be used as integral items during the trackback process. This allows traceback to be much faster. This cascade can also be operated at reduced lengths; in particular, as three stages, as two stages, or as a single stage. The corresponding incorporated register exchange likewise then produces pretraceback results of the same lengths in bits.

The traceback unit operates in the traditional manner for backwards traversing. This involves the repeated cycle of reading traceback memory to obtain the required word segment reflecting prior path decisions, shifting this word segment into the state index register to form the next state index needed for traceback, and using this data to form the next memory address. However, our design can move backwards up to four stages at a time due to the pretraceback mentioned above.

Flexibility was a key goal in the design of the VCP. It can operate on single shift register convolutional codes with constraint lengths of $K = 9, 8, 7, 6, 5$; and code rates of 1/2, 1/3, and 1/4. The defining polynomials for the desired code are taken as input versus hardwiring only a select few. The VCP also allows any puncturing pattern, has parameterized methods for partitioning frames for traceback, so that frame size essentially does not matter. And the convergence distance can be specified for partitioned frames. Thus, the VCP implementation can decode virtually any desired convolutional code found in the 2G, 2.5G and 3G wireless standards.

Efficient operation with a DSP was also achieved by memory mapping the device, by allowing block data transfers for input and/or output to be simultaneous with decoding, and by providing various signal lines for DSP/DMA synchronization such as input FIFO low and frame decode finished.

This decoder is very small and because of its very high throughput it is much more cost effective than a software approach. This frees the DSP to handle more channels and/or implement more advanced communication algorithms.

4.4.2 Turbo Decoder Coprocessor

Turbo coders are used in both the 3GPP and IS-2000 wireless standards. The turbo encoder shown in Figure 4.4 can deliver 10^{-6} BER performance at an SNR of 1.5 dB. The turbo encoder consists of two Recursive Systematic Convolution Coders (RSCC) that are

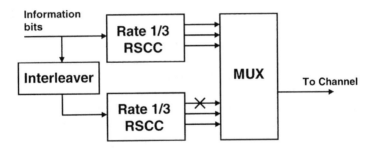

Figure 4.4   Turbo Encoder

connected in parallel as shown in Figure 4.4. The information bits are sent to both RSCC's. The lower RSCC information bits are interleaved prior to the coder. The output of both RSCC's is 3 bits, which are combined serially and later transmitted over the channel. The interleaved systematic bit from the lower RSCC is not transmitted because it is redundant. This leaves 5 bits that are punctured to make either a rate 1/4, 1/3, or 1/2 code.

The turbo decoder is an iterative decoder that uses the Maximum A Posteriori (MAP) algorithm. Each iteration of the decoder executes the MAP decoder twice. The first MAP decoder uses the non-interleaved data and the second MAP decoder uses the interleaved data. In each iteration, each MAP decoder feeds the other MAP decoder a new set of *a priori* estimates of the information bits, typically called *extrinsics*. In this way the MAP decoder pair can converge to a solution.

The data received from the channel needs to be scaled by $2/\sigma^2$ (where σ^2 is the signal noise variance) prior to use by the MAP decoders. This scaling is performed by the DSP.

The basic TCP architecture is shown in Figure 4.5. Flexible control allows the TCP to be configured to work in several modes. In the conceptually simplest mode the DSP loads an entire block of data to the TCP and it performs a single MAP decode on the data. The results are sent back to the DSP. This means the DSP will interleave the data between MAP decodes and is therefore involved in every iteration of the turbo decode. The data transfers are efficiently controlled by automation in the DSP's Enhanced DMA (EDMA) unit. This parti-

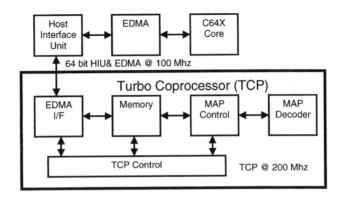

Figure 4.5   Turbo Coprocessor Architecture

cular operational mode allows the TCP to operate on a larger variety of codes than those in 3G, provided they use the same component RSCCs.

The TCP can also be set up to perform several iterations without DSP intervention. This greatly decreases the required bus bandwidth since the intermediate results are not being passed back and forth. In this mode, the TCP uses a look-up table to perform interleaving and can therefore perform as many iterations as required to converge. The TCP controller is in charge of writing the correct systematic, parity, and *a priori* data to the MAP decoder. After successful decode the DSP will retrieve the corrected data, typically via the EDMA.

To minimize power consumption it is common to use a stopping criterion that is a function of the MAP decoder outputs and is used to decide when convergence has occurred. It turns out that even though a maximum of 8–10 iterations is required to obtain best performance of the turbo decode, most of the time only 3–4 iterations are required for convergence. Therefore, a stopping criterion can have a significant impact on average MIPs requirements and therefore the power level. The TCP has a hardwired, proprietary stopping criterion for use in the multi-iteration mode. Of course, in the single MAP decode mode, the DSP is free to apply any stopping criterion.

For very large block sizes (in IS-2000 the turbo block can be as large as 20 kbits) the turbo decoder can perform a partial MAP decode using a sliding window technique. In this case the EDMA supplies the TCP with data, parity and *a priori* for a portion of the data block (codeword) with which to perform a portion of one MAP decode.

The MAP decode function is shown in Figure 4.6. The MAP controller can configure this block to perform alpha and beta updates as well as the output update from the extrinsic block. As is usual in turbo decoders, the iterative beta calculation is performed first and then the iterative alpha calculation is performed at the same time as the extrinsic output is performed using the latest alpha output as well as the previously derived betas. Therefore, we need beta storage but no alpha storage. A pipelined architecture allows four beta blocks to be generated in parallel with four alpha and output blocks. By this technique the maximum benefit of the circuit speed is obtained. The final design is capable of processing 16 channels at 384 kbps. Although this is more than the capacity of most base stations, it allows the turbo decoding to occur with low latency, which is a desirable requirement in the overall system.

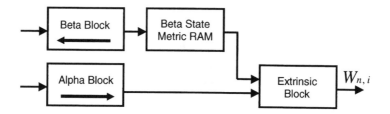

Figure 4.6 MAP Decoder Architecture

4.4.3 Correlator Coprocessor

The CCP is a programmable, highly flexible, vector based correlation machine that performs CDMA base station RAKE receiver operations for multiple channels. Because most RAKE receiver functions involve correlations and accumulations, regardless of the particular wireless protocol, a generic correlation machine can be used for various RAKE receiver tasks like finger despreading and search. However, though they are based on the same despreading core architecture, finger tasks and search tasks are processed on separate physical machines.

In addition to performing despreading functions (complex valued), which consist of code-sequence multiply and coherent accumulation, the CCP also accumulates "symbol" energy values (called non-coherent accumulations). For example, it accumulates the *early*, *on-time*, and *late* samples of a RAKE finger; these measurements are used for the finger's code-tracking loop (typically a DLL). For search operations, the CCP returns the accumulated energy values for a specified window of offsets.

The CCP performs all CR processing and energy accumulations according to the tasks that the DSP writes to the CCP's task buffers to control all CCP operations. The CCP does not perform SR receiver operations such as channel estimation, MRC, and de-interleaving, nor feedback loops such as AGC, AFC, and DLL. (For DLL, the CCP supplies the energy values to the feedback loop, but it does not operate on the loop itself.) All these symbol operations are performed on a TMS320C64x™ DSP. The first version of the CCP is tailored to support the IS-2000 3G standard but will be enhanced to support all future 3G standards.

Figure 4.7 shows an example of implementation using the CCP and shows how the CCP could be interfaced to the other components of the receive chain of a Digital Base Band (DBB) hardware configuration.

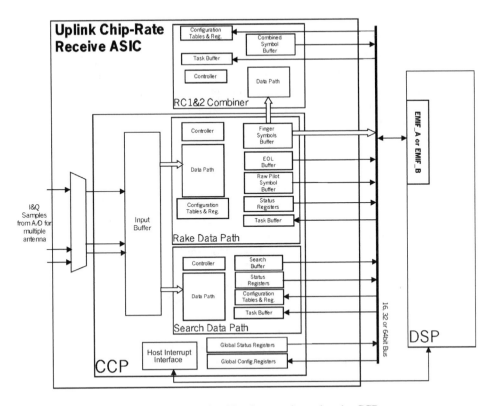

Figure 4.7   Example of Implementation using the CCP

The CCP is responsible for:

• Performing the despreading to provide data symbols per finger to the entity in charge of the MRC processing (may be either directly the DSP or another ASIC sub-block)
• Performing Early/On Time/Late (EOL) energy/IQ measurements for DLL
• Performing on-chip and 1/2-chip correlations and energy/IQ measurements for search purposes
• Providing raw pilot symbols per finger to the DSP

In IS-2000 RC 1&2, the FHT data path directly accesses the finger symbol buffer (output buffer of the RAKE data path) and performs the combining. Outputs of the combiner are written to the Combined Symbol Buffer (CSB). The DSP directly accesses that output buffer to get the combined symbols.

In RC 3&4, the DSP uses the computed raw pilot symbols to perform the channel estimation of each finger. Coefficients of the channel estimation are then sent to the entity in charge of the MRC processing. In this particular example, MRC processing is done in software, but it could also be processed by another hardware sub-block. Using those computed coefficients, the MRC multiplies despread symbols with the channel estimation coefficients and then sums the symbols coming from various fingers (paths) together to provide combined symbols. These combined symbols are then processed by further symbol processing stages in the base station receiver.

4.5 Summary and Conclusions

The goal of the work of this chapter is based upon the creation of a superior physical layer solution for the emerging 3G wireless base station market. Three key challenges that needed to be solved were: sufficient computational horsepower for large numbers of channels per unit, cost-effectiveness, and a high degree of flexibility. The approach discussed in this chapter achieves all three of these goals.

From the analysis presented in this chapter, for a typical situation with 64 users, the symbol rate processing requires 1100 MHz on a TMS320C64x™, and the CR processing requires 30 BOPS, assuming that only the DSP is used. Forward error correction decoding dominates the SR side while CDMA correlations dominate the CR side. To achieve a cost-effective solution it is clear that supplemental hardware support is needed.

The concept presented utilizes TI's newest DSP architecture, the TMSC32064x™, coupled with three FCPs: a Viterbi Decoder, a Turbo Decoder and a CCP. The FCPs are designed to be extremely efficient from a computation versus silicon area viewpoint, while at the same time being very flexible from an operational viewpoint. In addition, they were designed to interface with the DSP in an efficient manner to minimize DSP overhead in data and command interactions. Flexibility was achieved in the FCPs by building them in a parameterized, command driven manner, and for some, making them semi-programmable, such that they could be used for nearly every situation defined in the standards.

Overall, this flexibility is beneficial and needed for several reasons. In particular, flexibility allows multiple and/or changing standards to use the same device, improvements and changes to algorithms can be implemented quickly, enhancements can be realized gracefully, and more channels can be incorporated. Also, it allows for various approaches towards system partitioning onto processing devices. Examples include, separating versus combining uplink and downlink processing, or partitioning by functions versus numerous complete channels on a single unit. Lastly, flexibility provides a means for OEMs to differentiate their products.

Also, because of the power of TI's TMS320C64x™ DSP, this solution allows for future growth in approaches towards base station signal processing. Specifically, techniques such as adaptive beam forming, interference cancellation and multi-user detection, presently being developed, can be implemented on this architectural platform.

In addition, the TMS320C6416, which has recently been made available by Texas Instruments, incorporates the VCP and TCP coprocessors into the DSP.

Flexibility and a large amount of computational horsepower are achieved with the approach presented here because of the tremendous capabilities of the TI TMS32064x™ DSP together with the specific flexible coprocessors. A very competitive and cost-effective solution is the result.

5

The Use of Programmable DSPs in Antenna Array Processing

Matthew Bromberg and Donald R. Brown

5.1 Introduction

The increasing demand for communications services and the desire for increased data throughput in modern communications systems has fueled research and development into the use of adaptive antenna arrays. Since frequency bandwidth is in short supply and is expensive to acquire, the ability to separate users based on their spatial parameters is very attractive for wireless networks.

Adaptive antenna arrays offer the ability to increase the Signal-to-Noise Ratio (SNR) of a wireless communication link while at the same time permit the cancellation and removal of co-channel interference. Because of this an adaptive antenna array can be used to both dramatically increase the data rates of communication links as well as increase the number of users per cell that a wireless network can service. Some authors have reported well over an order of magnitude increase in network capacity [8].

As the computational power of modern Digital Signal Processors (DSPs) has increased, it has become possible to host adaptive array algorithms on these processors. Indeed the DSP has played a critical role in the feasibility of these systems. Many of the blind adaptive array algorithms require branching steps, iterative processing or require enough maintenance and flexibility to make hosting them in ASICs difficult. They are ideal however for a sufficiently powerful DSP. With the growing popularity and flexibility of software radios, DSPs will continue to enjoy a critical role in the design of these systems.

A conceptual block diagram of an adaptive array processor is shown in Figure 5.4. The DSP component of the processor includes much of the processing once the feeds from each antenna are digitized. This includes the application of the receiver weights, and the adaptation of those weights. In the simplest implementation, the weights applied to the data are fixed beforehand, and for sectorized antennas may be nothing more than choosing the antenna element with the largest gain in the direction of the Signal of Interest (SOI).

An example of the fixed beam approach is shown in Figure 5.1. Each of the five antenna elements have a cardiod gain pattern that maximizes the signal gain in one of five evenly spaced boresight angles. The antenna element with the largest gain is chosen. Unfortunately for the SOI emitter in this example, it falls in between the maximal response of two antennas and barely achieves a gain larger than one of the co-channel interferers.

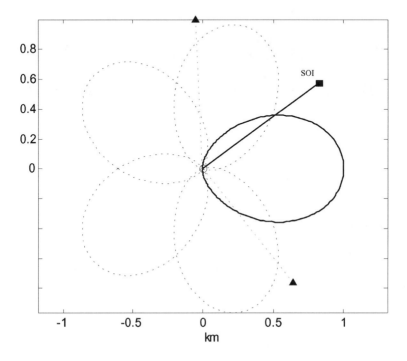

Figure 5.1   Choosing a fixed antenna pattern to enhance reception of the SOI

Although this approach can lead to performance gains, it falls short of the performance enhancements available with a fully adaptive array. An example of a beampattern for an adaptive array for the same antenna configuration is shown in Figure 5.6. In this case the interference is completely suppressed while at the same time the SOI SNR is enhanced by the receiver beamforming weights.

5.2 Antenna Array Signal Model

Most signal processing algorithms that exploit a multi-element antenna array are based on a simple signal model. Consider the transmitter and receiver geometry suggested in Figure 5.2.

The center of the coordinate system is chosen arbitrarily to be the geo-center of the receiver array. Each sensor has a coordinate in 3-space designated by $\mathbf{p}_1$ to $\mathbf{p}_M$. The location of the emitter is at $\mathbf{r}_0$. It is assumed that $\|\mathbf{r}_0\| \gg \|\mathbf{p}_k\|$, so that the received electromagnetic wave appears to be a plane wave. This will be become more precise shortly.

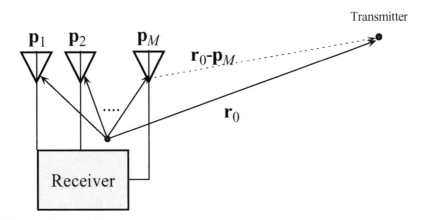

Figure 5.2 Array element and transmitter geometry

Assume the emitter emits an ideal spherical wave at a point in space and that the emitter is a baseband complex sinusoidal signal centered at frequency f and upconverted to the carrier frequency F_c. An ideal omni-directional antenna placed at some point in space $\mathbf{r}$ will observe a voltage due to the emitter of the form

$$V(t) = \frac{A}{\|\mathbf{r} - \mathbf{r}_0\|} \exp\left(2\pi j(f + F_c)\left(t - \frac{\|\mathbf{r} - \mathbf{r}_0\|}{c}\right)\right) \tag{1}$$

where c is the speed of light and A is a complex gain.

The response seen at antenna element k due to the emitter is therefore

$$V_k(t) = \frac{A}{\|\mathbf{p}_k - \mathbf{r}_0\|} \exp\left(2\pi j(f + F_c)\left(t - \frac{\|\mathbf{p}_k - \mathbf{r}_0\|}{c}\right)\right) \tag{2}$$

After downconversion the baseband signal becomes

$$V_k(t) = \frac{A}{\|\mathbf{p}_k - \mathbf{r}_0\|} \exp(2\pi jft) \exp\left(-2\pi j(f + F_c)\frac{\|\mathbf{p}_k - \mathbf{r}_0\|}{c}\right) \tag{3}$$

Because $\|\mathbf{r}_0\| \gg \|\mathbf{p}_k\|$ the following approximation holds

$$V_k(t) \approx \frac{A}{\|\mathbf{r}_0\|} \exp(2\pi jft) \exp\left(-2\pi j(f + F_c)\frac{\|\mathbf{r}_0\|}{c}\sqrt{1 - 2\frac{\mathbf{p}_k^T \mathbf{r}_0}{\|\mathbf{r}_0\|^2} + \frac{\|\mathbf{p}_k\|^2}{\|\mathbf{r}_0\|^2}}\right) \tag{4}$$

The term $\|\mathbf{p}_k\|^2/\|\mathbf{r}_0\|^2$ can be neglected and one can approximate, $\sqrt{1 - 2x} \approx 1 - x$ where $x = \mathbf{p}_k^T \mathbf{r}_0/\|\mathbf{r}_0\|^2$, because the higher order terms in the Taylor series are of order $o(1/\|\mathbf{r}_0\|^2)$ and can also be neglected.

This permits the complex baseband approximation

$$V_k(t) \approx \frac{A}{\|\mathbf{r}_0\|} \exp\left(-2\pi j\frac{\|\mathbf{r}_0\|}{\lambda}\right) \exp\left(2\pi j\frac{\mathbf{p}_k^T \mathbf{r}_0}{\|\mathbf{r}_0\|}\left(\frac{F_c + f}{c}\right)\right) \exp(2\pi jf(t - \tau_0)) \tag{5}$$

where $\lambda \triangleq c/F_c$ is the wavelength and $\tau_0 \triangleq \|\mathbf{r}_0\|/c$ is the emitter to array propagation delay. The term on the right in (5) is the original transmitted sinusoid delayed by τ_0. The rest of the array response at frequency f is nearly constant, independent of f, provided that

$$0 \approx \frac{f \mathbf{p}_k^T \mathbf{r}_0}{c \|\mathbf{r}_0\|} \ll 1 \tag{6}$$

Using the fact that $\mathbf{p}_k^T \mathbf{r}_0 = \|\mathbf{p}_k\| \|\mathbf{r}_0\| \cos(\theta_k)$, where θ_k is angle between $\mathbf{r}_0$ and $\mathbf{p}_k$, implies the following requirement for a flat antenna frequency response:

$$f \ll \frac{c}{\|\mathbf{p}_k\| \cos \theta_k} \tag{7}$$

The right hand side of (7) can be in the 1 GHz range or larger for a 1 foot radius antenna array for use in a cellular band. This seems to suggest initially that the array response is flat over a bandwidth that is 10–50 MHz. Unfortunately there are a number of issues that limit this result. There are non-ideal near field effects and multipath that dramatically reduce the flatness of the antenna response.

Assume however, that the array response is flat within the complex baseband $[0, B]$. Let $d(t)$ be a narrow band signal whose spectral support lies within this bandwidth. From (5) the response of antenna k to $d(t)$ can be written as,

$$V_k(t) \approx \frac{A}{\|\mathbf{r}_0\|} \exp\left(-2\pi j \frac{\|\mathbf{r}_0\|}{\lambda}\right) \exp\left(2\pi j \frac{\|\mathbf{p}_k\| \cos(\theta_k)}{\lambda}\right) d(t - \tau_0) \tag{8}$$

The receiver will typically contain a synchronization circuit that will, in conjunction with the transmitter, remove the delay τ_0 by either advancing the start of transmission, or delaying the receive gate. Also there is the presence of background radiation due to cosmic rays and noise introduced by the receiver. After digitizing the received baseband signal, one can therefore write the received signal at sensor k by,

$$x_k(n) = V_k(T_s n + \tau_0) + \varepsilon_k(n) \tag{9}$$

for sampling period T_s and noise process $\varepsilon_k(n)$, which includes the effect of the aforementioned background radiation.

Let $\mathbf{x}(n) \triangleq [x_1(n), x_2(n), ..., x_M(n)]^T$. This now yields the narrow-band antenna equation,

$$\mathbf{x}(n) = \mathbf{a}(\boldsymbol{\theta}) d(n) + \varepsilon(n) \tag{10}$$

where $\mathbf{a}(\boldsymbol{\theta})$ is the $M \times 1$ spatial signature vector, or array aperture vector at the direction cosine angle vector $\boldsymbol{\theta}$ induced by the emitter wavefront and where $d(n) \equiv d(T_s n)$. The kth element of the aperture vector is given by the gain factor that multiplies $d(t - \tau_0)$ on the right hand side of (8).

In order to understand how multipath restricts the maximum bandwidth size B consider the effect of K reflectors in the environment as shown in Figure 5.3. Each reflector excites its own array vector yielding the baseband continuous time signal model

$$\mathbf{x}(t) = \sum_{k=0}^{K} \mathbf{a}_k \xi_k d(t - \tau_k) + \varepsilon(t) \tag{11}$$

where $\mathbf{a}_k$ is the spatial signature vector due to the kth reflected path and ξ_k is the complex reflection coefficient. One also assumes here that the receiver has been synchronized to the

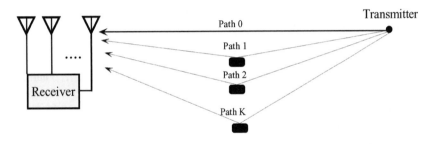

Figure 5.3   Linear beamformer processing at the receiver

direct path (path 0) so that τ_k is the differential multipath delay between the reflected path and the direct path. (This forces $\tau_0 = 0$.)

To make linear beamforming possible it is desirable to keep the array underloaded. That means that the number of significant emitters received by the array should be smaller than the number of sensors, M. If there is significant multipath, however, each emitter will load the array by a factor of K. Thus in a five-element antenna array if there are three emitters, each with two significant multipaths, the array will be overloaded, seeing the equivalent of six emitters and preventing the application of conventional beamforming techniques.

To reduce this loading effect on the array, it is desirable that the signal be narrow-band enough so that the approximation

$$d(t - \tau_k) \approx \alpha_k d(t) \tag{12}$$

is true. If it is, then (11) can be written as,

$$\mathbf{x}(t) = \sum_{k=0}^{K} \mathbf{a}_k \xi_k \alpha_k d(t) + \varepsilon(t) \tag{13}$$

$$\equiv \tilde{\mathbf{a}} d(t) + \varepsilon(t)$$

which consolidates the effect of the multipath into a single spatial signature vector $\tilde{\mathbf{a}}$, preserving the basic structure of the narrow-band antenna assumption.

A simple analysis of the bandwidth required to validate this considers the frequency representation of the narrow-band signal $d(t - \tau_k)$.

$$d(t - \tau_k) = \int_0^B \exp(2\pi jft)d(f)\exp(-2\pi jf\tau_k)df \tag{14}$$

It is apparent that (12) is true if $\exp(-2\pi jf\tau_k) \approx \alpha_k$. A simple approximation of this type is

$$\exp(-2\pi jf\tau_k) \approx \exp\left(-2\pi j\frac{B}{2}\tau_k\right) \tag{15}$$

This approximation holds provided that

$$\frac{B}{2}\tau_k \ll 1, \qquad B \ll \frac{2}{\tau_k} \tag{16}$$

In [34] for a particular challenging suburban cellular environment, the RMS delay spread is measured to be of the order of 2 μs. Assuming a coherence bandwidth a factor of 20 smaller than the right hand side of (16) yields,

$$B \approx 50\,\text{kHz} \tag{17}$$

Wideband signals, that exceed this design specification, can be channelized and processed over several sub-channels of bandwidth B. Of course the actual choice of B will depend on the application, the size of the array and the amount of multipath observed.

5.3 Linear Beamforming Techniques

Because of its simplicity, linear beamforming plays a key role in adaptive array processing. Significant performance gains can be achieved by enhancing the Signal-to-Interference Noise power Ratio (SINR) and by removing co-channel interference. Linear beamforming is also highly amenable to DSP solutions. In many implementations the application of the linear combining weights as well as the computation of those weights are computed in real time in dedicated DSP hardware.

A block diagram of a generic linear beamformer at a receiver is shown in Figure 5.4. The receiver consists of M antennas, and M RF chains that are digitized to complex baseband. The beamformer consists of a simple complex linear combiner with linear combining weights, $\mathbf{w} \equiv [w_1, w_2, ..., w_M]^T$. Each feed from the antenna is multiplied by a complex weight and added to obtain the overall response.

The output of the beamformer will be an estimate of the desired symbol $\hat{d}(n)$ at time sample n. The symbols are extracted from a receive environment that may contain unwanted noise and co-channel interference. The weights are adapted to suppress the interference and enhance the SOI. Multiple weight vectors can be used if there is more than one SOI.

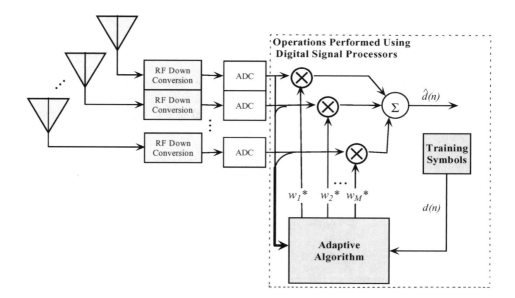

Figure 5.4   Linear beamformer processing at the receiver

The adaptation of the receiver weights typically requires the use of statistics derived from the received data vector $\mathbf{x}(n)$, whose elements are the complex digitized feeds from each antenna, and a reference or training signal $d(n)$. The training signal is either known, due to the transmission of a known training sequence, or estimated from the output of a CODEC or determined implicitly by exploiting properties of the SOI.

Many weight adaptation algorithms in use can be derived by optimizing an objective function that measures the performance of the system. The most commonly used objective function is the mean square error

$$\mu \equiv \langle |d(n) - \hat{d}(n)|^2 \rangle_n \equiv \frac{1}{N} \sum_{n=1}^{N} |d(n) - \mathbf{w}^H \mathbf{x}(n)|^2 \tag{18}$$

The mean square error performance function can be justified in part by considering the maximum likelihood estimator of the SOI given the narrow-band antenna model discussed previously and some simplifying statistical assumptions.

5.3.1. Maximum Likelihood Derivation

To formulate a statistical model for the estimation of unknown signal parameters, it is assumed that the signals are received over a wireless channel, downconverted and presented to the DSP hardware at complex baseband. The likelihood function itself is derived from the following signal model

$$\mathbf{x}(n) = \mathbf{a}d(n) + \mathbf{i}(n) \tag{19}$$

where $\mathbf{x}(n)$ is an $M \times 1$ complex received data vector at time sample n, $\mathbf{a}$ is an $M \times 1$ complex, received spatial signature or aperture vector, $d(n)$ is the transmitted information symbol at time n and $\mathbf{i}(n)$ is an interference vector due to environmental noise and other signals in the environment.

The received data is an $M \times 1$ vector due to M digitized feeds from multiple antennas and possibly multiple spreading frequencies, and/or polarizations. The bandwidth of the received waveform is assumed to be small relative to the coherence bandwidth of the wireless channel so that over this bandwidth the channel has flat fading characteristics and can be treated as a complex constant multiply.

The sample index n may include adjacent frequency bins if the signal has been channelized consistent with Discrete Multitone Modulation (DMT) or Orthogonal Frequency Division Multiplexing (OFDM). If so it is assumed that bulk signal delays have been removed either through signal processing, or through an adjustment of the transmitter start times.

The mathematical analysis is simplified if a block of data of N samples is processed at a given time and the received vectors are stacked into a matrix. Therefore, the conjugate received data and signal matrices are defined by

$$\mathbf{X} \triangleq [\mathbf{x}(1), \mathbf{x}(2), ..., \mathbf{x}(N)]^H \tag{20}$$

$$\mathbf{s} \triangleq [d(1), d(2), ..., d(N)]^H \tag{21}$$

$$\mathbf{I} \triangleq [\mathbf{i}(1), \mathbf{i}(2), ..., \mathbf{i}(N)]^H \tag{22}$$

This permits (19) to be written as

$$\mathbf{X} = \mathbf{sa}^H + \mathbf{I} \tag{23}$$

If we assume that $\mathbf{i}(n)$ is a complex, circularly symmetric, Gaussian random vector, with unknown covariance matrix $\mathbf{R_{ii}}$, and that the aperture vector $\mathbf{a}$ is unknown and deterministic, we can write the log-likelihood function for this signal model [22]

$$\rho_{ML}(\mathbf{R_{ii}}, \mathbf{a}) = -NM \ln(\pi) - N \ln|\mathbf{R_{ii}}| - \mathrm{tr}\Big\{\mathbf{R_{ii}}^{-1}(\mathbf{X} - \mathbf{sa}^H)^H(\mathbf{X} - \mathbf{sa}^H)\Big\} \tag{24}$$

Maximizing this expression over $\mathbf{a}$ yields $\mathbf{a} = \mathbf{X}^H \mathbf{s}/\mathbf{s}^H \mathbf{s}$, which we substitute into (24) to get

$$\rho_{ML}(\mathbf{R_{ii}}) = -NM \ln(\pi) - N \ln|\mathbf{R_{ii}}| - \mathrm{tr}\Big\{\mathbf{R_{ii}}^{-1}\mathbf{X}^H P_\perp(\mathbf{s})\mathbf{X}\Big\} \tag{25}$$

where $P_\perp(\mathbf{s}) \equiv \mathbf{I} - (\mathbf{ss}^H/\mathbf{s}^H\mathbf{s})$.

To optimize over the unknown interference covariance matrix, substitute $\mathbf{J} \triangleq \mathbf{R_{ii}}^{-1}$ in (25) so that

$$\rho_{ML}(\mathbf{J}) = -NM \ln(\pi) + N \ln|\mathbf{J}| - \mathrm{tr}\Big\{\mathbf{J}\mathbf{X}^H P_\perp(\mathbf{s})\mathbf{X}\Big\} \tag{26}$$

From matrix calculus one notes that $\partial \mathrm{tr}(\mathbf{JY})/\partial \mathbf{J}^* = \mathbf{Y}$ and $\partial|\mathbf{J}|/\partial \mathbf{J}^* = \mathbf{J}^{-1}$ for a positive definite matrix $\mathbf{J}$ and an arbitrary positive definite matrix $\mathbf{Y}$. Therefore after differentiation of the maximum likelihood function in (26) with respect to $\mathbf{J}^*$ and setting the result to $\mathbf{0}$ the optimal $\mathbf{J}$ can be written as

$$\mathbf{J} = \Big(\frac{1}{N}\mathbf{X}^H P_\perp(\mathbf{s})\mathbf{X}\Big)^{-1}$$

Substitution of the optimal $\mathbf{J}$ into the likelihood function (26) then yields

$$\rho_{ML} = NM \ln\Big(\frac{N}{e\,\pi}\Big) - N \ln|\mathbf{X}^H P_\perp(\mathbf{s})\mathbf{X}| \tag{27}$$

Using the definition $P(\mathbf{X}) \triangleq \mathbf{X}(\mathbf{X}^H\mathbf{X})^{-1}\mathbf{X}^H$, this can be written as

$$\rho_{ML} = NM \ln\Big(\frac{N}{e\,\pi}\Big) - N \ln|\mathbf{X}^H\mathbf{X}| - N \ln\Big(1 - \frac{\mathbf{s}^H P(\mathbf{X})\mathbf{s}}{\mathbf{s}^H\mathbf{s}}\Big) \tag{28}$$

Maximizing (28) is now seen to be equivalent to maximizing the following quantity

$$\max_{\mathbf{s}\in\mathbb{C}} \rho(\mathbf{s}, \mathbf{X}) \tag{29}$$

where

$$\rho(\mathbf{s}, \mathbf{X}) \equiv \frac{\mathbf{s}^H P(\mathbf{X})\mathbf{s}}{\mathbf{s}^H\mathbf{s}} \tag{30}$$

and $\mathbb{C}$ is a signal constraint set appropriate for the given application.

Examples of constraint sets might include a set of known waveforms or symbols chosen from a class of known constellations, or constant modulus signals, or signals constrained to be in a known subspace. Each constraint set type yields a potentially unique algorithm.

The likelihood function in (30) can be related to the minimum Mean Square Error (MSE) objective function by noting that

$$\min_{\mathbf{w}} \frac{\|\mathbf{Xw} - \mathbf{s}\|^2}{\|\mathbf{s}\|^2} = 1 - \rho(\mathbf{s}, \mathbf{X}) \tag{31}$$

This can be seen by differentiating the Normalized Mean Square Error (NMSE) function

$$\tilde{\mu}(\mathbf{w}, \mathbf{s}, \mathbf{X}) \triangleq \frac{\|\mathbf{Xw} - \mathbf{s}\|^2}{\|\mathbf{s}\|^2} \tag{32}$$

with respect to $\mathbf{w}^*$ and setting the result to $\mathbf{0}$ to solve for $\mathbf{w}$. This results in an optimal $\mathbf{w}$ determined by the normal equations,

$$\hat{\mathbf{w}} \equiv \left(\mathbf{X}^H \mathbf{X}\right)^{-1} \mathbf{X}^H \mathbf{s} \tag{33}$$

Solving (33) is referred to as the Least Squares (LS) algorithm.
The NMSE is related to the MSE in (18) by

$$\tilde{\mu} = \frac{\mu}{\langle |d(n)|^2 \rangle} \tag{34}$$

The MSE and NMSE objective function formulations are equivalent in the case of known training signals or signals that have a constant mean square over the adaptation block (e.g. constant modulus signals). For more general constraints, that lead to other types of blind adaptive beamforming, the NMSE formulation is preferred.

It is also possible to formulate the likelihood function in terms of the SINR at the output of the beamformer. The time averaged SINR is defined by

$$\gamma \triangleq \frac{\langle |\mathbf{w}^H \mathbf{a}d(n)|^2 \rangle}{\langle |\mathbf{w}^H \mathbf{i}(n)| \rangle} \tag{35}$$

Minimizing the NMSE objective function over $\mathbf{w}$ can be shown to be equivalent to maximizing the time averaged SINR, wherein the spatial signature vector $\mathbf{a}$ is replaced with its maximum likelihood estimate $\hat{\mathbf{a}} = \mathbf{X}^H \mathbf{s}/\mathbf{s}^H \mathbf{s}$. The estimated SINR in this case can be written as

$$\hat{\gamma}(\mathbf{w}, \mathbf{s}, \mathbf{X}) = \frac{\|P(\mathbf{s})\mathbf{Xw}\|^2}{\|P_\perp(\mathbf{s})\mathbf{Xw}\|^2} \tag{36}$$

where $P(\mathbf{s}) \triangleq \mathbf{s}(\mathbf{s}^H \mathbf{s})^{-1}\mathbf{s}^H$ and $P_\perp(\mathbf{s}) \triangleq \mathbf{I} - P(\mathbf{s})$.

A performance bound for the best linear beamformer can be found by optimizing (35) over $\mathbf{w}$ and taking the limit as the collect time N approaches infinity. This results in a simple formula for the maximum obtainable SINR

$$\gamma_\infty = \mathbf{a}^H \mathbf{R}_{\mathbf{ii}}^{-1} \mathbf{a} R_{dd} \tag{37}$$

where $R_{dd} \equiv E(|d(n)|^2)$ is the mean power of the transmitted information symbols. The maximum obtainable SINR serves as a yardstick to measure the performance of any given linear beamforming algorithm. A beamforming algorithm derived from the likelihood function formulated in (36) or (32) will often adhere to the performance predicted by the maximum obtainable SINR as the collect time gets large. There is also a minimum NMSE that is achievable and it is related to the maximum obtainable SINR by the formula

$$\tilde{\mu}_{\infty} = \frac{1}{\gamma_{\infty} + 1} \tag{38}$$

Note that the signal model for the likelihood function in (30), (32) or (36) is based on a single received SOI waveform, with all other emitters in the environment treated as interferers. If multiple overlapped signals are present, this model does not achieve the best possible performance. The problem of estimating multiple overlapped SOI waveforms is known as multi-user detection and is deferred until a later section.

5.3.2 Least Mean Square Adaptation

The complexity of inverting an $M \times M$ matrix when solving (33) for $\mathbf{w}$ has led to the consideration of algorithms that employ simple gradient descent techniques. By perturbing the weights in the direction of the negative gradient of the MSE objective function, the MSE can be reduced at each iteration, without requiring the inverse of a matrix.

The gradient of the MSE objective function in (18) can be written as

$$\frac{\partial \mu}{\partial \mathbf{w}^*} = -\langle x(n)e^*(n) \rangle \tag{39}$$

where $e(n) \triangleq d(n) - \hat{d}(n) = d(n) - \mathbf{w}^H x(n)$.

Updating the beamforming weights using a gradient descent technique would take the form

$$\hat{\mathbf{w}}(n+1) = \hat{\mathbf{w}}(n) - \mu \frac{\partial \mu}{\partial \mathbf{w}^*}(n) \tag{40}$$

where μ is the step size, $\hat{\mathbf{w}}(n)$ is the estimated weight vector at time sample n and $(\partial \mu / \partial \mathbf{w}^*)(n)$ is an estimate of the gradient at time sample n.

The LMS algorithm [44] approximates the gradient by averaging over the most recent sample, so that

$$\frac{\partial \mu}{\partial \mathbf{w}^*}(n) \approx -x(n)e^*(n) \tag{41}$$

The LMS algorithm update can therefore be written as

$$\hat{\mathbf{w}}(n+1) = \hat{\mathbf{w}}(n) + \mu x(n)e^*(n) \tag{42}$$

This update requires only M complex multiplies and M complex additions per sample. Note also that $e(n)$ is easily computed once $\hat{d}(n)$, the output of the beamformer is made available.

The simplicity of this algorithm is achieved at the expense of a slow convergence rate for the weights and a reduction in performance from that of the optimal weights in (33). The misadjustment from optimality is controlled by μ. The smaller μ is, the less the misadjustment, but unfortunately the slower the convergence.

A necessary condition for convergence of the algorithm is for

$$\mu < \frac{2}{\lambda_{\max}} \tag{43}$$

where λ_{max} is the maximum eigenvalue of the received data autocorrelation matrix $\mathbf{R_{xx}} \equiv E(\mathbf{x}(n)\mathbf{x}(n)^H)$. In practice one can set

$$\mu < \frac{2}{\mathrm{tr}(\mathbf{R}_{xx})} \tag{44}$$

The denominator on the right hand side of (44) is the received mean power summed over all the antenna elements.

The LMS algorithm converges at a rate roughly proportional to $(1 - \mu\lambda_{min})^n$, where λ_{min} is the minimum eigenvalue of the received data autocorrelation matrix $\mathbf{R_{xx}}$. At convergence, the MSE due to the LMS algorithm exceeds that of the MSE of the optimal filter by an amount $J_{ex}(\infty)$. For sufficiently small μ the excess mean square error is approximately

$$J_{ex}(\infty) \approx J_{min} \frac{\mu}{2} \mathrm{tr}(\mathbf{R_{xx}}) \tag{45}$$

where J_{min} is the MSE of the optimal weight vector [44].

There are a number of DSP implementation issues associated with the LMS algorithm. An issue of great importance is to choose a μ small enough to guarantee algorithm convergence and a small excess MSE, but also that allows the algorithm to converge sufficiently quickly. There is a limit to how small μ can be however. For fixed point processors a small μ may cause the gradient step to be strongly affected by quantization error, which can cause the algorithm to diverge.

Another implementation issue occurs when the received data autocorrelation matrix has eigenvalues that are almost zero so that $\mathbf{R_{xx}}$ is ill-conditioned. After scaling this can happen commonly enough if the environment contains a strong dominant emitter, forcing $\mathbf{R}_{xx}$ to be nearly rank 1. It is possible in this case for the LMS algorithm to excite modes that diverge, causing hardware overflow or underflow. This occurs in part to undamped modes that when coupled with non-linear effects such as round-off error, cause the LMS algorithm to diverge. A simple fix for this is to use the leaky LMS update defined by

$$\hat{\mathbf{w}}(n + 1) = (1 - \mu\eta)\hat{\mathbf{w}}(n) + \mu\mathbf{x}(n)e^*(n) \tag{46}$$

where η is a small positive constant. The leaky LMS has additional bias, but prevents the algorithm from diverging for ill-conditioned $\mathbf{R}_{xx}$.

5.3.3 Least Squares Processing

Although the LMS is computationally simple, it is possible to significantly reduce the computations required to implement (33), which can asymptotically approach the optimal beamformer performance. It is not necessary to include every sample in the computation of the autocorrelation estimates, $\hat{\mathbf{R}}_{xx} \equiv (1/N)\mathbf{X}^H\mathbf{X}$, and $\hat{\mathbf{R}}_{xd} \equiv (1/N)\mathbf{X}^H\mathbf{s}$. Instead the time averages can be decimated by a factor of P, so that only every Pth sample is used. The optimal weights are then computed by

$$\hat{\mathbf{w}} = \hat{\mathbf{R}}_{xx}^{-1}\hat{\mathbf{R}}_{xd} \tag{47}$$

or its equivalent. It is also possible to reduce the dimensionality of the autocovariance matrix by beamforming after the application of a set of M_1 fixed beams contained in the columns of an $M \times M_1$ matrix $\mathbf{T}$, wherein $M_1 < M$. One therefore performs linear beamforming on the transformed receive data vector [30]

$$\tilde{x}(n) \equiv \mathbf{T}^H x(n) \tag{48}$$

Since the computation of the weights in (47) has a complexity of order M^2N, this can result in considerable savings. The performance is also superior to simply reducing the number of antennas, since the processing gain of the full M-element array is still available to the beamformer. These techniques may allow LS processing to become feasible in situations where DSP cycle counts are limited.

A practical implementation of (47) will not actually compute the estimated autocorrelation matrix $\hat{\mathbf{R}}_{xx}$ directly, but rather work with its Cholesky factor, $\mathbf{R}_x$. The Cholesky factor is the unique upper triangular matrix that has the property

$$\mathbf{R}_x^H \mathbf{R}_x = \hat{\mathbf{R}}_{xx} \tag{49}$$

with real diagonal elements. The Cholesky factor can be computed directly from the received data matrix $\mathbf{X}$ defined in (20). Not only is this more numerically stable than (47), but the number of bits of numerical precision, required to implement the algorithm is half that of (47). This is because in (47), the computation of $\hat{\mathbf{R}}_{xx}$ involves computing the square of the input data.

Given the $N \times M$ conjugated, received data matrix $\mathbf{X}$, the Cholesky factor is computed by taking the QR-decomposition of $\mathbf{X}$. This operation factors $\mathbf{X}$ into a product of two matrices

$$\mathbf{Q}\mathbf{R}_x = \frac{1}{\sqrt{N}}\mathbf{X} \tag{50}$$

where $\mathbf{Q}$ is an $N \times M$ orthonormal matrix, wherein $\mathbf{Q}^H\mathbf{Q} = \mathbf{I}$ and $\mathbf{R}_x$ is the upper triangular Cholesky factor of $\hat{\mathbf{R}}_{xx}$.

The QR decomposition can be computed efficiently by use of the Modified Gram–Schmidt Orthogonalization technique (MGSO). The MGSO algorithm implements the following steps:

Set $\mathbf{p}_m = \mathbf{x}_m$
For $m = 1$ to M
    $r_{mm} = \|\mathbf{p}_m\|^2$
    $\mathbf{q}_m = \mathbf{p}_m/r_{mm}$
For $i = m + 1$ to M
    $r_{mi} = \mathbf{q}_m^H\mathbf{p}_i$
    $\mathbf{p}_i = \mathbf{x}_i - r_{mi}\mathbf{q}_m$
end i
end m

where x_m is the mth column of $\mathbf{X}$, $\mathbf{q}_m$ is the mth column of $\mathbf{Q}$, and r_{mi} is the element in the mth row and ith column of the Cholesky factor $\mathbf{R}_x$. The MGSO algorithm obtains the desired matrix factorization by projecting the unprocessed columns of $\mathbf{X}$ onto a subspace that is orthogonal to the columns of $\mathbf{q}_k$ found previously. This means that

$$\mathbf{q}_m = r_{mm}\mathrm{P}_\perp(\mathbf{q}_{m-1})\mathrm{P}_\perp(\mathbf{q}_{m-2})...\mathrm{P}_\perp(\mathbf{q}_1)x_m \tag{51}$$

where $\mathrm{P}_\perp(\mathbf{V}) \equiv \mathbf{I} - \mathbf{V}(\mathbf{V}^H\mathbf{V})^{-1}\mathbf{V}^H$. Since the vectors $\mathbf{q}_1...\mathbf{q}_{m-1}$ already form an orthonormal basis, this implies that $\mathbf{q}_m$ will be orthogonal to $\mathbf{q}_k$ for $k = 1$ to $m - 1$, thereby extending the orthonormal basis to include $\mathbf{q}_m$.

Normally the QR-decomposition is performed on an augmented matrix that allows $\mathbf{R}_x$ to be updated in a sample recursive or a block recursive manner. Recursive least squares is a modification of least squares that applies an exponential forgetting factor to previous samples or blocks of data. Mathematically the statistical averages are computed using the following recursive formulas

$$\hat{\mathbf{R}}'_{xx} = \lambda\hat{\mathbf{R}}_{xx} + \frac{(1-\lambda)}{N_1}\mathbf{X}_1^H\mathbf{X}_1 \tag{52}$$

$$\hat{\mathbf{R}}'_{xd} = \lambda\hat{\mathbf{R}}_{xd} + \frac{(1-\lambda)}{N_1}\mathbf{X}_1^H\mathbf{s}_1 \tag{53}$$

where in this context $\hat{\mathbf{R}}_{xx}$, and $\hat{\mathbf{R}}_{xd}$ denote the exponential averaged statistics up to the previous time block and $\hat{\mathbf{R}}'_{xx}$, and $\hat{\mathbf{R}}'_{xd}$ denote the statistics averaged up to the current time block. The new block of data is an $N_1 \times M$ matrix denoted by $\mathbf{X}_1$. Each row of $\mathbf{X}_1$ is the Hermitian transpose of a received data vector. Similarly $\mathbf{s}_1$ is the new $N_1 \times 1$ training symbol vector containing the conjugated complex training symbols.

Recursive least squares can be conceptualized as the minimization over $\mathbf{w}$ of the weighted least squares problem

$$\mu_W \equiv (1-\lambda)\|\Lambda^{1/2}\mathbf{X}\mathbf{w} - \Lambda^{1/2}\mathbf{s}\|^2 \tag{54}$$

where the diagonal weighting matrix is defined by

$$\Lambda \equiv \frac{1}{\sqrt{N_1}}\begin{bmatrix} \lambda^L\mathbf{I} & \mathbf{0} & \mathbf{0} & \mathbf{0} \\ \mathbf{0} & \ddots & \mathbf{0} & \mathbf{0} \\ \mathbf{0} & \mathbf{0} & \lambda\mathbf{I} & \mathbf{0} \\ \mathbf{0} & \mathbf{0} & \mathbf{0} & \mathbf{I} \end{bmatrix} \tag{55}$$

and where $\mathbf{X}$ is an $N_1L \times M$ matrix consisting of L blocks of data and $\mathbf{s}$ is an $N_1L \times 1$ training symbol vector.

The parameters that need to be updated recursively are $\mathbf{R}_x$ the Cholesky factor of the autocorrelation matrix, $\mathbf{w}$ the beamforming weights, and the whitened cross correlation $\mathbf{k} \equiv \mathbf{R}_x^{-H}\mathbf{R}_{xd} = \mathbf{Q}^H\mathbf{s}$. The whitened received data vector $\mathbf{Q}_1 \equiv \mathbf{R}_x^{\prime-H}\mathbf{X}_1$, which is simply the bottom part of the updated Q-matrix $\mathbf{Q}'$, can be useful as well. If $\mathbf{QR}_x$ is the QR-decomposition of the weighted received data matrix $\Lambda^{1/2}\mathbf{X}$, then $\mathbf{Q}'\mathbf{R}'_x$ is defined to be the QR-decomposition of the matrix

$$\mathbf{X}' \equiv \begin{bmatrix} \sqrt{\lambda}\Lambda^{1/2}\mathbf{X} \\ \sqrt{\frac{(1-\lambda)}{N_1}}\mathbf{X}_1 \end{bmatrix} \tag{56}$$

which is the weighted data matrix augmented by the new data block $\mathbf{X}_1$. Similarly $\mathbf{s}$ is updated by

$$\mathbf{s}' \equiv \begin{bmatrix} \sqrt{\lambda}\Lambda^{1/2}\mathbf{s} \\ \sqrt{\frac{(1-\lambda)}{N_1}}\mathbf{s}_1 \end{bmatrix} \tag{57}$$

By the definition of the QR-decomposition and the definition of $\mathbf{k}$ one can write

$$\mathbf{Q}'^H[\mathbf{X}' \ \mathbf{s}'] = [\mathbf{R}'_x \ \mathbf{k}'] \tag{58}$$

If the old $\mathbf{Q}$ is applied to the top half of the data matrix

$$\mathbf{A} \triangleq \begin{bmatrix} \mathbf{Q} & \mathbf{0} \\ \mathbf{0}^H & 1 \end{bmatrix}^H \begin{bmatrix} \sqrt{\lambda}\mathbf{\Lambda}^{1/2}\mathbf{X} & \sqrt{\lambda}\mathbf{\Lambda}^{1/2}\mathbf{s} \\ \sqrt{\frac{(1-\lambda)}{N_1}}\mathbf{X}_1 & \sqrt{\frac{(1-\lambda)}{N_1}}\mathbf{s}_1 \end{bmatrix} \tag{59}$$

it yields

$$\mathbf{A} = \begin{bmatrix} \sqrt{\lambda}\mathbf{R}_x & \sqrt{\lambda}\mathbf{k} \\ \sqrt{\frac{(1-\lambda)}{N_1}}\mathbf{X}_1 & \sqrt{\frac{(1-\lambda)}{N_1}}\mathbf{s}_1 \end{bmatrix} \tag{60}$$

The new QR-decomposition can now be achieved by finding the QR-decomposition of the $(M + N_1) \times M$ matrix

$$\mathbf{A}_1 \triangleq \begin{bmatrix} \sqrt{\lambda}\mathbf{R}_x \\ \sqrt{\frac{(1-\lambda)}{N_1}}\mathbf{X}_1 \end{bmatrix} \tag{61}$$

so that

$$\tilde{\mathbf{Q}}\mathbf{R}'_x = \mathbf{A}_1 \tag{62}$$

The Cholesky factor obtained in this way must be the required updated Cholesky factor by the uniqueness of the QR-decomposition. This is because an orthonormal matrix

$$\mathbf{Q}' \equiv \begin{bmatrix} \mathbf{Q} & \mathbf{0} \\ \mathbf{0}^H & 1 \end{bmatrix}\tilde{\mathbf{Q}}$$

has been exhibited such that

$$\mathbf{X}' = \mathbf{Q}'\mathbf{R}'_x \tag{63}$$

Equations (58), (59) and (63) also reveal that

$$\tilde{\mathbf{Q}}^H\mathbf{A} = [\mathbf{R}'_x \ \ \mathbf{k}'] \tag{64}$$

Thus a full update of the Cholesky factor and the weights can take place by forming the matrix $\mathbf{A}$ and performing a QR-decomposition [29].

The QR-decomposition of $\mathbf{A}_1$ is performed in a matter that exploits the sparsity of $\mathbf{A}_1$ and permits a simultaneous application of $\tilde{\mathbf{Q}}^H$ to

$$\begin{bmatrix} \sqrt{\lambda}\mathbf{k}^H & \sqrt{\frac{(1-\lambda)}{N_1}}\mathbf{s}_1^H \end{bmatrix}^H$$

This can be performed by a modification of MGSO that fully exploits the sparsity of $\mathbf{A}$ [26], or by successively applying elementary orthonormal matrices to the left hand side of $\mathbf{A}$ so that $\sqrt{[(1-\lambda)/N_1]}\mathbf{X}_1$ is annihilated column by column. Either the Householder reflections [25] or the Givens rotations [20] are capable of performing this task. If the latter two techniques are used, $\mathbf{A}$ should be augmented with the block column matrix $[\mathbf{0} \ \ \mathbf{I}]^H$ if it is desired to retain the whitened received data $\mathbf{Q}_1$.

The updated beamformer weights are obtained from

$$\mathbf{w}' = \mathbf{R}'^{-1}_x\mathbf{k}' \tag{65}$$

For upper triangular matrices, this is performed in a straightforward manner by back-substitution. The last component of $\mathbf{w}'$, w'_M, is solved first, followed by w'_{M-1}, etc.

$$w'_m = \frac{1}{r'_{mm}}\left(k'_m - \sum_{j=m+1}^{M} r'_{mj}w'_j\right) \tag{66}$$

The optimal weights obtained from least squares processing are only optimal for the block of data for which they were trained. In general there will be a misadjustment in MSE that deviates from the minimum MSE obtainable. Equivalently the SINR at the output of the beamformer, will be smaller than the maximum obtainable SINR.

For time averaged least squares processing this misadjustment reduces to zero as the block size N approaches infinity. In this case it can be shown that [29]

$$J_{ex}(N) = \frac{M}{N-M-1}J_{\min} \tag{67}$$

when $N > M$. This should be contrasted with the LMS case in (45), which has a lower bound for its misadjustment.

The exponentially weighted least squares also has a lower bound for its misadjustment given by the approximate formula

$$J_{ex}(\infty) \approx \frac{M(1-\lambda)}{1+\lambda}J_{\min} \tag{68}$$

5.3.4 Blind Signal Adaptation

In order to obtain an optimal linear beamformer, it is necessary to train the beamforming weights using a reference signal. In many applications, however, a reference signal is either unavailable, or would consume an excessive amount of system resources. Time samples that are devoted to preambles or mid-ambles, etc. necessarily reduce the available bandwidth for carrying the communication symbol payload.

It is also possible that a receiver either does not know or cannot use a signal's training sequence due to either synchronization issues or because the actual modulation format is initially unknown.

For these situations, instead of using an actual reference signal for the purpose of least squares processing, only known properties of the signal are exploited. The more that is actually known about the signal the better the beamforming performance. Nevertheless, very weak signal properties, such as constant modulus, cyclostationary, or bursty signals are often sufficient to enable the training of good beamformers.

A very general procedure for property exploitation employs alternating directions optimization on (32) [6]. Given a signal constraint set $\mathbb{C}$, repeatedly perform the following until the weights have converged:

1. find $\mathbf{s}$ as the minimizer to

$$\mathbf{s} = \arg\min_{\mathbf{s}\in\mathbb{C}} \frac{\|\mathbf{Xw}-\mathbf{s}\|^2}{\|\mathbf{s}\|^2} \tag{69}$$

2. find the optimal $\mathbf{w}$ given $\mathbf{s}$,

$$\mathbf{w} = (\mathbf{X}^H\mathbf{X})^{-1}\mathbf{X}^H\mathbf{s} \tag{70}$$

3. return to step 1.

Recall that (70) can also be written as

$$\mathbf{w} = \mathbf{R}_x^{-1}\mathbf{Q}^H\mathbf{s} \tag{71}$$

This can be facilitated by computing the QR-decomposition of the received data matrix $\mathbf{X}$ in advance. For many signal constraint sets, we can either solve (69) exactly or in some cases obtain an approximate solution, such as finding a nearby waveform in the contraint set, or a closest constellation point. This procedure in general is referred to as property restoral and can include a large number of procedures such restoring the modulus of a signal, or demodulating the signal and then remodulating it.

A block diagram of a generic blind beamformer is shown in Figure 5.5. The reference signal is determined from the output of a non-linear property restoral step that approximates a solution to (69). In some cases it is appropriate to make more than one pass through the data in accordance with the alternating direction optimization suggested in (69) and (70).

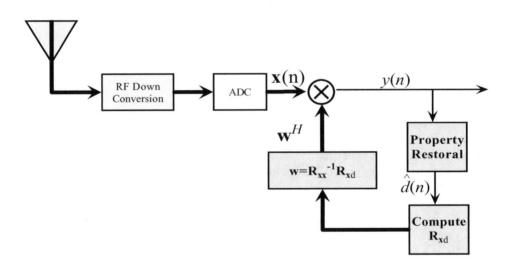

Figure 5.5   Generic processor for a blind beamformer

5.3.4.1 Constant Modulus Algorithms

The situation where the signal constraint set is the set of constant modulus signals encompasses a wide variety of important modulation types. Some of these signal types include, FM, CDMA, PSK, MSK, and FSK. Constant modulus algorithms have even been used to successfully beamform QAM signals. Although there are gradient descent forms of the constant modulus algorithm, they converge slowly and are subject to misadjustment.

A least squares version of the constant modulus algorithm can be derived from (69) as follows [1]

$$\hat{d}(n) = \arg \min_{|d(n)|=1} \tilde{\mu}(\mathbf{w}, \mathbf{s}, \mathbf{X})$$

$$\tilde{\mu}(\mathbf{w}, \mathbf{s}, \mathbf{X}) \equiv \frac{1}{N} \sum_n |\mathbf{w}^H x(n) - d(n)|^2 \tag{72}$$

Using the fact that $|d(n)| = 1$ one can write the NMSE in (72) as

$$\tilde{\mu}(\mathbf{w}, \mathbf{s}, \mathbf{X}) = \frac{1}{N} \sum_n \left\{ |y(n)|^2 - 2\mathrm{Re}(d^*(n)y(n)) + 1 \right\} \tag{73}$$

where $y(n) \equiv \mathbf{w}^H x(n)$. The optimal constant modulus $d(n)$ that minimizes $\tilde{\mu}(\mathbf{w}, \mathbf{s}, \mathbf{X})$ can be written as

$$\hat{d}(n) = \frac{y(n)}{|y(n)|} \tag{74}$$

The result in (74) coupled with (70) constitutes the single target constant modulus algorithm. These equations can be performed iteratively over the same data block, or can be performed recursively from one data block to the next using the results of Section 5.3.3. In the latter case one can consider setting λ smaller at the beginning, and gradually increasing λ towards unity as the algorithm converges.

If there are multiple constant modulus signals in the environment, the single target algorithm will typically lock onto the signal with the largest received SINR. At this point, in theory one could subtract this signal from the received data and apply the algorithm to the altered data. This constitutes a projection onto a subspace orthogonal to the signals copied so far. It is possible to apply this change directly to the weights which are of dimension $M \times 1$ as opposed to the full dimensionality of the received data matrix $M \times N$. However it is appropriate to relax a strict orthogonality constraint once a new weight vector begins to converge in response to a different signal. Such a procedure is referred to as a soft orthogonality constraint, and if set up correctly it can permit the simultaneous copy of all constant modulus signals received at the array [1].

5.3.5 Subspace Constraints

One of the most general signal properties that can be efficiently exploited, is when a signal belongs to a known subspace. Mathematically this can be expressed as

$$\mathbf{s} = \mathbf{Tu} \tag{75}$$

For this to be effective the dimension of the linear combining vector $\mathbf{u}$, $N_u \times 1$, must be smaller than the dimension of $\mathbf{s}$, $N \times 1$. The columns of the $N \times N_u$ matrix $\mathbf{T}$ form a subspace that contain all the signals of interest. In most cases $\mathbf{T}$ is sparse or easily analyzed.

The subspace constraint encompasses signals that have finite time, or frequency support, i.e. that are zero outside a known specified region. This would include impulsive or bursty signals, band-limited signals, frequency hopped signals, TDMA signals, etc. This constraint also encompasses signals that are real (e.g. BPSK), or signals that were transmitted as a linear combination from a known basis function (e.g. OFDM or CDMA).

The subspace constraint based beamformer can be derived from (36). After some rearranging this can be written as

$$\hat{\gamma}(\mathbf{w}, \mathbf{s}, \mathbf{X}) = \frac{\mathbf{s}^H P(\mathbf{Xw})\mathbf{s}}{\mathbf{s}^H P_\perp (\mathbf{Xw})\mathbf{s}} \qquad (76)$$

Substitution of (75) into (76) yields

$$\hat{\gamma}(\mathbf{w}, \mathbf{u}, \mathbf{X}) = \frac{\mathbf{u}^H \mathbf{T}^H P(\mathbf{Xw})\mathbf{Tu}}{\mathbf{u}^H \mathbf{T}^H P_\perp (\mathbf{Xw})\mathbf{Tu}} \qquad (77)$$

Because the numerator of the Rayleigh quotient in (77) is a rank 1 matrix, the eigenvector $\mathbf{u}$ that maximizes (77) can be written as

$$\mathbf{u} = (\mathbf{T}^H \mathbf{T})^{-1} \mathbf{T}^H \mathbf{Xw} \qquad (78)$$

This implies that the optimal SOI waveform is found from

$$\hat{\mathbf{s}} = P(\mathbf{T})\mathbf{Xw} \qquad (79)$$

where $P(\mathbf{T}) \equiv \mathbf{T}(\mathbf{T}^H \mathbf{T})^{-1} \mathbf{T}^H$.

Substitution of this into (36) yields

$$\hat{\gamma}(\mathbf{w}, \mathbf{X}) = \frac{\mathbf{w}^H \mathbf{X}^H P(\mathbf{T})\mathbf{Xw}}{\mathbf{w}^H \mathbf{X}^H P_\perp (\mathbf{T})\mathbf{Xw}} \qquad (80)$$

where $P_\perp (\mathbf{T}) \equiv \mathbf{I} - P(\mathbf{T})$. Note that the SINR estimate in (80) is asymptotically biased in that $\hat{\gamma}(\mathbf{w}, \mathbf{X}) \rightarrow \text{SINR} + 1$ as the collect time goes to infinity.

The numerator of the Rayleigh quotient in (80) contains the matrix $\mathbf{X}^H P(\mathbf{T})\mathbf{X}$ which is an autocorrelation matrix that contains the SOI when it is on, or present in the environment, whereas the denominator matrix $\mathbf{X}^H P_\perp (\mathbf{T})\mathbf{X}$ is the data autocorrelation when the signal is off or not present. This interpretation is exact when the SOI is confined to a known subset of data samples (in time or frequency), and the background interference is present in the remaining samples.

Following the aforementioned interpretation, (80) can be written as the following eigenvector problem [2]

$$\mathbf{X}_{\text{on}}^H \mathbf{X}_{\text{on}} \mathbf{w} = (\gamma + 1) \mathbf{X}_{\text{off}}^H \mathbf{X}_{\text{off}} \mathbf{w} \qquad (81)$$

where $\mathbf{X}_{\text{on}} \triangleq P(\mathbf{T})\mathbf{X}$ and $\mathbf{X}_{\text{off}} \triangleq P_\perp (\mathbf{T})\mathbf{X}$. As the collect time, N, goes to infinity, the following approximations hold

$$\mathbf{X}_{\text{off}}^H \mathbf{X}_{\text{off}} \approx N_{\text{off}} \mathbf{R}_{\mathbf{ii}}$$

$$\mathbf{X}_{\text{on}}^H \mathbf{X}_{\text{on}} \approx N_{\text{on}} \left(\mathbf{R}_{\mathbf{ii}} + \mathbf{aa}^H R_{dd} \right) \qquad (82)$$

where $\mathbf{R}_{\mathbf{ii}}$ is the autocovariance of the background interference and R_{dd} is the autocovariance of the SOI symbol vector.

The approximation in (82) leads to the following relationships

$$\frac{N_{\text{off}}}{N_{\text{on}}} \mathbf{R}_{\text{off}}^{-H} \mathbf{R}_{\text{on}}^{H} \mathbf{R}_{\text{on}} \mathbf{R}_{\text{off}}^{-1} - \mathbf{I} \approx R_{dd} \mathbf{R}_{\text{off}}^{-H} \mathbf{a} \mathbf{a}^{H} \mathbf{R}_{\text{off}}^{-1} \tag{83}$$

$$\left(\frac{N_{\text{off}}}{N_{\text{on}}} \mathbf{R}_{\text{off}}^{-H} \mathbf{R}_{\text{on}}^{H} \mathbf{R}_{\text{on}} \mathbf{R}_{\text{off}}^{-1} - \mathbf{I} \right) \mathbf{v} = \gamma \mathbf{v} \tag{84}$$

where $\mathbf{R}_{\text{on}}$ is the Cholesky factor, the upper triangular matrix from the QR-decomposition of $\mathbf{X}_{\text{on}}$, $\mathbf{R}_{\text{off}}$ is the Cholesky factor associated with $\mathbf{X}_{\text{off}}$ and $\mathbf{v} = \mathbf{R}_{\text{off}} \mathbf{w}$.

The matrix

$$\mathbf{U} \triangleq \frac{N_{\text{off}}}{N_{\text{on}}} \mathbf{R}_{\text{off}}^{-H} \mathbf{R}_{\text{on}}^{H} \mathbf{R}_{\text{on}} \mathbf{R}_{\text{off}}^{-1} - \mathbf{I}$$

from the left hand side of (83) is approximately rank 1. This means that a simple power method technique will rapidly find its largest eigenvalue and associated eigenvector. The DSP can implement the power method by applying $\mathbf{U}$ multiple times to a randomly chosen initial $\mathbf{v}$. A procedure that applies $\mathbf{U}$ three times to $\mathbf{v}$ might look like

1. $\mathbf{v} = [1 \ 0 \ 0 \ \dots \ 0]^{T}$
2. $\mathbf{R}_{u} = \mathbf{R}_{\text{on}} \mathbf{R}_{\text{off}}^{-1}$
3. $\mathbf{z} = \mathbf{R}_{u} \mathbf{v}$
4. $\mathbf{v} = (N_{\text{off}}/N_{\text{on}}) \mathbf{R}_{u}^{H} \mathbf{z} - \mathbf{v}$
5. $\mathbf{v} = \mathbf{v}/\| \mathbf{v} \|$
6. While loop count < 4, return to step 3
7. $\mathbf{w} = \mathbf{R}_{\text{off}}^{-1} \mathbf{v}$.

5.3.6 Exploiting Cyclostationarity

There is a class of beamforming algorithms that exploits more subtle statistical properties of communications waveforms. In particular many bauded communication signals exhibit a correlation with copies of the original signal that have been frequency shifted by multiples of $1/T$ where T is the baud period. Signals that have this property are said to exhibit cyclostationarity [17]. More generally a cyclostationary signal is a signal whose autocorrelation function is periodic in time with period T. A signal that exhibits cyclostationarity has a component signal that is cyclostationary. Thus a signal that exhibits cyclostationarity will have a non-zero correlation with frequency shifted and possibly time shifted versions of itself. This correlation is referred to as the cyclic spectrum.

To see how this arises, consider the communication waveform represented by

$$d(t) = \sum_{k=-N}^{N} c_{k} p(t - kT) \tag{85}$$

where $p(t)$ is a finite duration pulse and c_{k} is a complex symbol transmitted at time kT. Taking the Fourier transform of both sides of (85) yields

$$D(f) = \sum_{k=-N}^{N} c_{k} \exp(-2\pi jfkT) P(f) \tag{86}$$

$$\equiv C(f)P(f) \tag{87}$$

where $P(f)$ is the Fourier transform of $p(t)$, and $C(f) \triangleq \sum_{k=-N}^{N} c_k \exp(-2\pi jfkT)$.

Cyclostationarity occurs due to the fact that $C(f - m/T) = C(f)$ for all m. From this one can write

$$\int_{-\infty}^{\infty} D(f)D^*\left(f - \frac{m}{T}\right)df = \int_{-\infty}^{\infty} |C(f)|^2 P(f)P\left(f - \frac{m}{T}\right)^* df \tag{88}$$

From Parseval's relation, we have

$$\langle d(t)d^*(t)\exp(-2\pi jmF_b t)\rangle_t = \frac{F_b}{2N+1}\int_{-\infty}^{\infty} d(t)d^*(t)\exp(-2\pi jmF_b t)dt \tag{89}$$

$$= \int_{-\infty}^{\infty} \frac{F_b}{2N+1}|C(f)|^2 P(f)P(f - mF_b)^* df$$

where $F_b \triangleq 1/T$ is defined to be the baud frequency. If we take the statistical expectation of both sides of (89) and if we assume the complex symbol sequence c_k is white, with unit power, then we have

$$R_{dd}^{F_b} \triangleq E\big(d(t)d^*(t)\exp(-2\pi jmF_b t)\big)$$

$$= F_b \int_{-\infty}^{\infty} P(f)P(f - mF_b)^* df \tag{90}$$

Thus the degree of correlation depends on the spectral efficiency of the pulse waveform. The spectral efficiency is a measure of how much spectral energy $P(f)$ has outside frequency band $f \in [-F_b/2, F_b/2]$. Since no pulse shaping is perfect, there are usually some non-zero cyclic spectral components.

To derive beamforming weights that can exploit this correlation one can attempt to maximize the time averaged cross-correlation function at cycle frequency α,

$$\rho_c(\alpha, \mathbf{w}, \mathbf{v}) \triangleq \frac{|\langle \hat{d}(T_s n)\tilde{d}^*(T_s n)\rangle_n|}{\sqrt{\langle|\hat{d}(T_s n)|^2\rangle_n \langle|\tilde{d}(T_s n)|^2\rangle_n}} \tag{91}$$

where T_s is the sampling period of the receiver, $\hat{d}(T_s n) \triangleq \mathbf{w}^H \mathbf{x}(n)$ is a symbol estimate at the output of the beamformer and $\tilde{d}(T_s n) \triangleq \mathbf{v}^H \mathbf{x}(n)\exp(2\pi j\alpha n)$ is a frequency shifted version of the symbol waveform estimated by the application of another beamforming vector $\mathbf{v}$.

This is a deviation from the maximum likelihood approach recommended so far, but it is closely related. The correlation coefficient $\rho_c(\alpha, \mathbf{w}, \mathbf{v})$ is positive and less than one. It achieves unity only if $\hat{d}(Tsn)$ is perfectly correlated with $\tilde{d}(Tsn)$, which has been frequency shifted by α. It is similar in form to the NMSE and SINR objective functions described earlier in (32) and (36).

Substituting $\mathbf{w}^H \mathbf{x}(n)$ for $\hat{d}(T_s n)$, $\mathbf{v}^H \mathbf{x}(n)\exp(2\pi j\alpha n)$ for $\tilde{d}(T_s n)$ in (91) and maximizing over $\mathbf{v}$ yields

$$\rho_c^2(F_b, \mathbf{w}) = \frac{\mathbf{w}^H \mathbf{R}_{xx}^{\alpha} \mathbf{R}_{xx}^{-1} \mathbf{R}_{xx}^{\alpha H} \mathbf{w}}{\mathbf{w}^H \mathbf{R}_{xx} \mathbf{w}} \tag{92}$$

where

$$\mathbf{R}_{xx}^{\alpha} \triangleq \langle \mathbf{x}(n)\mathbf{x}(n)^{H} \exp(-2\pi j \alpha n) \rangle_{n} \tag{93}$$

$\mathbf{R}_{xx} \equiv \langle \mathbf{x}(n)\mathbf{x}(n)^{H} \rangle_{n}$ and the optimal $\mathbf{v}$ is given by

$$\hat{\mathbf{v}} = \mathbf{R}_{xx}^{-1}\mathbf{R}_{xx}^{\alpha H}\mathbf{w}. \tag{94}$$

One could also have maximized initially over $\mathbf{w}$ to obtain

$$\hat{\mathbf{w}} = \mathbf{R}_{xx}^{-1}\mathbf{R}_{xx}^{\alpha}\mathbf{v} \tag{95}$$

Typically $\rho_{c}(\alpha, \mathbf{w})$ would be optimized over both α and $\mathbf{w}$ and would obtain a maximum at $\alpha = T_{s}F_{b}$ the baud rate. From the Rayleigh quotient in (92) a necessary condition for optimality is that $\mathbf{w}$ is an eigenvector in the following eigenvalue problem

$$\mathbf{R}_{xx}^{\alpha}\mathbf{R}_{xx}^{-1}\mathbf{R}_{xx}^{\alpha H}\mathbf{w} = \lambda\mathbf{R}_{xx}\mathbf{w} \tag{96}$$

The solution that yields the largest λ is chosen. The optimization can be efficiently obtained by alternating directions, solving (94) and (95) repeatedly.

This technique is called Self-Coherence Restoral in [3] or SCORE. The SCORE technique is a very general technique that works for a large class of communication waveforms. Moreover it does not necessarily require baud synchronization or an exact carrier synchronization. Its disadvantage is that it may require a large number of time samples to converge, especially for spectrally efficient pulse shaping. For these reasons it is highly attractive as an initial step, both to help in the synchronization process of determining an unknown baud rate and to provide an initial set of beamforming weights.

5.3.7 Transmit Beamformer Techniques

For many cellular applications it is impractical to put a lot of antennas in the remote units. This is because there may be thousands of remotes in each cell making it critical to minimize the remotes' cost and complexity. Moreover the remote power consumption is typically severely limited due both to FCC safety regulations and the practical demands of a portable unit.

The base station on the other hand may realistically be able to support many antennas since cost and complexity are amortized by the number of users serviced by each base station. Also the power constraints on the base station are much less severe than for the remotes.

Therefore an important and practical problem is how to determine complex gains for each antenna element that can be used to transmit information from the base station to the remote. Transmit beamforming can offer similar performance enhancements as receive beamforming. The difficulty lies in the fact that in the most general case, insufficient information is available at the transmitter to adapt its weights.

In networks that contain more than one cell, or in an ad-hoc network, the transmit beamforming problem becomes more complex, because all the transmitters that are in any given receivers field of view must cooperate so as not to cause co-channel interference that a receiver cannot remove. For this reason we must consider the effects of other transceivers in the network for an adequate analysis.

Divide all the transceivers in a network into two groups for a given narrow band channel:

those that are transmitting and those that are receiving. For the uplink transmission case, the users in group 1 are transmitting and the users in group 2 are receiving. This can be described by the following signal model

$$\mathbf{x}_2(n; q) = \mathbf{i}_2(n; q) + \mathbf{H}_{21}(q, q)\mathbf{s}_1(n; q) \tag{97}$$

$$\mathbf{i}_2(n; q) = \boldsymbol{\varepsilon}_2(n; q) + \sum_{p \neq q} \mathbf{H}_{21}(q, p)\mathbf{s}_1(n; p) \tag{98}$$

where q is a link number from 1 to Q links that is associated with a communication channel between a network node in group 1, $\ell_1(q)$, and a network node in group 2, $\ell_2(q)$, $\mathbf{x}_2(n; q)$ is the $M_2 \times 1$ received data vector at time sample n and node $\ell_2(q)$ and where M_1 is the number of antenna elements used by the transceivers in group 1 and M_2 is the number of antenna elements used by the transceivers in group 2. The $M_2 \times M_1$ matrix $\mathbf{H}_{21}(q, p)$ represents the channel response from node $\ell_1(p)$ to node $\ell_2(q)$, $\mathbf{i}_2(n; q)$ is the interference vector due to all the other emitters in the network and $\boldsymbol{\varepsilon}_2(n; q)$ is due to the background and receiver noise seen at node $\ell_2(q)$. The transmitted waveform at node $\ell_1(q)$, $\mathbf{s}_1(n; p)$ is given by the equation

$$\mathbf{s}_1(n; q) = \sum_{k=1}^{M_c} \lambda_1(k, q)\mathbf{g}_1(k, q) \, d_1(n; k, q) = \mathbf{G}_1(q)\boldsymbol{\Lambda}_1(q)\mathbf{d}_1(n; q) \tag{99}$$

where $\mathbf{d}_1(n; q)$ is an $M_c \times 1$ vector of independent, unit-power, uplink source data transmitted over link q and time slot n, $\mathbf{G}_1(q)$ is an $M_1 \times M_c$ matrix of transmit weights, and $\boldsymbol{\Lambda}(q)$ is an $M_c \times M_c$ diagonal matrix with diagonal entries $\lambda_1(k, q)$ representing the transmit gains for each link q and transmission mode k, and where $M_c \leq \text{rank } \{\mathbf{H}_{21}(q, q)\}$.

This model permits a node to transmit over multiple transmit beams (up to M_c). This can be advantageous when the channel matrix $\mathbf{H}_{21}(q, q)$ is full rank due to multipath, polarization diversity or because the channel model incorporates multiple independent channels, which would be represented as block diagonal terms within $\mathbf{H}_{21}(q, q)$.

The downlink transmission case can be described by (97) after the group indices 1 and 2 and have been swapped in all the subscripts.

A useful tool for obtaining transmit beamformer weights is to optimize a network performance criterion that can guarantee all the links in the network meet some quality objective. Since the information-theoretic channel capacity represents a bound on the maximum bit rate achievable in a given channel it is attractive as a network performance criterion.

Associated with every transmit weight vector $\mathbf{g}_t(k, q)$ is a corresponding receive weight vector $\mathbf{w}_r(k, q)$, where the index pair $(t, r) = (1, 2)$, if group 1 is transmitting and $(t, r) = (2, 1)$ if group 2 is transmitting. Therefore after the application of the receive weights one has

$$\hat{d}_t(n; k, q) = \mathbf{w}_r^H(k, q)\tilde{\mathbf{i}}_r(n; q) + \mathbf{w}_r^H(k, q)\mathbf{H}_{rt}(q, q)\lambda_t(k, q)\mathbf{g}_t(k, q)d_t(n; k, q) \tag{100}$$

where

$$\tilde{\mathbf{i}}_r(n; q) \triangleq \mathbf{i}_r(n; q) + \sum_{m \neq k} \mathbf{H}_{rt}(q, q)\lambda_t(m, q)\mathbf{g}_t(m, q)d_t(n; m, q) \tag{101}$$

and $\hat{d}_t(n; k, q)$ is an estimate of the transmitted symbol $d_t(n; k, q)$. Assuming that the co-channel interference produced by the other transmitters in the network, $\tilde{\mathbf{i}}_r(n; q)$ is a complex, circularly symmetric Gaussian random vector, then $\mathbf{w}_r^H(k, q)\tilde{\mathbf{i}}_r(n; q)$ is Gaussian and the optimal source statistics are Gaussian. As a result of this, the channel capacity of the channel model suggested by (100), which maximizes the mutual information $I(\hat{d}_t(n; k, q); d_t(n; k, q))$, is given by

$$D_{rt}(k, q) \equiv \log(1 + \gamma_r(k, q)) \tag{102}$$

where $\gamma_r(k, q)$ is the SINR observed for transmission mode k at the receive end of link q. This SINR is seen to be

$$\gamma_r(k, q) \triangleq \frac{\left| \mathbf{w}_r^H(k, q) \mathbf{H}_{rt}(q, q) \mathbf{g}_t(k, q) \right|^2 \pi_t(k, q)}{\mathbf{w}_r^H(k, q) \mathbf{R}_{\tilde{\mathbf{i}}_r \tilde{\mathbf{i}}_r}(k, q)} \mathbf{w}_r^H(k, q) \tag{103}$$

where the interference covariance statistics are given by

$$\mathbf{R}_{\tilde{\mathbf{i}}_r \tilde{\mathbf{i}}_r}(k, q) \sum_{(m,p) \neq (k,q)} \pi_t(m, p) \mathbf{H}_{rt}(q, p) \mathbf{g}_t(m, p) \mathbf{g}_t^H(m, p) \mathbf{H}_{rt}^H(q, p) + \mathbf{R}_{\varepsilon_r \varepsilon_r}(q) \tag{104}$$

$\pi_t(m, p) \triangleq \lambda_t^2(m, p)$, $\mathbf{R}_{\varepsilon_r \varepsilon_r}(q)$ is the background noise covariance matrix and where the source symbols $d_t(n; m, p)$ are assumed to be mutually independent with zero mean and unit variance.

The decoupled capacity for link q is defined to be the sum of the capacities over all the transmission modes

$$D_{rt}(q) \equiv \sum_k \log(1 + \gamma_r(k, q)) \tag{105}$$

Since $D_{rt}(k, q)$ assumes all other transmission modes are interference one would expect $D_{rt}(q)$ to be smaller than the maximal joint mutual information $I(\mathbf{x}_r(q); \mathbf{d}_t(q))$. It can be shown, however, that there exist transmit weights $\mathbf{g}_t(k, q)$ and receive weights $\mathbf{w}_r(k, q)$ which allow $D_{rt}(k, q)$ to acheive this upper bound. It follows that little is lost by assuming the decoupled capacity as an objective function for a given link.

To formulate an objective function suitable for the entire network, it is desirable to be able to provide a minimal QoS for the entire network. An attempt is made therefore, to maximize the worst case capacity. This is achieved by the *decoupled network capacity* metric given by

$$D_{rt}(\mathbf{W}_r, \mathbf{G}_t) \triangleq \max_{\pi_t(k,q)} \min_q D_{rt}(q) \tag{106}$$

where the dependency on all the transmit weights $\{\mathbf{g}_t(k, q)\} \leftrightarrow \mathbf{G}_t$ and all the receive weights $\{\mathbf{w}_r(k, q)\} \leftrightarrow \mathbf{W}_r$ is explicitly shown.

To simplify the analysis of the decoupled network capacity, assume that the network is interference limited so that the background noise covariance is negligible compared to the interference covariance, permitting the approximation $\mathbf{R}_{\varepsilon_r \varepsilon_r}(q) \approx \mathbf{0}$. The max-min formulation of the decoupled network capacity metric has an optimal solution, wherein all of the decoupled capacities at each link q are equal after maximization over the transmit powers $\pi_t(k, q)$.

To see this, assume that a solution with optimal transmit powers requires a link q with a decoupled capacity that is larger than the maximum worst case value of the decoupled network capacity, D_{rt}^+ (it certainly cannot be smaller). Replace the transmit powers for that link with a scaled version that is slightly smaller, $\pi_t(k, q) \leftarrow \varepsilon \pi_t(k, q)$, for all k, and $0 < \varepsilon < 1$. Decrease ε until $D_{rt}(q)$ is equal to the previous D_{rt}^+. The performance of all the other links in the network must either improve or stay the same since they are seeing reduced co-channel interference. The decoupled network capacity, therefore, remains unchanged. This contradicts the assumption that link q must have a capacity that is larger than the optimum. It follows that there is an optimal solution with all the decoupled capacities at the same value. If the optimization occurs in a channel that does not have independent or isolated subchannels, so that all $\mathbf{w}_r(k,q)^H \mathbf{H}_{rt}(q,p)\mathbf{g}_t(k,p)$ are non-zero, then equal decoupled capacities is the only possibility.

It is instructive to reformulate the network metrics in terms of a transfer power matrix. Define the $M_cQ \times M_cQ$ matrix $\mathbf{P}_{rt}$ by first defining

$$\mathbf{P}_{rt}(k,q,m,p) \triangleq \left|\mathbf{w}_r^H(\text{k},\text{q})\mathbf{H}_{rt}(\text{q},\text{p})\mathbf{g}_t(\text{m},\text{p})\right|^2 \tag{107}$$

The matrix $\mathbf{P}_{rt}$ is formed from $\mathbf{P}(k,q,m,p)$ by flattening the index pair (k,q) into a single row index of length M_cQ, and flattening the index pair (m,p) into a single column index of the same length. Flatten also the output SINRs $\gamma_r(k,q)$ in a similar fashion into a vector $\boldsymbol{\gamma}_r$ and the transmit powers $\pi_t(k,q)$ into a vector $\boldsymbol{\pi}_t$. This permits (103) with (104) to be written as

$$\delta(\boldsymbol{\gamma}_r)(\mathbf{P}_{rt} - \delta(\mathbf{P}_{rt}))\boldsymbol{\pi}_t = \delta(\mathbf{P}_{rt})\boldsymbol{\pi}_t \tag{108}$$

after both sides are multiplied by the denominator. The operator $\delta(\mathbf{v})$ applied to a vector $\mathbf{v}$, indicates that the elements of the vector are placed in the diagonal of a square diagonal matrix of appropriate size. For a matrix, the convention is adopted that $\delta(\mathbf{M})$ is a diagonal matrix that has the same diagonal elements as the matrix $\mathbf{M}$.

In the case of a single transmit mode, $M_c = 1$, one can write

$$D_{rt}(\mathbf{W}_r, \mathbf{G}_t) = \log(1 + \gamma_r) \tag{109}$$

where γ_r is found from (108) as a solution to the eigenvalue problem

$$\left(\delta(\mathbf{P}_{rt})^{-1}\mathbf{P}_{rt} - \mathbf{I}\right)\boldsymbol{\pi}_t = \frac{1}{\gamma_r}\boldsymbol{\pi}_t \tag{110}$$

Therefore $1/\gamma_r$ is the Perron root of the non-negative matrix $\delta(\mathbf{P}_{rt})^{-1}\mathbf{P}_{rt} - \mathbf{I}$, [24], and can be written as

$$\gamma_r = \frac{1}{\rho\left(\delta(\mathbf{P}_{rt})^{-1}\mathbf{P}_{rt}\right) - 1} \tag{111}$$

The transmit power vector, $\boldsymbol{\pi}_t$ is the Perron vector of $\delta(\mathbf{P}_{rt})^{-1}\mathbf{P}_{rt} - \mathbf{I}$.

In theory, to obtain transmit weights one would attempt to optimize (106) with respect to the transmit weight vectors $\mathbf{g}_t(k,q)$. This initially appears to be a daunting task, because every $\mathbf{g}_t(k,q)$ can potentially affect the interference in the denominator of $\gamma_r(m,p)$ in every capacity term $D_r(m,p)$. In [8] this problem is overcome in networks that possess channel reciprocity, i.e. wherein one can assume that $\mathbf{H}_{12}(q,p) = \mathbf{H}_{21}^T(p,q)$. Time Division Duplex (TDD) networks can have this property, because both transmission and reception occur over the same frequency channels, but are multiplexed in time.

If channel reciprocity is in effect then it can be shown under a wide variety of conditions [7,8] that the downlink decoupled network capacity equals the uplink decoupled network capacity

$$D_{12}(\mathbf{W}_1, \mathbf{G}_2) = D_{21}(\mathbf{G}_2^*, \mathbf{W}_1^*). \tag{112}$$

This suggests that if the transmit weights at a given receiver are set to a scalar multiple of the conjugate of its optimal receive weights, those transmit weights will be optimal for the entire network.

$$\mathbf{g}_t(k, q) = \mathbf{w}_t^*(k, q). \tag{113}$$

The advantage of this, is that the receive weights can be easily optimized at each receiver using only local information. In fact it is clear from (103) and (105) that the optimal $\mathbf{w}_r(k, q)$ is obtained from maximizing the output SINR. This can be done using any of the linear beamforming techniques suggested so far and clearly only require an estimate of the received data covariance matrix and a training signal (or a property to exploit).

Optimization of the transmit gains is performed numerically by solving the following problem, [8], for each link q in the network

$$\min \sum_k \pi_t(k, q) \ s.t. \ D_{rt}(q) = \beta \tag{114}$$

If the denominator of (103) is assumed nearly constant then $\gamma_r(k, q)$ can be approximated by

$$\gamma_r(k, q) \approx \frac{\left| \mathbf{w}_r^H(k, q) \mathbf{H}_{rt}(q, q) \mathbf{g}_t(k, q) \right|^2 \pi_t(k, q)}{\hat{R}_r(k, q)} \tag{115}$$

where $\hat{R}_r(k, q)$ is the estimated, post beamforming, interference power over transmit mode k and link q. This approximation permits (114) to be solved in closed form using water-filling arguments [13]. The network is then optimized by maximizing β subject to a total transmit power constraint. In practice this means increasing β until a node is at its maximum allowable power. This approximation for the interference power tends to hold because the receive weights put nulls in the direction of interfering emitters, making $\hat{R}_r(k, q)$ insensitive to changes in the transmit powers of the interfering emitters. For the case where one can assume channel reciprocity both $\hat{R}_r(k, q)$ and $\mathbf{w}_r^H(k, q) \mathbf{H}_{rt}(q, q) \mathbf{g}_t(k, q)$ are easily estimated at the receiver, without any additional estimation of the channels in the network. This makes it amenable to real time DSP implementation at each transceiver.

In [41], the case where $M_c = 1$ is addressed for an objective function that minimizes the total transmitted power subject to an SINR constraint. There it is shown that the procedure in (114) for this special case converges to a local minimum and in the one cell case to a global minimum. It is also shown in [41] that the objective functions for the uplink and downlink are the same when the transmit weights are set equal to the conjugate of the receive weights.

It is not difficult to see that (112) holds in the case where the background noise covariance is neglected. Indeed if the channel matrices are reciprocal and (113) holds, then the transfer matrix will be symmetric

$$\mathbf{P}_{12} = \mathbf{P}_{21}^T \tag{116}$$

For the $M_c = 1$ case this immediately implies (112). Using the fact that the spectral radius is invariant with respect to the transpose operation as well as to changing the order of a matrix multiplication, one can write

$$\rho\left(\delta(\mathbf{P}_{12})^{-1}\mathbf{P}_{12}\right) = \rho\left(\delta(\mathbf{P}_{21})^{-1}\mathbf{P}_{21}^T\right)$$

$$= \rho\left(\mathbf{P}_{21}^T\delta(\mathbf{P}_{21})^{-1}\right)$$

$$= \rho\left(\delta(\mathbf{P}_{21})^{-1}\mathbf{P}_{21}\right).$$

This fact then leads to

$$\gamma_1 = \frac{1}{\rho\left(\delta(\mathbf{P}_{12})^{-1}\mathbf{P}_{12}\right) - 1}$$

$$= \frac{1}{\rho\left(\delta(\mathbf{P}_{21})^{-1}\mathbf{P}_{21}\right) - 1}$$

$$= \gamma_2$$

which implies (112) because of (109).

For the general case where $M_c > 1$, one notes that a necessary condition for $\gamma \equiv \gamma_r$ to be a solution of the eigen-equation (108) is that

$$\left|\left(\mathbf{I} + \delta(\gamma)^{-1}\right)^{-1}\delta(\mathbf{P}_{21})^{-1}\mathbf{P}_{21} - \mathbf{I}\right| = 0 \qquad (117)$$

The network capacity objective function for the uplink can therefore be written as

$$D_{21} \triangleq \max_{\gamma} \min_{m} \sum_{q \in U(m)} \sum_{k} \log(1 + \gamma(k, q)), \quad \text{such that} \qquad (118)$$

γ satisfies (117) and $\gamma \geq 0$ where $\gamma = [\gamma(1,1), \gamma(2,1), ..., \gamma(M_c, 1), \gamma(1,2), ..., \gamma(M_c, Q)]^T$.

The previous characterization can now be repeated for D_{12}. In (117) replace $\mathbf{P}_{21}$ with $\mathbf{P}_{12}$ to establish constraints on the output SINRs. It is not hard to see that the constraints are identical when the channels are reciprocal so that $\mathbf{P}_{21}^T = \mathbf{P}_{21}$. It follows that $D_{12} = D_{21}$, which establishes (112).

In many cases, however, the channels are not reciprocal, such as in Frequency Division Duplex (FDD), where each transceiver transmits and receives in possibly widely separated frequency bands. The concept of reciprocity can still play a role in helping to determine optimal transmit weights however. If downlink transmission weights are needed, one can postulate the existence of a virtual uplink channel that has reciprocal channel responses. This permits the optimal transmit weights to be found as the optimal receive weights of the virtual uplink channel [41]. The drawback of course is the need for all of the MIMO channel responses seen by each remote to be transmitted back to the base station. The burden can be reduced somewhat if the channel responses have low rank and if the channels are restricted only to those in the base stations field of view. The concept of using a feedback channel for adapting transmit weights can be found in [18] and [19].

To see how the exploitation of reciprocity can result in optimal transmit weights, consider the beampattern in Figure 5.6. If the center circle represents a base station and the SOI is at the 'x', then it is clear that the received beampattern will do a good job of enhancing the SOI SINR and removing the two interferers. If one transmits with the same weights, then one also enhances the link gain to SOI, while minimizing the interference presented to the other nodes in the network. This is precisely what is desired for a set of transmit weights.

The performance advantage obtained when transmit beamforming is exploited can be substantial. In [8] numerical experiments for 19 cell wireless networks, with eight antenna element basestations, demonstrate an increase of channel capacity approaching a factor of 30 or more over the single antenna basestation.

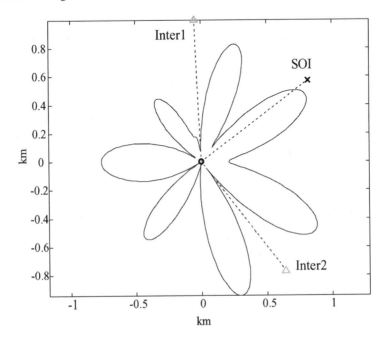

Figure 5.6   Receive and transmit beampattern

5.4 Multiple Input Multiple Output (MIMO) Signal Extraction

5.4.1 MIMO Linear System Model

In this section we consider the case of a discrete time, linearly modulated, time invariant communication system with t inputs and r outputs. Such a system can be represented in general by the linear input–output relationship

$$x = Hu + n \qquad (119)$$

where $H \in \mathbb{C}^{r \times t}$ represents the effective MIMO *channel matrix* mapping a vector of random channel inputs u of arbitrary joint distribution to the vector observation x and where $n \sim \mathcal{N}(0, R_{nn})$ represents additive Gaussian channel noise. Without loss of generality, we assume in this section that the channel input vector u is zero mean and that it may be partitioned into desired and auxiliary elements which we denote as $u = [u_d^H, u_a^H]^H$. The problem in this case is to estimate $\mathbf{u}_d$ given the observation x. This task is complicated by the fact that the desired signals may interfere with each other and that the auxiliary signals may also corrupt the observation. This general formulation allows for desired and auxiliary portions of the channel input vector to be generated by random processes with arbitrary joint distribution. The following examples illustrate the generality of this formulation.

5.4.1.1 MIMO Communication System Examples

Example 1:   Consider first the simple case of a single user with a single antenna transmitting a burst of N symbols through a multipath communications channel to a receiver also using a single antenna. The "channel" in this case represents the combined effects of the transmitter pulse shaping filter, the actual propagation channel between the transmit and receive antennas, and the receiver input filter prior to sampling. Assuming that the channel's delay spread may be upper bounded by $L < \infty$ symbol periods and that that the receiver samples the entire received signal at the baud rate, we have a MIMO system model with $t = N$ inputs and $r = N + L - 1$ outputs. The single-user channel matrix has the well known Toeplitz structure

$$H = H_{su} = \begin{bmatrix} q_0 & & & \\ q_1 & q_0 & & \\ \vdots & q_1 & & \\ q_{L-1} & \vdots & \ddots & q_0 \\ & q_{L-1} & & q_1 \\ & & & \vdots \\ & & & q_{L-1} \end{bmatrix} \quad \text{(complete observation)} \qquad (120)$$

where $\{q_n\}_{n=0}^{L-1}$ denotes the baud-rate sampled impulse response of the FIR channel.

An implicit assumption in the prior formulation is that the receiver is able to form an observation of the entire transmission. This may not be practical when N is large and is impossible for communication systems transmitting a continuous stream of data. In this case, the receiver is faced with the task of estimating the desired signal(s) with a partial observation. The MIMO model remains valid in this case but it is necessary to construct the partial-observation single user channel matrix by extracting a band of r rows from the the complete-observation single user channel matrix given in (120). The partial-observation single user channel matrix also has the Toeplitz structure and is given as

$$H = H_{su} = \begin{bmatrix} q_{L-1} & q_{L-2} & \cdots & q_0 & & \\ & q_{L-1} & q_{L-2} & \cdots & q_0 & \\ & & \ddots & & & \\ & & & q_{L-1} & q_{L-2} & \cdots & q_0 \end{bmatrix} \quad \text{(partial observation)} \quad (121)$$

In this case, the MIMO system has r outputs and $t = r + L - 1$ inputs.

Note that in the complete-observation case, it is possible for the MIMO channel matrix to have full column rank since it is a "tall" matrix. In the partial-observation case, full column rank is not possible since the matrix is not "tall". The full column rank condition has significant implications for the linear estimation techniques to be discussed later in this chapter.

Example 2: Here, we extend the first example by considering the case of multiple receive antennas. Let M_r denote the number of receive antennas. In this case, the number of observations at the output is multiplied by a factor of M_r but the number of inputs to the MIMO channel matrix remains unchanged. Assuming that the maximum delay spread over all M_r propagation channels is upper bounded by $L < \infty$, the MIMO channel matrix may be written as

$$H = \begin{bmatrix} H_{su}^1 \\ \vdots \\ H_{su}^{(M_r)} \end{bmatrix}$$

where $H_{su}^{(m)}$ is the single-user channel matrix representing the channel between desired users transmission and the mth receive antenna.

   Note that the presence of additional receive antennas may result in a MIMO channel matrix with full column rank even in the case when none of the individual subchannel matrices have full column rank. This is important for MIMO channel capacity and the existence of zero-forcing solutions for all of the channel inputs as discussed later in this section. Clearly, it is possible to select M_r large enough such that H becomes "tall" in this case. However, being "tall" is only necessary and not sufficient for having full column rank. Full column rank is achieved when there is enough subchannel disparity between the M_r receive signals to extract all t MIMO channel inputs perfectly in the noise-free case.

   We also point out that oversampling (sampling faster than the baud rate) at the receiver input has the same dimensional effect on the MIMO channel matrix as multiple receive antennas. In the context of the MIMO framework, antenna arrays and oversampling are quite similar in that they both attempt to exploit diversity in the communication system by creating multiple subchannels through which the user transmission may be observed. However, these methods differ physically in how these subchannels are created. Antenna arrays create subchannels spatially whereas oversampling creates subchannels temporally. Antenna arrays may fail to produce diversity in the case when the receive antennas are not sufficiently separated in space. Oversampling may fail to produce diversity when the transmission does not have sufficient excess bandwidth. In both cases, a failure to produce exploitable diversity occurs when the subchannels are too similar to provide any benefits to the receiver.

Example 3: Here, we extend the prior example by considering the case of multiple transmit antennas. This example applies to the case where one or more users, each with one or more transmit antennas, transmit symbols to a receiver also using one or more antennas. Each transmit antenna is assumed to have its own set of symbols of arbitrary distribution which may or may not be independent of the symbols of the other transmit antennas. We also note that, as in the prior examples, symbols may be classified as desired or auxiliary.

   Denoting the number of transmit antennas as M_t and assuming that the maximum delay spread over all $M_t M_r$ propagation channels is upper bounded by $L < \infty$, the MIMO channel matrix may be written as

$$H = \begin{bmatrix} H_{su}^{(1,1)} & \cdots & H_{su}^{(1,M_t)} \\ \vdots & & \vdots \\ H_{su}^{(M_r,1)} & \cdots & H_{su}^{(M_r,M_t)} \end{bmatrix}$$

where $\boldsymbol{H}^{(n,m)}$ is the single-user channel matrix representing the channel between the nth transmit antenna and the mth receive antenna (or oversampling phase). We note that there is no implicit assumption of synchronism among the M_t transmitters here since the MIMO communication system model includes the effects of multipath. We also note that additional transmit antennas have the effect of widening the MIMO channel matrix which may result in linearly dependent columns and no increase in the the rank of $\boldsymbol{H}$. This may affect the channel capacity calculations as well as the existence of zero-forcing solutions as discussed later in this section.

5.4.2 Capacity of MIMO Communication Channels

In this section we summarize some of the recent results in the literature regarding the capacity calculation of MIMO channels. In contrast to the MIMO signal extraction techniques discussed in the following section which specify all signal processing to be performed by the receiver, techniques used to achieve the theoretical channel capacity require participation by both the transmitter and the receiver.

The information capacity of the general MIMO channel, under the assumption of AWGN with variance σ^2, was derived in [32]. Denoting C as the information capacity, K as the rank of $\boldsymbol{H}$, $\lambda_{H,n}$ as the nth singular value of $\boldsymbol{H}$, and $\lambda_{u,n}$ as the nth singular value of the channel input vector covariance matrix, we summarize the primary theorem below.

Theorem 1 *The information capacity of the discrete time MIMO channel $\boldsymbol{H} \in \mathbb{C}^{r \times t}$ is equal to the sum of the information capacities for the K effective SISO subchannels of $\boldsymbol{H}$ corresponding to the non-zero singular values of $\boldsymbol{H}$. Specifically, the MIMO channel capacity may be written as*

$$C = \sum_{n=1}^{K} \log_2 \left(\frac{1 + \lambda_{u,n} |\lambda_{H,n}|^2}{\sigma^2} \right) \text{ bps/Hz}$$

A justification for this theorem is given below by considering a clever transmission technique that directly reveals the K independent SISO subchannels available for communication. Since each of these subchannels is unrelated to the others, the total MIMO channel capacity is then equal to the sum of the channel capacities for each of the SISO subchannels.

Denote the Singular Value Decomposition (SVD) of $\boldsymbol{H}$ as $\boldsymbol{U}_H \boldsymbol{\Lambda}_H \boldsymbol{V}_H^H$. Exposure of the K available SISO subchannels involves two steps:

1. At the transmitter, all inputs to the MIMO channel matrix must be constrained to lie in the subspace of $\mathbb{C}^t$ spanned by the first K columns of $\boldsymbol{V}_H$ corresponding to the non-zero singular values. These inputs are denoted as $\boldsymbol{u} = \boldsymbol{V}_H \boldsymbol{v}$ where the first K elements of $\boldsymbol{v}$ are non-zero and the remaining elements are all equal to zero.
2. At the receiver, the linear operator $\boldsymbol{U}_H^H$ is applied to the observation vector.

Performing these steps results in a new observation vector $\boldsymbol{y}$ given as

$$\boldsymbol{y} = \boldsymbol{U}_H^H \boldsymbol{x}$$

$$= \boldsymbol{U}_H^H \boldsymbol{U}_H \boldsymbol{\Lambda}_H \boldsymbol{V}_H^H \boldsymbol{V}_H \boldsymbol{v} + \boldsymbol{U}_H^H \boldsymbol{n}$$

$$= \boldsymbol{\Lambda}_H \boldsymbol{v} + \boldsymbol{U}_H^H \boldsymbol{n}$$

This last expression reveals the independent subchannels of the MIMO system. Since n is assumed to be white and U_H^H is an orthogonal linear transformation, then the distribution of $U_H^H n$ is also white with variance σ^2 and zero mean. Moreover, since Λ_H is diagonal, this new system clearly possesses K isolated SISO channels. The channel capacity of this system is identical to the channel capacity of the original MIMO system since the linear transformations at the transmitter and the receiver are both orthogonal.

Maximization of the channel capacity using the water-pouring algorithm under transmit power constraints is also discussed in [32]. Extensions to these results for the case when the MIMO channel itself is a random parameter are given in [33].

Another recent body of work the explores the capacity of MIMO channels is the BLAST (Bell-labs-LAyered-Space-Time) communication system. This work originated in [15] and extensions to it were presented in [12] and [27]. A fundamental contribution is a derivation of the information capacity for a particular case of MIMO channels. Under a narrowband channel assumption where there is no multipath present, the MIMO channel matrix may be written such that the ijth entry is a complex number representing the gain between the jth transmit antenna and the ith receive antenna. If the number of transmit antennas, denoted by M_t, is equal to the number of receive antennas, then the MIMO channel matrix is square. Assuming that the channels are time invariant and known, that the channels are impaired only by AWGN, and that the transmitted signal vector is composed of statistically independent equal power components, each with Gaussian distribution, the Shannon capacity of this MIMO system can be written as

$$C = \log_2\left(\det\left(I + \frac{\rho}{M_t} HH^H\right) \right) \text{ bps/Hz}$$

where ρ is the average received SNR. Numerical results have shown remarkable spectral efficiencies in uncorrelated environments. The spectral efficiency of the BLAST system is degraded when the signals arriving at the receive antennas are correlated and similar expressions for channel capacity with correlated channels is given in [27].

5.4.3 Linear Estimation of Desired Signals in MIMO Communication Systems

This section considers linear receiver-based techniques for estimating a desired signal in the context of the MIMO system model of Section 5.4. The basic MIMO model given in (119) is assumed here. All of the linear techniques considered here form an estimate of the desired signal u_d by applying a matrix or vector operator F to the observation

$$\hat{u}_d = F^H(Hu + n)$$

where $F \in \mathbb{C}^{r \times t_d}$ and where t_d is the number of desired signals requiring estimation.

5.4.3.1 Zero-Forcing Detection

The zero-forcing detector generates an estimate of the desired symbol vector by linearly canceling, or zeroing all structured interference sources in the MIMO system. Specifically, the zero forcing detector F is selected such that $F^H H = [I_{t_d}, 0_{t-t_d}]$. The zero-forcing detector is not guaranteed to exist in every MIMO system and when it does exits it is not guaranteed to be unique. The following theorem given in [9] explicitly describes a set of sufficient and necessary existence conditions for the zero-forcing detector in the context of the MIMO system model.

Theorem 2   *Denote V_d as the subspace of $\mathbb{C}^r$ spanned by the t_d columns of H corresponding to the desired input symbols. Denote V_a as the subspace of $\mathbb{C}^r$ spanned by the remaining $t - t_d$ columns of H. The zero-forcing solution exists if and only if $dim(V_d) = t_d$ and $V_d \cap V_a = 0$.*

Note that this theorem reduces to a full column rank condition on H when $t_d = t$. Full column rank is not required for the existence of the zero-forcing detector when $t_d < t$. Intuitively, this theorem states that each column of H associated with a desired symbol must have some component orthogonal to the span of the subspace generated by all of the other columns in H for the zero-forcing detector to exist. The zero-forcing detector linearly combines the observation vector x such that only the orthogonal component of the desired symbols appears at its output.

If the zero-forcing detector exists, the resulting estimate may be expressed as

$$\hat{u}_d = u_d + F^H n$$

The zero-forcing detector requires full knowledge of the MIMO channel matrix H in order to compute the appropriate left inverse. No knowledge of the joint distribution of u is needed.

5.4.3.2 Linear Minimum Mean Squared Error Detection

The Linear Minimum Mean Squared Error (LMMSE) detector generates an optimum estimate of the desired symbol vector in the sense that the resulting estimates will have the least mean squared error over the class of all possible linear detectors. MSE has been shown to be a good proxy for bit error rate in many practical cases yet is much more amenable to analysis than the bit error rate criterion. Specifically, the LMMSE detector F is selected such that the MSE is minimized for each element in the desired symbol vector, e.g.

$$f_\ell = \arg \min_{g \in \mathbb{C}^{r \times 1}} \mathrm{E}\left[\|g^H x - u_d(\ell)\|^2\right]$$

where $u_d(\ell)$ is the ℓth desired symbol. The vector LMSSE detector is then formed as $F = [f_1, ..., f_{t_d}]$. Unlike the zero-forcing detector, the LMMSE detector accounts for additive channel noise is its design. The LMMSE detector also has a well known closed form solution that may be expressed as

$$F = \left(\mathrm{E}[xx^H]\right)^{-1} \mathrm{E}[xu_d^H]$$

It can be shown that if the zero-forcing detector exists for a particular MIMO channel matrix H then the LMMSE detector must also exist. Moreover, the LMMSE detector may also exist in cases where the zero-forcing detector does not exist since $\mathrm{E}[xx^H]$ will have full column rank when AWGN is present in the MIMO communication system irrespective of the rank of H. If we assume that the AWGN and the channel input vector are independent, we can write

$$\mathrm{E}[xx^H] = H\mathrm{E}[uu^H]H^H + \sigma I$$

where σ^2 is the variance of the AWGN. This expression shows that, even though $\mathrm{E}[xx^H]$ will have full column rank when $\sigma > 0$, the inverse of $E[xx^H]$ may be ill-conditioned when σ is small and $H E[uu^H] H^H$ does not have full rank.

The LMMSE detector, like the zero-forcing detector, needs full knowledge of the MIMO channel matrix for direct computation. The LMMSE detector also needs knowledge of the input signal covariances. However, it is possible to avoid requiring this knowledge if the communication system is designed to allow for a training signal. If a training signal is present then u_d is known and it is possible to to estimate the LMMSE detector by computing the sample covariances of the received signal and the training signal. Adaptive approaches including Least-Mean-Squares (LMS) and Recursive-Least-Squares (RLS) also provide a method to approximate the LMMSE detector by exploiting the training signal in the absence of knowledge of exact channel parameters and input distributions.

5.4.3.3 Blind-Adaptive Linear Estimation

Several researchers have recently considered the question ''How close can we get to LMMSE performance with a blind linear estimator?'' Clearly, training an adaptive LMS or RLS estimator is undesirable since it requires use of bandwidth that could otherwise be used for the transmission of useful data. If the channel is time-varying, periodic training will be required at additional expense and complexity. Blind adaptive approaches use the data itself to adapt a receiver such that it produces high quality estimates without the need for training. This section summarizes some of the recent results in this field.

One popular blind adaptive estimation technique for MIMO channels, called Constrained Minimum Output Energy (CMOE), was first described in [28]. The CMOE detector is not truly blind in that it does require knowledge of the column of H corresponding to the desired symbol. However, knowledge of the other columns of H channels is not required. For clarity, we describe the single parameter CMOE detector here. Vector parameter CMOE detectors can be realized by building a bank of single parameter CMOE detectors in parallel.

The criterion behind the CMOE detector for the ℓth desired symbol is to find a vector $f_\ell \in \mathbb{C}^r$ such that the average output energy $E[|f_\ell^H x|^2]$ is minimized subject to the constraint $f_\ell^H h_\ell = 1$ where $h_\ell \in \mathbb{C}^r$ is the column of H corresponding to the ℓth desired symbol. If the communication system is such that a large number or continuous stream of observations are available at the receiver then the CMOE detector can be computed without any training sequence or knowledge of H other than its ℓth column.

The CMOE detector can be better understood by geometric analysis. Specifically, the constraint portion of the CMOE criterion causes the detector to always have a unit projection on the line given by the ℓth column of the MIMO channel matrix. This implies that a constant amount of the desired user's signal will be passed by the CMOE detector. Interfering signals may also appear at the output of the CMOE detector but the minimization portion of the CMOE criterion forces the CMOE detector to seek a solution such that the amount of interference energy passed through the detector is minimized. The amount of desired energy remains fixed due to the constraint, hence it is easy to show that the CMOE detector maximizes the SINR of its output. If the interference energy passed by the detector can be reduced to zero, then the CMOE detector has found a solution with a unit projection in the desired direction and also orthogonal to the interfering columns of H.

It has been proven analytically in [23] that the CMOE detector produces estimates proportional to the LMMSE detector. The CMOE detector may be computed directly from the sample covariances of the received signal. Extensions to the CMOE detector have also been developed in [28] that relax the requirements for full knowledge of the ℓth column of $\boldsymbol{H}$.

Another blind adaptive approach with close connections to the LMMSE detector is the Constant Modulus (CM) estimator. Unlike the CMOE detector, the CM estimator does not require knowledge of any channel parameters and is considered to be a truly blind estimator. Since the first development of the CM criterion in [21] and [37], many researchers noticed that CM estimators tended to be very similar to LMMSE estimators. Recently, bounds on the MSE performance of CM estimators were derived in [35].

The CM estimator satisfies the criterion

$$\boldsymbol{f} = \arg \min_{\boldsymbol{g} \in \mathbb{R}^r} \mathrm{E}\left[\left(|\boldsymbol{g}^H \boldsymbol{x}|^2 - \gamma \right)^2 \right]$$

where γ is a real valued scalar design parameter that specifies the desired modulus. The intuition behind this approach is clear for signals that possess the CM property, e.g. M-PSK. Selecting a linear estimator that is able to restore the received signal to a CM often tends to also cancel interference and produce high quality estimates. Remarkably, the CM criterion is also quite effective on signals that do not possess the CM property, e.g. M-QAM.

CM estimators are often computed in practice via a stochastic gradient descent of the CM cost surface. Unlike the LMMSE criterion which has a quadratic cost surface with a unique minimum, the CM criterion has a cost surface with multiple minima. Roughly speaking, each minima of the CM cost surface is associated with a CM estimator for a particular desired channel input. It turns out that the initialization of the stochastic gradient descent essentially selects the CM estimator to which the stochastic gradient descent algorithm will converge. This implies that, if initialized poorly, there exists the possibility that the CM estimator may converge to an undesired solution yielding estimates for an undesired signal. In [36], sufficient conditions are derived for the initialization of the stochastic gradient descent algorithm that guarantee local convergence of the CM estimator to a desired solution.

5.4.4 Non-linear Estimation of Desired Signals in MIMO Communication Systems

This section considers non-linear receiver-based techniques for estimating a desired signal in the context of the MIMO system model. The basic MIMO model given in (119) is assumed here. Unlike the prior section where all estimates are formed as a linear combination of the observation vector, the estimators considered in this section use non-linear techniques to achieve performance that is often superior to linear detection. It has been noted in [40] that even the best linear detectors often significantly underperform the optimum (non-linear) maximum likelihood detector. Non-linear detectors that can bridge this "yawning gap" between the best linear and unconstrained optimum performance are essential to provide practical, high performance options for real world applications. We explore some of the recent results in this field below.

5.4.4.1 Maximum Likelihood Detection

Given an observation $\boldsymbol{x}$ from the MIMO system model of (119) and assuming that the channel and joint distribution of $\boldsymbol{u}$ are known to the receiver, it is possible to formulate the *joint* maximum likelihood estimate as

$$\hat{\boldsymbol{u}}_d = \arg \max_{v \in \Omega u_d} P(\boldsymbol{u}_d = \boldsymbol{v} \mid \boldsymbol{x})$$

where Ω_{u_d} denotes the set of all possible realizations for $\boldsymbol{u}_d$. It is clear from this formulation that the joint maximum likelihood detector has the desirable property that it also minimizes the probability of joint decision error, e.g. $P(\hat{\boldsymbol{u}}_d \neq \boldsymbol{u}_d)$. Minimum joint decision error however, does not always imply that the individual decisions will also have minimum error probability [43]. To minimize the individual error probabilities for desired symbols, we must consider the *individual* maximum likelihood detector. The individual maximum likelihood detector also seeks to maximize the conditional probability but only for a single desired channel input rather than multiple channel inputs. The individual maximum likelihood detector for the ℓth desired input may be formulated as

$$\hat{u}_d(\ell) = \arg \max_{v \in \Omega_{u_d(\ell)}} P(u_d(\ell) = v \mid \boldsymbol{x})$$

where $\Omega_{u_d(\ell)}$ denotes the set of all possible realizations for $u_d(\ell)$. It is clear from this formulation that the individual maximum likelihood detector minimizes the probability of decision error for $u_d(\ell)$, e.g. $P(\hat{u}_d(\ell) \neq u_d(\ell))$.

The optimal performance of maximum likelihood detection does not come without a price. First, full knowledge of the MIMO channel $\boldsymbol{H}$ and the joint distribution of $\boldsymbol{u}$ is required to compute the joint and/or individual maximum likelihood estimates. Second, symbols in $\boldsymbol{u}$ chosen from finite alphabets cause the joint distribution of $\boldsymbol{u}$ to be non-differentiable in general. This means that maximization of the likelihood functions typically requires an exhaustive search through the set of admissable values for $\boldsymbol{u}_d$ which may be computationally expensive. For example, in a case where one desires to compute a maximum likelihood estimate for 100 binary symbols, Ω_{u_d} will have 2^{100} elements. In any case, the maximum likelihood detector, joint or individual, is an important detector in that it sets the baseline by which all other detectors (linear or non-linear) are measured. Detectors that exhibit near-maximum-likelihood performance with low computational complexity over a variety of common operating conditions are receiving significant research attention today. Two such detectors are discussed below.

5.4.4.2 Successive Interference Cancellation

The Successive Interference Cancellation (SIC) detector is essentially a low-complexity intuitive algorithm for extracting desired signals from MIMO channel observations. One of the first performance studies of the SIC detector was presented in [31] where it was shown that SIC may offer good performance in several cases. The SIC detector operates on the observation vector $\boldsymbol{x}$ as follows:

1. The SIC detector first decides on the order in which it will estimate the t channel inputs. There are $t!$ possibilities but a common choice is to set the detection order such the most reliable signals are detected first and the least reliable are detected last. The detection order is denoted as $\ell_1, ..., \ell_t$.
2. Set $k = 1$ and initialize the residual observation to the actual observation, i.e. $\boldsymbol{x}_{r_1} = \boldsymbol{x}$.
3. The ℓ_kth channel input is estimated *from the residual observation* using one of the prior

techniques, usually a low-complexity technique such as matched filtering. The resulting estimate is denoted as $\hat{u}_d(\ell_k)$.

4. This estimate is then multiplied by the ℓ_k column of $\boldsymbol{H}$ and substracted from the observation to generate a new residual observation $\boldsymbol{x}_{r_k} = \boldsymbol{x}_{r_{k-1}} - \boldsymbol{h}_{\ell_k}\hat{u}_d(\ell_k)$.

5. If there are more signals to estimate, increment k and loop back to step 3.

The SIC algorithm is sometimes called the "onion-peeling" algorithm in the sense that layers of the received signal are peeled away much like the layers of an onion. Clearly, if the cancellation at step 4 is effective, the amount of intereference on the other signals in the observation is reduced. The problem with SIC detection occurs when the cancellation at step 4 is not effective. It is possible in this case to actually increase the amount interference generated on the other desired signals and reduce the reliability of detecting them correctly.

The SIC detector, in combination with very low rate convolutional codes, was shown to achieve the capacity limit of spread sprectrum multiple access channels with background Gaussian noise in [42]. Achieving this capacity limit does require coordinated processing of all transmissions in as much as each simultaneous transmission is required to have a particular symbol energy. These symbol energies are computed as functions of the interfering symbol energies and the background noise power. The analysis in [42] shows that as the number of transmitters becomes large, this communication system asymptotically approaches the Shannon limit.

The SIC detector requires full knowledge of the MIMO channel matrix but does not explicitly require knowledge of the joint input distribution. The SIC detector algorithm may be modified such that not all signals are estimated (partial-SIC) or such that multiple passes through the SIC algorithm are performed. Multiple passes are often advantageous in practical systems since the first signals are decided without the benefit of any intereference cancellation on the first pass whereas these signals will now be decided on a residual observation with less expected interference on subsequent passes.

5.4.4.3 Parallel Interference Cancellation

The Parallel Interference Cancellation (PIC) detector is another low complexity non-linear detection algorithm useful for extracting desired signals from MIMO channel observations. Like SIC, the PIC detector attempts to estimate and cancel interference from the original observation to create a residual observation with less expected interference. The PIC detector was first suggested for digital cellular communication systems in [38] and [39] where it was called the "multistage detector". The basic algorithm for the PIC detector is as follows:

1. The PIC detector forms an estimate of some or all of the MIMO channel input vector (including any combination of desired and/or undesired symbols) using a simple technique such as matched filtering or zero-forcing detection. These estimates are then multiplied against their associated columns in the MIMO channel matrix $\boldsymbol{H}$ and subtracted from the original observation $\boldsymbol{x}$ to form a residual observation $\boldsymbol{x}_r$. Unlike the SIC detector, all of these operations are performed simultaneously rather than serially.

2. Each of the desired channel channel inputs is then estimated *from the residual observation* using a simple technique such as matched filtering. Again this is done simultaneously rather than serially.

The PIC detector is an intuitively satisfying algorithm. If the vector estimation at step 1 is accurate, then the interference sources will be cancelled and the reliability of the estimates for the desired signals will be improved in step 2. However, the PIC detector suffers from the same interference cancellation sensitivities as the SIC detector. If the interference is poorly estimated in step 1, it is possible to corrupt the residual observation to the point where the estimates of step 2 are worse than if no interference cancellation was attempted at all.

The PIC detector has several advantages over the SIC detector including the fact that its parallel structure has less inherent detection delays than the serial structure of the SIC detector. Also, for signals received with equal powers, the PIC detector provides equal performance estimates for all of the channel inputs whereas the SIC detector provides asymmetric performance in this case. An analytical comparison of PIC and SIC detector SINR performance was given for digital cellular systems in [10].

Multiple stages of the PIC detector are also simple to realize. In this case, if each stage improves the quality of the desired symbol estimates then multiple stages of PIC detection will converge to good quality symbol estimates. It is possible however, that the PIC detector will not converge to a good solution and may in fact diverge. An analysis of this behavior for a particular form of PIC detection was presented in [11] where it was shown that even in cases where a few stages of PIC detection dramatically improved the estimation quality, more stages of PIC detection actually caused the estimates to have less quality than the first stage.

Recent research has considered estimation techniques within the PIC framework that have less tendancy to exhibit this behavior [14]. Here the authors develop a hybrid linear–non-linear detector with the PIC structure that is robust to unreliable interference estimation in step 1. This new PIC detector was shown to provide significant performance advantages over the conventional PIC detector with little additional complexity.

5.4.5 Conclusions

In this section we have summarized some of the recent results in the field of MIMO communication channels. MIMO communication channels have been shown to be quite general and to represent a variety of operating scenarios including one or more transmitters, one or more receivers, and multiple propogation paths. The capacity of MIMO channels was also discussed and techniques for achieving this capacity were described. Several linear and non-linear signal extraction techniques were discussed for MIMO channels. These technniques differ widely in several factors including the required knowledge at the receiver, the computational complexity, and the resulting performance. Looking forward, it seems that MIMO communication systems will remain an active and important research topic for some time.

References

[1] Agee, B., 'Blind Separation and Capture of Communication Signals Using a Multitarget Constant Modulus Beamformer,' in *Proceedings of the 1989 IEEE Military Communications Conference*, Boston, MA, October 1989.

[2] Agee, B., 'Fast Acquisition of Burst and Transient Signals Using a Predictive Adaptive Beamformer,' in *Proceedings of the 1989 IEEE Military Communications Conference*, Boston MA, October 1989.

[3] Agee, B., Schell, S. and Gardner, W., 'Self-coherence Restoral: a New Approach to Blind Adaptive Signal Extraction Using Antenna Arrays', *IEEE Proceedings* 78(4), April 1990, 753–767.

[4] Agee, B., Kelly, P. and Gerlach, D., 'The Backtalk Airlink for Full Exploitation of Spectral and Spatial Diversity in Wireless Communication Systems', in *Proceedings of the Fourth Workshop on Smart Antennas in Wireless Mobile Communications*, July 1997.

[5] Agee, B., Bromberg, M., Gerlach, D., Ho, M., Jesse, M., Mechaley, R., Stephenson, D., Golden, T., Nix, D., Naish, R., Gibbons, D., Maxwell, R., Hoole, E. and Ryan, D., Highly Bandwidth-Efficient Communications, *PCT Application No. WO 98/37,638*, 27 August 1998.

[6] Agee, B., Bruzzone, S., Bromberg, M., 'Exploitation of Signal Structure in Array-Based Blind Copy and Copy-Aided DF Systems', *International Conference on Acoustics, Speech and Signal Processing*, May 1998.

[7] Agee, B., 'Exploitation of Internode Mimo Channel Diversity in Spatially-Distributed Multipoint Communication Networks', in *Proceedings of the Tenth Annual VA Technology Symposium on Wireless Personal Communication*, June 2000.

[8] M. Bromberg and B. Agee, 'The LEGO approach for achieving max-min capacity in reciprocal multipoint networks', in Proceedings of the Thirty Fourth Asilomar Conference on Signals, Systems, and Computers, Oct. 2000.

[9] D. Brown, D. Anair, and C. Johnson, 'Linear detector length conditions for DS-CDMA perfect symbol recovery', in Proceedings of the 1999 Signal Processing Advances in Wireless Communications Conference, Annapolis, MD, pp. 178–81, May 9-12 1999.

[10] Brown, D. and Johnson, C., 'SINR, Power Efficiency, and Theoretical System Capacity of Parallel Interference Cancellation', in *Proceedings of the 2000 Conference on Information Sciences and Systems*, Vol. 1, Princeton, NJ, 15–17 March 2000, pp. TA2.1–TA2.6.

[11] Brown, D., Motani, M., Veeravalli, V., Poor, H. and Johnson, C., 'On the Performance of Linear Parallel Interference Cancellation', *IEEE Transactions on Information Theory*, July 2001, in press.

[12] Chizhik, D., Rashid-Farrokhi, F., Ling, J. and Lozano, A., 'Effect of Antenna Separation on the Capacity of BLAST in Correlated Channels', *IEEE Communications Letters*, 4, November 2000, 337–339.

[13] Cover, T. and Thomas, J., *Elements of Information Theory*, Wiley, New York, 1991.

[14] Divsalar, D., Simon, M. and Raphaeli, D., 'Improved Parallel Interference Cancellation for CDMA', *IEEE Transactions on Communications*, 46, February 1998, 258–268.

[15] Foschini, G. and Gans, M., 'On Limits of Wireless Communications in a Fading Environment When Using Multiple Antennas', *Wireless Personal Communications*, 6, March 1998, 311–335.

[16] Gardner, W., 'Multiplication of Cellular Radio Capacity by Blind Adaptive Spatial Filtering', in *Proceedings of the IEEE International Conference on Selected Topics in Wireless Communication, ICWC*, June 1992, pp. 102–106.

[17] Gardner, W., *Introduction to Random Processes*, McGraw-Hill, New York, 1990.

[18] Gerlach, D. and Paulraj, A., 'Adaptive Transmitting Antenna Methods for Multipath Environments', in *Proceedings of the 1994 Global Telecommunications Conference*, Vol. 1, 1994, pp. 425–429.

[19] Gerlach, D., Adaptive Transmitting Antenna Arrays at the Base Station in Mobile Radio Networks, Ph.D. Thesis, Stanford University, June 1995, pp. 84–85.

[20] Givens, W., 'Computation of Plane Unitary Rotations Transforming a General Matrix to Triangular Form', *SIAM Journal of Applied Math*, 6, 26–50.

[21] Godard, D, 'Self Recovering Equalization and Carrier Tracking in Two-Dimensional Data Communication Systems', *IEEE Transactions on Communications*, 28, November 1980, 1867–1875.

[22] Goodman, N.R., 'Statistical Analysis Based on a Certain Multivariate Complex Gaussian Distribution (an Introduction)', *Annals of Mathematical Statistics*, 34(1), March 1963, 152–177.

[23] Honig, M., Madhow, U. and Verdu, S., 'Blind Adaptive Multiuser Detection', *IEEE Transactions on Information Theory*, 41, July 1995, 944–960.

[24] Horn, R. and Johnson, C., *Matrix Analysis*, Cambridge University Press, New York, 1985.
Householder, J., 'Unitary Triangularization of a Non-Symmetric Matrix', *Journal of ACM*, 5, 339–342.

[26] Ling, F., Manolakis, D. and Proakis, J., 'A Recursive Modified Gram–Schmidt Algorithm for Least Squares Estimation', *IEEE Transactions Acoustics, Speech, and Signal Processing*, ASSP-34, 829–836.

[27] Loyka, S. and Mosig, J., 'Channel Capacity of n-Antenna BLAST Architecture', *Electronics Letters*, 26, March 2000, 660–661.

[28] Madow, U., 'Blind Adaptive Interference Suppression for Direct-Sequence CDMA', *Proceedings of the IEEE*, 86, October 1998, 2049–2069.

[29] Manolakis, D., Ingle, V. and Kogon, S., *Statistical and Adaptive Signal Processing*, McGraw Hill, New York, 2000.

[30] Morgan, D.R., 'Partially Adaptive Array Techniques', *IEEE Transaction on Antennas and Propagation*, 26(6), 1978, 823–833.

[31] Patel, P. and Holtzman, J., 'Analysis of a Simple Successive Interference Cancellation Scheme in a DS-CDMA System', *IEEE Journal on Selected Areas in Communication*, 12, June 1994, 796–807.

[32] Raleigh, G. and Cioffi, J., 'Spatio-Temporal Coding for Wireless Communication', *IEEE Transactions on Communications*, 46, March 1998, 357–366.

[33] Raleigh, G. and Jones, V., 'Multivariate Modulation and Coding for Wireless Communication', *IEEE Journal on Selected Areas in Communication*, 17, May 1999, 851–866.

[34] Rappaport, T., *Wireless Communications*, Prentice Hall, Englewood Cliffs, NJ, 1996.

[35] Schniter, P. and Johnson, C., 'Bounds for the MSE Performance of Constant Modulus Estimators', *IEEE Transactions on Information Theory*, 46, November 2000, 2544–2560.

[36] Schniter, P. and Johnson, C., 'Sufficient Conditions for the Local Convergence of Constant Modulus Algorithms', *IEEE Transactions on Signal Processing*, 48, October 2000, 2785–2796.

[37] Treichler, J. and Agee, B., 'A New Approach to Multipath Correction of Constant Modulus Signals', *IEEE Transactions on Acoustics, Speech, and Signal Processing*, 31, April 1983, 459–472.

[38] Varanasi, M. and Aazhang, B., 'Near-Optimum Demodulation for Coherent Communications in Asynchronous Gaussian CDMA Channels', in *Proceedings of the 22nd Conference on Information Sciences and Systems*, Princeton, NJ, March 1988, pp. 832–839.

[39] Varanasi, M. and Aazhang, B., 'Multistage Detection in Asynchronous Code-Division Multiple-Access Communications', *IEEE Transactions on Communications*, 38, April 1990, 509–519.

[40] Varanasi, M., 'Group Sequence Detection: Bridging the Yawning Gap Between Linear and Optimum Multiuser Detection', in *Proceedings of the 1995 Workshop on Information Theory Multiple Access and Queueing*, 1995.

[41] Visotsky, E. and Madhow, U., 'Optimum Beamforming using Transmit Antenna Arrays', in *Proceedings of the IEEE 49th Conference on Vehicular Technology*, Vol. 1, 1999, pp. 851–856.

[42] Viterbi, A., 'Very Low Rate Convolutional Codes for Maximum Theoretical Performance of Spread-Spectrum Multiple-Access Channels', *IEEE Journal on Selected Areas in Communication*, 8, May 1990, 641–649.

[43] Verdú, S., *Multiuser Detection*, Cambridge University Press, New York, 1998.

[44] Widrow, B. and Stearns, S., *Adaptive Signal Processing*, Prentice Hall, Englewood Cliffs, NJ, 1985.

6

The Challenges of Software-Defined Radio

Carl Panasik and Chaitali Sengupta

The current move to create and adopt third generation (3G) wireless communications standards has raised tremendous expectations among engineers. To some extent, there is a perception that adopting the new standards will result, almost instantaneously, in being able to design a plethora of multi-purpose wireless Internet appliances with features and capabilities far beyond those found in today's wireless telephones and palmtop organizers.

Recent discussion in the industry has suggested that the coming months will bring a kind of super-communications/entertainment appliance. With form factors and battery life similar to today's wireless phones, this system will deliver high-fidelity audio and full-motion video from the Internet, while it also provides voice communications capabilities and serves such as a Bluetooth transceiver. It will access the wireless LAN at the office, serve as a cellular phone during the commute, and connect to another wireless LAN at home. It will recognize local wireless infrastructures such as high-speed data kiosks in airports. Most important, it will communicate flawlessly with the infrastructure anywhere in the world where users choose to take it.

Embedded in this set of expectations is a reasonable amount of fact. 3G appliances certainly will deliver features and capabilities beyond those currently available (2G) and recent introductions of 2.5G. Recent advances in Digital Signal Processor (DSP) technology will enable high-fidelity audio and full-motion video in handheld wireless appliances in the near future. Bluetooth technology is ready to hit the streets.

But the idea that a super wireless communications appliance will arrive any time soon is pure fiction. Current technology does not permit the union of all 3G possibilities in a single handheld appliance. It cannot accommodate, in one small system, the vastly different requirements of communications standards found in different parts of the world.

Still, heightened expectations exist and OEMs hope to be the first to market with a super wireless appliance, which will provide all the capabilities currently imagined – and more. Crucial technological limitations exist although research and development may serve to overcome them. One avenue is a Software-Defined Radio (SDR) approach to the handheld appliance as a means to surmount impediments to multi-mode communications and to mitigate key difficulties posed by varying communications standards.

6.1 Cellular Communications Standards

Operators desire that the users are drawn away from their PCs and televisions to the cellular phones as their personal (possibly worn as closely as a watch) source of information and entertainment. Unlike somewhat portable television receivers, cellular providers desire that the mobile phones be used worldwide. Given this scenario, it is interesting to explore the solution to all these goals, using technological feasibility today and in the near future.

The ideal world for cellular providers would be one in which there was a single standard and single RF band of communication frequencies, so that they only needed to support one phone. But today, one is faced with a variety of wireless standards in different parts of the world, with different capabilities. Figure 6.1 shows the various wireless data standards that will be available in the next few years. The figure presents data rate as a function of range, parameterized by standard. Note that longer ranges imply higher mobility (more accurately, velocity), while the higher data rates are for stationary objects. The accommodation of various standards on one network requires a handset, which supports the various standards. Ideally, this is accomplished with a radio that has minimal fixed functionality, and analog processing. Flexible, powerful digital processing could provide software defined functions and re-configuration.

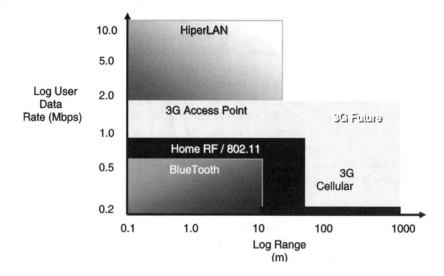

Figure 6.1 Wireless standards: data rates and ranges. Note, 802.11 is presented as the original IEEE standard. Later modifications increased the data rate to 2 Mbps and now 11 Mbps.

6.2 What is SDR?

Recently, SDR [2–5], has been suggested as the solution to meet a wide variety of requirements. Clearly SDR has advantages to the customer of providing one information appliance, which is always connected, no matter the location, communicating voice and data. In addition, the handset manufacturer only has to support one "chassis" in the handset. This simplifies worldwide support. Going back to the user's ultimate desire for information, SDRs bring the ability to access various forms of communication through a compact transceiver. As previously

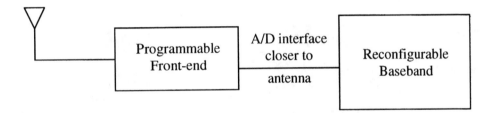

Figure 6.2   The SDR for multi-mode, multi-band transceivers. The concept utilizes a multi-band RF front-end and a baseband primarily re-configurable in software

mentioned, a single PDA which can access the wireless LAN in the office, connect to the 3G cellular system during commute and connect to a home-based wireless LAN at home would offer ubiquitous anywhere, anytime information. There may be other opportunities to direct the information appliance to recognize local wireless infrastructures, such as high-speed data kiosks at an airport, or displaying movie trailers in the parking lot of the 28-plex cinema!

The functionality of an SDR is realized by software performing signal processing tasks in the digital domain as shown in Figure 6.2. Instead of processing analog RF signals to isolate channels or bands and eliminate noise from adjacent bands, the SDR converts the wireless signal to digital data streams at the earliest opportunity.[1] Powerful digital signal processing provides more flexible software-defined functions and hardware reconfigurations.

For the handheld 3G wireless communications appliance, this kind of digital signal processing is essential to accommodate varying standards and multiple modes. It permits the transceiver hardware to be reconfigured for a variety of purposes, significantly reducing system size and cost. In a cellular telephone, for example, SDR would use the same hardware for communications anywhere in the world, despite the variety of standards now in place. Spectrum regulation bodies in different parts of the world have defined region-specific uplink and downlink frequencies for 2G and 3G communications. This requires that the ideal universal handheld change both the radio frequencies and the filtering needed to separate transmit from receive. Figure 6.3 shows the worldwide spectrum for IMT-2000, the 3G wireless standard. Although Europe, China and Japan have nearly common receive and transmit frequencies, the US has chosen a compacted band with a duplex of only 90 MHz. This not only affects changes in radio frequencies, but in the filtering used to separate transmit from receive. In addition, there is no agreed-upon single worldwide RF band for uplink and downlink within any standard. For instance, today's standard GSM phone purchased in Europe will not operate in the US's GSM infrastructure, as the RF frequency bands differ.

As functions are added to the appliance, this duplication of hardware results in a system that is unacceptably large, complex and expensive. If the idea is to create a super appliance nearly as small and portable as today's wireless telephones, the only available solution is an

---

[1] Note that the discussion of transceivers centers upon the receiver. The receiver must separate the various signals from the base station (adjacent users), from other base stations (co-channel users) and from other services (out of band users) which may have significantly larger signals than those in-band. The receiver must be very linear throughout its large dynamic range and work with feeble signals. Synchronization is often a part of the detection/demodulation process. On the other hand, the transmitter must create a signal, which has very low out-of-band components. Most transmitters are implemented with variable center frequency and use direct conversion from digital.

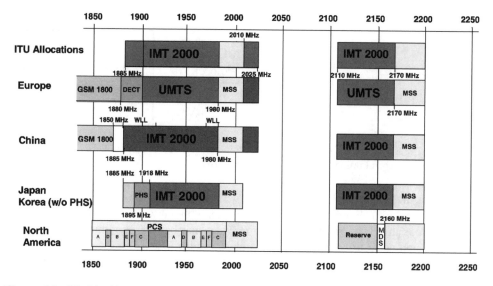

Figure 6.3 World-wide spectrum allocation for third generation cellular. Reproduced courtesy of REF. [10]

ever-increasing amount of digital signal processing, which permits functional reconfiguration of a system for the purpose at hand.

Most wireless data systems use the same frequency channel for both transmit and receive, a method called Time Domain Duplex (TDD), as shown in Figure 6.4. Bluetooth and 802.11 share the 2.4-GHz ISM band, while HiperLAN and several proprietary systems operate in the 5.2-GHz ISM band. In these systems the mobile transmits to the base (or access point) for a time slot, then the base transmits information to the mobile for another time slot. The number of time slots for each transmission can be re-allocated (between uplink and downlink) as the data traffic requires. In contrast, most voice-dominated cellular systems utilize separate bands for transmit and receive. This method, known as Frequency Domain Duplex (FDD), enables simultaneous two-way communication. A device that operates in FDD and TDD modes would enable universal communication.

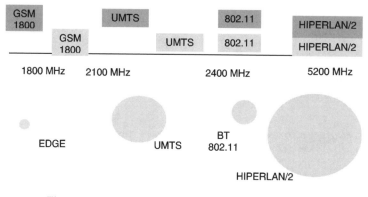

Figure 6.4 Wireless standards: spectrum and bandwidth

Various modulation techniques currently in use further complicate the problem of developing a single multi-band, multi-mode, multi-function appliance. Most commercial communications systems use phase shift keyed modulation. However, some wireless data systems can invoke higher ordered modulation (i.e. 16-QAM) for very high data rate communications in situations in which optimal propagation paths exist. In all cases, the channel bandwidths are proportional to the data rates of each system. Changes in data rate require a variable filter bandwidth in the SDR.

An interesting evolution in communications is the move from frequency channel-defined systems to code channel-defined systems. The most common cellular standards today, such as GSM, define narrowband frequency channels that are time-shared by many users, a system called Time Division Multiple Access (TDMA). Code channel-defined systems, on the other hand, utilize spreading codes to separate several users on a common frequency channel. In the US, this Code Division Multiple Access (CDMA) technology [1] is commonly deployed in the Personal Communication Systems (PCS) band at 1900 MHz. The digitized voice signals are multiplied by a chipping code that identifies the user. At the receiver, all the codes are demodulated and separated by a digital baseband correlator to isolate the specific user.

Moving the imaginary super-wireless appliance out of the realm of fiction and into fact requires the development of a handset capable of dealing with all attributes of multiple standards. Such an SDR-enabled appliance must accommodate not only the variations that are common today, but also those that may emerge as individual standards or modes evolve to meet growing needs and circumvent problems discovered after implementation.

SDR research is proceeding along two vectors. The first approach entails moving towards hardware and software that can be reconfigured to provide different functions. This aspect involves the ideas of reconfiguration and the extensive use of software in all layers of the protocol stack including the physical layer. The first step is re-configuration of parts that are implemented in software even in current systems. These parts include three basic components: signal-processing algorithms in the physical layer, the protocol stack, and applications. The SDR concept advocates reconfiguration of all three parts from the network or from an application, based on demand. Reconfiguration also encompasses the idea of re-configurable hardware. This reconfiguration again can be at a system level or at a function level in a particular system. The first implies using the same piece of hardware with different parameters in different modes, whereas the second implies using the same piece of hardware for different types of processing, at different times.

The second, and arguably more effective approach, creates a smaller and perhaps more useful wireless appliance, which involves moving more analog functions into the digital domain.

6.3 Digitizing Today's Analog Operations

Digitizing operations currently performed in analog requires moving the Analog to Digital Converter (ADC) ever closer to the radio's antenna. In the ideal SDR, the ADC would immediately follow the antenna and all filtering, mixing and amplification would take place in the digital domain. However, the actualization of this ideal is far in the future. For the time being, it is more practical to think about radio designs closer to those found in today's cellular phones: transceivers that include both an analog stage and a digital stage. The

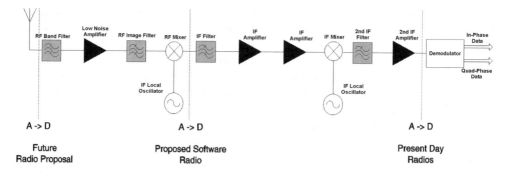

Figure 6.5 Heterodyne receiver. SDR moves the ADC toward the antenna

ADC's location may move, but a certain amount of high frequency analog processing will remain essential.

Any movement of the ADC toward the antenna increases the complexity of the digital circuitry. Consider the operation of a typical heterodyne receiver (see Figure 6.5). At the antenna, all signals in a given environment are present, including short-wave radio, broadcast radio and television, satellite, cellular, PCS, point-to-point microwave and military radar. In a cellular system, the signals transmitted from the base station are arranged in a band with many tightly packed, narrow channels. The receiver's first stage separates the desired communication band from all other signals. The band is chosen by the RF band filter, amplified by the low noise amplifier, and down-converted to a useful intermediate frequency by the first mixer. The channel filter removes mixer spurious responses and adjacent channel users. The intermediate frequency signal is then amplified and down-converted to a much lower frequency, nearly the same as the channel bandwidth. Digital conversion takes place at this point.

Each subscriber will receive a signal level determined by the path loss between its location and the base station. Adjacent channels are used by adjacent base stations, which may have signals that are much larger than the desired base station. This can happen in the following scenario. At the edge of a cell a building may block the path to the desired base station, while the adjacent cell has a clear line-of-sight path [6]. All cellular systems have receiver specifications that enable reception in this scenario. For instance, the GSM system with 200-kHz channels [7] specifies the in-band blocking levels presented in Table 6.1 for a reference sensitivity level of -102 dBm.

Therefore, the complete channel filter function must reject signals beyond 600 kHz by 59 dB.

If the radio is redesigned so that the last filter stage is implemented in the digital domain, the ADC must be able to handle signals that previously were defined as in-band. At the same

Table 6.1 In-band blocking levels for the GSM mobile station

Frequency range from carrier (kHz)	In-band blocking levels (dBm)
600–800	-43
800–1600	-43
1600–3000	-33

time it must have enough dynamic range to accurately sample signals (from adjacent users) that may be an order of magnitude larger. This is because the channel is now the entire down-converted band of the base station, some 35 MHz wide.

6.4 Implementation Challenges

The ideal handset SDR mentioned above, in which the ADC immediately follows the antenna, would require digitizing the entire cellular downlink band and then applying digital filtering to select appropriate data streams for particular applications. Such a receiver could be easily reconfigured for various standards, modes and functions at the flick of a switch. A user could carry on a cell phone conversation at one moment and electronically purchase a snack from a vending machine the next. But without extensive additional engineering, these functions could not occur simultaneously. Still, a reconfigurable handset would be significantly more desirable than the situation of carrying a different appliance or a different plug-in card for each application.

The key challenges that must be met in order to match the analog and digital partitioning to the SDR paradigm, are high power dissipation, low immunity to adjacent channel interference, A/D sampling rates, A/D dynamic range and sensitivity to timing errors (i.e. phase noise).

6.5 Analog and ADC Issues

A realistic, workable SDR model requires moving the ADC closer to the antenna to achieve increased digital processing. This requirement leads to the need for ADCs that can cope with signals of large bandwidth and high dynamic range. This, in turn, leads to increased power consumption in the digital baseband due to high sample rates and increased word length.

The RF and analog sections of the SDR system are subject to problems related to antennas, filtering, power amplifiers, etc. For multi-mode operation, the simplest solution is to have separate transceivers for different modes (RF bands). This "velcro radio" leads to problems with handset size and cost. On the other hand, digital-based multimode components face the challenge of providing required performance in different cellular systems and conditions. Table 6.2 compares the ADC and filter requirements for GSM and the 3G standard, W-CDMA.

However, even if it is possible to design the above analog components that are necessary for any radio, irrespective of where the analog–digital division is, increased digital processing, will require ADCs that can cope with signals of large bandwidth and high dynamic range. Figure 6.5 shows how the ADC must move closer to the antenna in the SDR, as compared to its location today. Consequently, the digital signals have large word lengths

Table 6.2 Receiver component requirements for the dual-mode cellular phone

Standard	Sampling frequency (Msps)	Number of bits	Channel sidelobes (dB)
GSM	0.400	12	59
W-CDMA	32	5	45

and high sample rates. There are ADCs available that have high sample rates and large bandwidth but the supported dynamic ranges are limited, particularly at minimal power consumption levels. ADCs capable of digitizing the entire cellular band exist today, they are found largely in specialized equipment, such as high-speed sampling oscilloscopes. Unfortunately, these ADCs require enormous processing power and consume the battery charge quickly. Some of today's 2G cellular downlinks are 35 MHz wide; thus, the ADC would have to perform roughly 70 megasamples per second. The proposed 3G downlinks are even wider at 60 MHz. The only possibility in this area are advanced ADCs applying noise-shaping techniques that support high dynamic ranges, in specific frequency bands. This would allow the SDR concept to work for a specific mode, but the multi-mode scenario will require clearing several more technical hurdles.

The fact is that this level of sophistication is not really necessary in a handheld wireless appliances because the equipment itself serves only one individual at a time. Unlike a base station, in which processing in different modes for multiple users must occur simultaneously, the handheld device can be much more specialized within any configuration.

6.6 Channel Filter

In the receiver (Figure 6.5), the Intermediate Frequency (IF) filter is the most significant contributor to delay distortion (group delay). This filter, typically centered in the 70–120-MHz range, removes the mixer products and interferers outside the first adjacent channels. The matched, channel, demodulator filter is implemented using the DSP. It has two functions: to suppress the adjacent channel signals and, as a matched filter, to improve the ISI response.

Surface Acoustic Wave (SAW) filters are routinely used for IF filters, as they have many zeroes which can be used to provide the required narrow transition bandwidth and the desired constant group delay.

Novel, efficient methods of programmable digital filtering may be found in Ref. [8]. The author presents an IF filter stage which first down-converts the signal, sampled at four times the IF frequency, to baseband by multiplying by ± 1. The signal is then filtered with a high decimation rate, multiplier-less CIC filter, selecting a narrow channel from a large bandwidth signal. The CIC filter is followed by a conventional, symmetric, FIR filter, requiring fewer taps at the lower (down-sampled) sampling rate. Simulations show that an efficient channel filter can be realized in today's technology.

6.7 Delta-Sigma ADC

Overcoming these problems will require several engineering innovations, some of which are being developed in the world's major research laboratories. A useful ADC architecture uses a $\Delta\Sigma$ modulator to sample a narrow channel, thereby suppressing adjacent channel signals. This modulator acts as a filter to reduce the sampled bandwidth to a level manageable by the ADC within acceptable power consumption limits. Just such a modulator with spec-compliant GSM and W-CDMA performance was described at the IEEE International Solid-State Circuits Conference in San Francisco [9]. Various laboratories are currently developing novel, efficient methods of programmable digital channel filtering. Simulations show that an efficient channel filter can be realized in today's digital technology. Another needed innovation is the development of down-conversion techniques that are much more power efficient than those currently in use. Nevertheless, substantially more work is needed in this area to make SDR a practical reality for 3G wireless and related communications applications.

6.8 Conclusion

With the transition from 2G to 3G wireless communications producing a total of three 3G modes, in addition to Bluetooth and GPS, it is clear that terminals that can support multiple modes will be imperative. Whether such appliances will feature integrated (on-chip) RF will be determined by issues of cost, risk and technical feasibility. The first available multi-mode systems will rely on multiple RF front-ends and combined digital baseband systems.

Today's technology is near to the realization of a useful SDR at the low power consumption levels essential to handheld wireless appliances. The concept and efforts behind SDR will eventually provide subscribers with small personal units containing a plethora of information and entertainment functions that will not be restricted to use with one transmission standard or in a specific part of the world. For wireless industry operators and equipment manufacturers, that prospect alone makes the continuing development of SDR-enabled handhelds worthwhile and exciting.

References

[1] Viterbi, A.J., *CDMA Principles of Spread Spectrum Communications*, Addison-Wesley, Reading, MA, 1995.
[2] Mitola, J. and Maguire, G.Q., 'Cognitive radio: Making software radios more personal', *IEEE Personal Communications*, August 1999.
[3] Hentsche, T., Henker, M. and Fettweis, G. 'The digital front-end of software radio terminals', *IEEE Personal Communications*, August 1999.
[4] Mitola, J., 'Technical challenges in the globalization of software radio'. *IEEE Communications Magazine*, February 1999.
[5] Shepherd, R., 'Engineering the embedded software radio', *IEEE Communications Magazine*, November 1999.
[6] Rappaport, T.S., *Wireless Communications: Principles and Practice*, Prentice-Hall, Englewood Cliffs, NJ, 1996.
[7] European Telecom Standards Institute (ETSI), 05.05 Radio Transmission and Reception, January, 1994.
[8] Hinton, D., 'Efficient IF filter for dual-mode GSM/W-CDMA transceiver', *WTBU Technical Activity Report*, August 2000.
[9] Burger, T. and Huang, Q., 'A 13.5 mW, 185 M sample/s delta/sigma modulator for UMTS/SGSM dual-standard IF reception, *International Solid State Circuits Conference*, 2001, Vol. 427, pp. 44–45.
[10] UMTS Forum Report No. 5, 'Minimum Spectrum Demand per Public Terrestrial UMTS Operator in the Initial Phase', 8 September 1998.

7

Enabling Multimedia Applications in 2.5G and 3G Wireless Terminals: Challenges and Solutions

Edgar Auslander, Madhukar Budagavi, Jamil Chaoui, Ken Cyr, Jean-Pierre Giacalone, Sebastien de Gregorio, Yves Masse, Yeshwant Muthusamy, Tiemen Spits and Jennifer Webb

7.1. Introduction

7.1.1. "DSPs take the RISC"

From the mid-1980s to the mid-1990s, we were in the "Personal Computer" era and CISC microprocessors fuelled the semiconductor market growth (Figure 7.1). We are now in a new era where people demand high personalized bandwidth, multimedia entertainment and information, anywhere, anytime: Digital Signal Processing (DSP) is the driver of the new era (Figure 7.2). There are many ways to implement DSP solutions; no matter what, the world surrounding us is analog; analog technology is therefore key. In this chapter, we will explore the different ways to implement DSP solutions and present the case of the dual core DSP + RISC, which will introduce the innovative OMAP™ hardware and software platform by Texas Instruments.

Whether it is a matter of cordless telephones or modems, hard disk controllers or TV decoders, applications integrating signal processing need to be as compact as possible. The tendency is of course to put more and more functions on a single chip. But in order to be really efficient, the combination of several processors in silicon demands that certain principles be respected.

In order to respond to the requirements of real-time DSP in an application, the implementation of a DSP solution involves the use of many ingredients: analog to digital converters, digital to analog converters, ASIC, memories, DSPs, microcontrollers, software and asso-

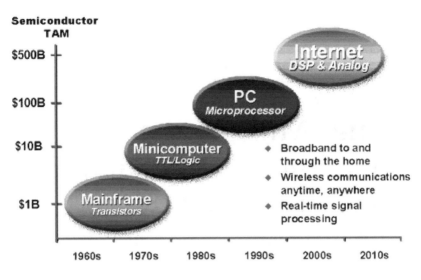

Figure 7.1.   DSP and analog drive the Internet age

ciated development tools. There is a general and steady trend towards increased hardware integration, which is where the advantage of offering "systems on one chip" comes in. Several types of DSP solutions are emerging; some using dedicated ASIC circuits, others integrating one or more DSPs and/or microcontrollers.

Since the constraints of software and development differentiation (even of standards!) demand flexibility and the ability to react rapidly to changes in specifications, the use of non-programmable dedicated circuits often causes problems. The programmable solution calls for the combination of processors, a memory and both signal processing instructions and control instructions, as well as consideration of the optimum division between hardware

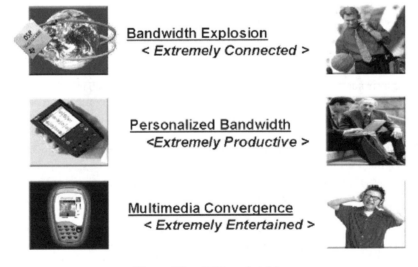

Figure 7.2.   DSP market drivers

and software. In general, it is necessary to achieve optimum management of flows of data and instructions, and in particular to monitor exchanges between memory banks so as not to be heavily penalized in terms of performance. Imagine, for example, two memory banks, one accessed from time to time and the other during each cycle: instead of giving both banks a common bus, it is undoubtedly preferable to create two separate buses so as to minimize power consumption and preserve the bandwidth. More than ever, developing a DSP solution requires in-depth knowledge of the system to be designed in order to find the best possible compromise between parameters which are often contradictory: cost, performance, consumption, risk, flexibility, time to market entry, etc.

As far as control-type functions are concerned, RISC processors occupy a large share of the embedded applications market, in particular for reasons of "useful performance" as compared to their cousins, CISC processors. As for signal processing functions, DSPs have established themselves "by definition". Whatever method is chosen to combine these two function styles in a single solution, system resources, tasks and inputs/outputs have to be managed in such a way that the computations carried out don't take more time than that allowed under the real-time constraint. The sequencing, the pre-empting of system resources and tasks as well as the communication between the two are ensured by a hard real-time kernel.

There is a choice of four scenarios to combine control functions (a natural fit for RISC) and signal processing functions (a natural fit for DSP): the use of a DSP plus a RISC, a RISC on its own or with a DSP co-processor, a DSP on its own or lastly a new integrated DSP/RISC component.

The first time that two processors, one a RISC and the other a DSP, were used in the industry on a single chip was by Texas Instruments in the field of wireless communications: this configuration is now very popular. It permits balanced division between control functions and DSP functions in applications that require a large amount of signal processing (speech encoding, modulation, demodulation, etc.) as well as a large amount of control (man–machine interface, communication protocols, etc.). A good DSP solution therefore requires judicious management of communications between processors (via a common RAM memory for example), development tools permitting co-emulation and parallel debugging and the use of RISC and DSP cores suitable for the intended application.

In the case of a RISC either with or without a DSP co-processor, it must be remembered that RISC processors generally have a simple set of instructions and an architecture based on the "Load/Store" principle. Furthermore, they have trouble digesting continuous data flows that need to be executed rapidly, special algorithms or programs with nested loops (often encountered in signal processing) because they have not been designed for that purpose. In fact, they have neither the appropriate addressing mode, nor a bit manipulation unit, nor dedicated multipliers or peripherals. So, although it is possible to perform signal processing functions with RISC processors with a reduced instruction set, the price to pay is the use of a large number of operations executed rapidly, which leads to over consumption linked to this use of "brute force". To avoid having a hardwired multiplier and thus "resembling a DSP too closely", some RISCs are equipped with multipliers of the "Booth" type based on successive additions. This type of multiplier is advantageous when the algorithms used only require a small number of rapid multiplications, which is not often the case in signal processing. The trends that are emerging are therefore centered more on "disguising a RISC processor as a DSP" or using small DSP coprocessors. In the case of the latter, the excess burden of the DSP activity – generation of addresses and intensive calculations – is too heavy in most applications and, in addition, this can limit the bandwidth of the buses.

It must be acknowledged that current DSPs are not suitable for performing protocol functions or human–machine interfaces, or for supporting most non-specialized DSP operating systems. These operating systems very often need a memory management unit to support memory visualization and regional protection that isn't found in conventional DSPs. However, the use of a DSP processor without microcontroller is suitable in embedded applications that either do not need a man–machine interface or have a host machine that is responsible for the control functions. These applications represent a sizeable market: most modern modems in particular fall within that category. Moreover, DSPs are advantageously replacing microcontrollers in many hard disk control systems and even in some electric motors.

A new breed of single core processor has recently emerged: the DSP/RISC (not to be confused with the dual core DSP + RISC single chip architecture). The main advantage of a DSP/RISC processor, combining DSP and control functions, lies in avoiding the need for communication between processors, that is to say, only using one instruction sequencing machine, thus making a potential saving on the overall memory used, consumption and the number of pins. It remains to be seen whether these benefits will be borne out in the applications, but system analysis is often complicated so it is possible to come out in favor of these new architectures. The main problems constituted by this approach arise at the level of application software development. In fact, the flexibility of designing the software separately according to type is lost: for example a man–machine interface on the one hand and speech processing on the other. Between a DSP and a microcontroller, the programs used are different in nature and the implementation or adaptation requirements are greater as far as the controller is concerned: contrary to what one might expect, having the software in two distinct parts can thus be advantageous. At least at first, this problem of "programming culture" should not be neglected; teams which were different and separated up to now should form just one, generating, over and above technical pitfalls, human and organizational difficulties. Furthermore, betting on a single processor flexible enough to respond to the increasing demands placed on both DSP power and control is a daring wager, but it could be taken up for some types of applications, a priori at the lower end of the range: it still remains to be seen whether it will all be worth the effort.

Let us focus now on wireless terminals. Wireless handsets contain two parts: The modem part and the applications part. The modem sends data to the network via the air interface and retrieves data from the air interface. The application part performs functions that the user wants to use: speech, audio, image and video, e-mail, e-commerce, fax transmission; some other applications enhance user interface: speech recognition and enhancement (name dialing, acoustic echo cancellation), keyboard input (T9), handwritten recognition; other applications entertain the user (games...), help him/her organize his/her time (PIM functionality, memo)... Since wireless bandwidth is limited and expensive, speech, audio image and video signals will be heavily compressed before transmission; this compression requires extensive signal processing.

The modem function required traditionally a DSP for signal processing of the Layer1 modem and a microcontroller for Layer 2 and 3. Similarly, some applications (speech, audio, video compression...) require extensive signal processing and therefore should be mapped to DSP in order to consume minimum power while other applications are better mapped to the microprocessor (Figure 7.3).

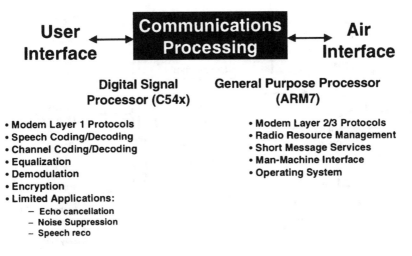

**Digital Signal   General Purpose Processor
Processor (C54x)   (ARM7)**

- **Modem Layer 1 Protocols**   **• Modem Layer 2/3 Protocols**
- **Speech Coding/Decoding**   **• Radio Resource Management**
- **Channel Coding/Decoding**   **• Short Message Services**
- **Equalization**   **• Man-Machine Interface**
- **Demodulation**   **• Operating System**
- **Encryption**
- **Limited Applications:**
  - **Echo cancellation**
  - **Noise Suppression**
  - **Speech reco**

Figure 7.3.   2G wireless architecture

Depending on the number of applications and on the processor performances, the DSP and/ or the microcontroller used for the modem can also be used for the application part. However, for phones which need to run media-rich applications enabled by the high bit rate of 2.5G and 3G, a separate DSP and a separate microcontroller will be required (Figure 7.4).

7.2. OMAP™ H/W Architecture

7.2.1. Architecture Description

The OMAP™ architecture, depicted in Figure 7.5, is designed to maximize the overall system performance of the 2.5G or 3G terminal while minimizing power consumption. This is achieved through the use of TI's state-of-the-art TMS320C55x DSP core and high perfor-

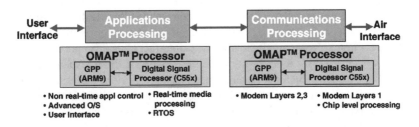

- **DSP and GPP provide the best architecture for low-power multimedia**
  - Optimum split of functions : real time DSP vs. control and OS
  - Minimum cycles/function and latency on real time media functions
  - Optimum OS support and non real time applications on GPP
- **An Open platform based on GPP & DSP needed to be introduced**

Figure 7.4.   3G wireless architecture

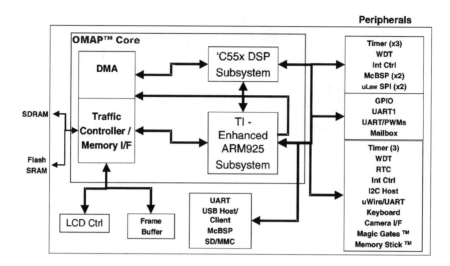

Figure 7.5.   OMAP1510 applications processor

mance ARM925T CPU. Both processors utilize a cached architecture to reduce the average
access time to instruction memory and eliminate power hungry external accesses. In addition
both cores have a Memory Management Unit (MMU) for virtual to physical memory translation and task to task protection.

OMAP™ also contains two external memory interfaces and one internal memory port. The
first supports a direct connection to synchronous DRAMs at up to 100 MHz. The second
external interface supports standard asynchronous memories such as SRAM, FLASH, or
burst FLASH devices. This interface is typically used for program storage and can be configured as 16 or 32 bits wide. The internal memory port allows direct connection to on-chip
memory such as SRAM or embedded FLASH and can be used for frequently accessed data
such as critical OS routines or the LCD frame buffer. This has the benefit of reducing the
access time and eliminating costly external accesses. All three interfaces are completely
independent and allow concurrent access from either processor or DMA unit.

OMAP™ also contains numerous interfaces to connect to peripherals or external devices.
Each processor has its own external peripheral interface that supports direct connection to
peripherals. To improve system efficiency these interfaces also support DMA from the
respective processor's DMA unit. In addition the design facilitates shared access to the
peripherals where needed. The local bus interface is a high speed bi-directional multi-master
bus that can be used to connect to external peripherals or additional OMAP™-based devices
in a multi-core product. Additionally, a high speed access bus is available to allow an external
device to share the main OMAP™ system memory (SDRAM, FLASH, internal memory).
This interface provides an efficient mechanism for data communication and also allows the
designer to reduce system cost by reducing the number of external memories required in the
system. In order to support common operating system requirements several peripherals are
included such as timers, general purpose input/output, a UART, and watchdog timers. These
peripherals are intended to be the minimum peripherals required in the system. Additional
peripherals can be added on the Rhea interfaces. A color LCD controller is also included to

support a direct connection to the LCD panel. The ARM™ DMA engine contains a dedicated channel that is used to transfer data from the frame buffer to the LCD controller where the frame buffer can be allocated in the SDRAM or internal SRAM.

7.2.2. Advantages of a Combined RISC/DSP Architecture

As depicted in the previous section, OMAP™ architecture is based on a combination of a RISC (ARM925) and a DSP (TMS320C55x). A RISC architecture, like ARM925, is best suited for control type code (OS, user interface, OS applications), whereas a DSP is best suited for signal processing applications, such as MPEG4 video, speech and audio applications.

A comparative benchmarking study (see Figure 7.5) has shown that executing a signal processing task would consume three times more cycles when executed on the latest RISC machine (StrongARM™, ARM9E, ARM10) compared to a TMS320C55x DSP. In terms of power consumption, it has been shown that a given signal processing task executed on such a RISC engine would consume more than twice the power required to execute the same task on a TMS320C55x architecture. The battery life, critical for mobile applications, will therefore be much higher in a combined architecture versus a RISC-only platform.

For instance, a single TMS320C55x DSP can process in real-time a full video-conferencing application (audio + video at 15 images/s), using only 40% of the total CPU computation capability. Sixty percent of the CPU is therefore still available to run other applications at the same time. Moreover, in a dual core architecture like OMAP™, the ARM™ processor is in that case fully available to run the operating system and its related applications. The mobile user can therefore still have access to his/her usual OS applications while processing a full videoconferencing application.

A single RISC architecture would have to use its full CPU computation capability to execute only the videoconferencing application, for twice the power consumption of the TMS320C55x. In addition, there is a gain because the two cores truly process in parallel. Therefore, the mobile user will not be able to execute any other application at the same time. Moreover, the battery life will be dramatically reduced.

7.2.3. TMS320C55x and Multimedia Extensions

The TMS320C55x DSP offers a highly optimized architecture for wireless modem and vocoding applications execution. Corresponding code size and power consumption are also optimized at the system level. These features also benefit a wider range of applications with some trade-offs in performance or power consumption.

The flexible architecture of the TI DSP hardware core allows extension of the core functions for multi-media specific operations. To facilitate the demands of the multi-media market for real-time low power processing of streaming video and audio, the TMS320C55x family device is the first DSP with such core level multi-media specific extensions. The software developer has access to the multi-media extensions using the *copr( )* instructions as described in Chapter 18.

One of the first application domains that will extend the functionality of wireless terminals is video processing. Motion estimation, Discrete Cosine Transform (DCT) and its inverse

Table 7.1 Video hardware accelerators characteristics

HWA type	Current consumption (at 1.5 V) (mA/MHz)	Speed-up factor versus software
Motion estimation	0.04	x5.2
DCT/iDCT	0.06	x4.1
Pixel interpolation	0.01	x7.3

function (iDCT) and pixel interpolation are the most consuming in terms of number of cycles for a pure software implementation using the TMS320C55x processor.

Table 7.1 summarizes the extensions' characteristics. The overall video codec application mentioned earlier is accelerated by a factor of 2 using the extensions versus a classic software implementation. By reducing cycle count, the DSP real-time operating frequency and, thus, the power consumption are also reduced. Table 7.2 summarizes performance and current consumption (at maximum and lowest possible supply voltage) of a TMS320C55x video MPEG4 coder/decoder using multimedia extensions, for various image rates and formats.

7.3. OMAP™ S/W Architecture

OMAP™ includes an open software infrastructure that is needed to support application development and provide a dynamic upgrade capability for a heterogeneous multiprocessor system design. This infrastructure includes a framework for developing software that targets the system design and Application Programmer Interfaces (APIs) for executing software on the target system.

Future 2.5G and 3G wireless systems will see a merge of the classical "voice centric" phone model with the data functionality of the Personal Digital Assistant (PDA). It is expected that non-voice multimedia applications (MPEG4 video, MP3 audio, etc.) will be downloaded to future phone platforms. These systems will also have to accommodate a variety of popular operating systems, such as WinCE, EPOC, Linux and others on the MCU side. Moreover, the dynamic, multi-tasking nature of these applications will require the use of operating systems on the DSP as well.

Table 7.2 MPEG4 video codec performance and power

Formats and rates	Millions of cycles/s	mA@1.5 V (0.1u Leff)	mA@0.9 V (0.1u Leff)
QCIF, 10 fps	18	12	7
QCIF, 15 fps	28	19	11
QCIF, 30 fps	55	37	22
CIF, 10 fps	73	49	29
CIF, 15 fps	110	74	44

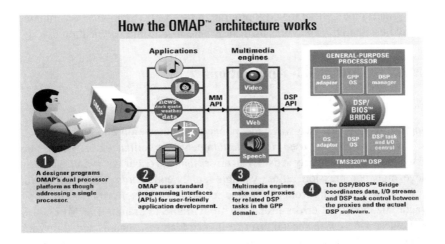

Figure 7.6. TI DSP/BIOS™ Bridge delivers seamless access to enhanced system performance

Thus the OMAP™ platform requires a software architecture that is generic to allow easy adaptation and expansion for future technology. At the same time, it needs to provide an I/O and processing performance that will allow it to be near the performance of a specific targeted architecture.

It is important to be able to abstract the implementation of the DSP software architecture from the General-Purpose Programming (GPP) environment. In the OMAP™ system, we do this by defining an interface architecture that allows the GPP to be the system master. The architecture of this "DSPBridge" consists of a set of APIs that includes device driver interfaces (Figure 7.6).

The most important function that DSPBridge provides is communications between GPP applications and DSP tasks. This communication enables GPP applications and device drivers to:

- Initiate and control tasks on the DSP
- Exchange messages with the DSP
- Stream data to and from the DSP
- Perform status queries

Standardization and re-use of existing APIs and application software are the main goals for the open platform architecture, allowing extensive re-use of previously developed software and a faster time to market of new software products.

On the GPP side, the API that interfaces to the DSP is called the Resource Manager (RM). The RM will be the singular path through which DSP applications are loaded, initiated and controlled. The RM keeps track of DSP resources such as MIPS, memory pool saturation, task load, etc., and controls starting and stopping tasks, controlling data streams between DSP and GPP, reserving and releasing shared system resources (e.g. memory), etc.

The RM projects the DSP in the GPP programming space and applications running in this space can address the DSP functions as if they were local to the application.

7.4. OMAP™ Multimedia Applications

7.4.1. Video

Video applications include two-way videophone communication and one-way decoding or encoding, which might be used for entertainment, surveillance, or video messaging. Compressed video is particularly sensitive to errors that can occur with wireless transmission. To achieve high compression ratios, variable-length codewords are used and motion is modeled by copying blocks from one frame to the next. When errors occur, the decoder loses synchronization, and errors propagate from frame to frame. The MPEG-4 standard supports wireless video with special error resilience features, such as added resynchronization markers and redundant header information. The MPEG-4 data-partitioning tool, originally proposed by TI, puts the most important data in the first partition of a video packet, which makes partial reconstruction possible for better error concealment.

TI's MPEG-4 video software for OMAP™ was developed based on reference C software, which was then converted to use ETSI C libraries, and then ported to TMS320C55x assembly code. The ETSI C libraries consist of routines representing all common DSP instructions. The ETSI routines perform the desired function, but also evaluate processing cycles and check for saturation, etc. Thus, the ETSI C, commonly used for testing speech codecs, provides a tool for benchmarking, and facilitates porting the C code to assembly.

As shown in Section 7.2.2, the video software runs very efficiently on OMAP™. The architecture is able to encode and decode in the same time as QCIF (176 × 144 pixels) images at 15 frames per second. The CPU loading for simultaneous encoding and decoding represents only 15% of the total DSP CPU capability. Therefore, 85% of the CPU is still available for running other tasks, such as graphic enhancements, audio playback (MP3), speech recognition.

The assembly encoder is under development, and typically requires about three times as much processing as the decoder. The main processing bottlenecks are motion estimation, DCT and IDCT. However, the OMAP™ hardware accelerators will improve the video encoding execution by a factor of two, through tight coupling of hardware and software.

OMAP™ provides not only the computational resources, but also the data-transfer capability needed for video applications. One QCIF frame requires 38016 bytes, for chrominance components down-sampled in 4:2:0 format, when transferring uncompressed data from a camera or to a display. The video decoder and encoder must access both the current frame and the previously decoded frame in order to do the motion compensation and estimation, respectively. Frame rates of 10–15 frames per second need to be supported for wireless applications.

3G standards for wireless communication, along with the new MPEG-4 video standard, and new low-power platforms like OMAP™, will make possible many new video applications. It is quite probable that video applications will differentiate between 2G and 3G devices, creating new markets and higher demand for wireless communicators.

7.4.2. Speech Applications

Continuous speech recognition is another resource-intensive algorithm. For example, commercial large-vocabulary dictation systems require more than 10 MB of disk space and 32 MB of RAM for execution. A typical embedded system, however, has constraints of low power, small memory size and little to no disk storage. Therefore, speech recognition algorithms designed for embedded systems such as wireless phones need to minimize resource usage while providing acceptable recognition performance.

We propose a dynamic vocabulary speech recognizer that is split between the DSP and the ARM™. The computation-intensive, small-footprint speech recognition engine runs on the DSP, while the computation non-intensive, larger footprint grammar, dictionary and acoustic model generation components reside on the ARM™. The interaction between the model generation and recognition modules is kept to a minimum and conducted via a hierarchy of APIs (described in Figure 7.4). The advantage of this architecture is that the application can handle new vocabularies in several (potentially unlimited) recognition contexts without pre-compilation or storage of the grammars and models.

For each new recognition context, the grammars and acoustic models are generated dynamically on the ARM™ and transferred to the recognizer on the DSP. For example, a voice-enabled web browser on the phone can now handle several different websites, each with its own different vocabulary. Similarly, for a voice-enabled stock quote retrieval application, company names can be dynamically added to or removed from the stock portfolio. The only caveat is that at any given time, the size of the active vocabulary is limited by the resource constraints on the DSP (e.g., size of the RAM available for the recognition search). However, given that different vocabularies can be swapped in and out depending on the recognition context, the application can be designed to give the user the perception of an unlimited vocabulary speech recognition system.

Similarly, the Text-To-Speech (TTS) system on the wireless device can be split between the ARM™ and DSP. The text analysis and linguistic processing modules of the TTS reside on the ARM™ along with the phonetic databases. The unit selection and waveform generation modules reside on the DSP. As with the speech recognizer, the interaction between the ARM™ and DSP modules is kept to a minimum and conducted via a hierarchy of APIs.

7.5. Conclusion

The OMAP™ multiprocessor architecture has been optimized to support heavy multimedia applications such as video and speech in 2.5G and 3G terminals. Such an architecture, combining two heterogeneous processors (RISC and DSP), several OS combinations and applications running on both the DSP and ARM™ can be made accessible seamlessly to application developers thanks to the DSPBridge concept. Moreover, this dual processor architecture is shown to be both more cost and power efficient than a single processor solution to the same problem.

Further Reading

[1] Auslander E., 'Le traitement du signal accepte le Risc', Electronique 73, September 1997.
[2] http://www.ti.com/sc/docs/apps/omap/overview.htm

8

A Flexible Distributed Java Environment for Wireless PDA Architectures Based on DSP Technology

Gilbert Cabillic, Jean-Philippe Lesot, Frédéric Parain, Michel Banâtre, Valérie Issarny, Teresa Higuera, Gérard Chauvel, Serge Lasserre and Dominique D'Inverno

8.1 Introduction

Java offers several benefits that could facilitate the use of wireless Personal Digital Assistants (WPDAs) for the user. First, Java is portable, and that means that it is independent of the hardware platform it runs on, which is very important for reducing the cost of application development. As Java can be run anywhere, the development of applications can be done on a desktop without the need of a real hardware platform. Second, Java supports dynamic loading of applications and can significantly contribute to extend the use of WPDA.

Nevertheless, even if Java has a very good potential, one of its main drawbacks is the need of resources for running a Java application. By resources, we mean memory volume, execution time and energy consumption, which are the resources examined for embedded system trade-off conception. It is clear that the success of Java is conditioned by the availability of a Java execution environment that will manage efficiently these resources.

Our goal is to offer a Java execution environment for WPDA architectures that enables a good trade-off between performance, energy and memory usage. This chapter is composed of three different parts. As energy consumption is very important for WPDAs, we first pose the problem of Java and energy. We propose a classification of opcodes depending on energy features, and then, using a set of representative WPDA applications, we analyze what the Java opcodes are that will influence significantly the energy consumption. In the second part we present our approach to construct a Java execution environment that is based on a modular decomposition. Using modularity it is possible to specialize some parts of the Java Virtual

Machine (Jvm) for one specific processor (for example to exploit low power features of the DSP). At last, as WPDA architectures are based on shared memory heterogeneous multi-processors such as the *Omap* platform [25], we present in the third part our ongoing work carried out around a distributed Jvm[1] in the context of multimedia applications. This distributed Jvm first permits the use of a DSP to a Java application and second exploits the energy consumption features of the DSP to minimize the overall energy consumption of the WPDA.

8.2 Java and Energy: Analyzing the Challenge

Let's first examine the energy aspects related to Java opcodes. Then let's give some approaches (hardware and software) to minimizing the energy consumption of opcodes.

8.2.1 Analysis of Java Opcodes

To understand the energetic problem connected with a Java execution, we describe in this section the hardware and software resources involved for a Java program, and in what way these resources affect the energy consumption. We propose a classification according to hardware and software components needed to realize them; the complexity of opcode's realization and energy consumption aspects.

Java code is composed of opcodes defined in the Java Virtual Machine Specification [14]. There are four different memory areas. The *constant pool* contains all the data constants needed to run the application (signature of methods, arithmetic or string constants, etc.). The *object heap* is the memory used by the Jvm to allocate objects. With each method there is associated a *stack* and a *local variable* memory space. For a method, local variables are the input parameters and the private data of the method. The stack is used to store variables needed to realize the opcodes (for example, an arithmetic addition opcode *iadd* implies the realization of two pops, an addition, and the push of the result on the stack), and to store the resulting parameters after an invocation of a method. Note that at the end and at the beginning of a method the stack is empty. At last, associated with the memory areas, a Pointer Counter (PC) identifies the next opcode to execute in the Java method.

8.2.1.1 Classification

There are 201 opcodes supported by a Jvm. We propose the following classification:

Arithmetic
Java offers three categories of arithmetic opcodes: integer arithmetic opcodes with low (integer, coded on 32 bits) and high (long, coded on 64 bits) representation, floating point arithmetic opcodes with simple precision (float, coded on 32 bits) and double precision (double, coded on 64 bits), and logic arithmetic (shift and back forwards, and elementary logic operations).

These opcodes involve the use of the processor's arithmetic unit and, if available, the floating-point unit coprocessor. All 64-bit based opcodes will take more energy than the

---

[1] See http://www.irisa.fr/solidor/work/scracthy.html for more details.

others because more pop and push on the stack are required for the opcode realization. Moreover, floating point opcodes require more energy than for other arithmetic opcodes.

PC

This category concerns Java opcodes that are used to realize a conditional jump, a direct jump, or an execution of a subroutine. Other opcodes (*tableswitch and lookupswitch*) realize multiple conditional jumps. All these opcodes manipulate the PC of the method. The penalty on energy here is linked to the cache architecture of the processor. For example a jump can possibly generate a cache miss to load the new byte-code sequence the execution of which will take a lot of energy.

Stack

The opcodes related to the stack realize quite simple operations: popping out one (*pop*) or two stack entries (*pop2*); duplicating one (*dup*) or two entries (*dup2*) (note that it is possible to insert a value before duplicating the entries using *dup*_x** opcodes); or swapping the two first data on the stack (*swap*). The energy connected to these opcodes depends on the memory transfers on the stack memory area.

Load-Store

We group in this category opcodes that make memory transfers between the stack, the local variables, the constant pool (by pushing a constant on the heap) and the object heap (by setting or getting a value for an object field, static or not). Opcodes between the stack and the local variables are very simple to implement and take less energy than other ones. Constant pool opcodes are also quite simple to realize, but require more tests to identify the constant value. The realization complexity for opcodes involving the stack and the object heap is complex, principally due on one hand to the huge number of tests needed to verify that the method has the right to modify or access a protected object field, and on the other hand due to the identification of the memory position of the object. It is important to see that this memory is not necessary in the cache, and solving the memory access can introduce an important penalty on the energy. At last, in Java, an array is a specific object and is allocated in the object heap. An array is composed of a fixed number either of object references, or either of basic Java types (boolean, byte, short, char, integer, long, float, and double). Every load-store of an array requires a dynamic born check of the array (to see if the index is valid). This is why array opcodes need more tests than other load-store heap opcodes.

Object Management

In this category, we group opcodes related to object management as: method invocation, creation of an object, throw of an exception, monitor management (used for mutual exclusion), and opcodes needed for type inclusion tests. All these opcodes require a lot of energy because they are very complex to realize and involve lots of Jvm steps (in fact, the invocation is the most complex). For example, the invocation of a method requires at least the creation of a local variable memory zone and a stack space, to identify the method and then to execute the method entry point.

8.2.2 Analyzing Application Behavior

The goal of our experiments is to understand precisely the opcodes that significantly influence the energy consumption. To achieve this goal, we first chose a set of representative applications that cover the range of use of a wireless PDA (multimedia and interactivity based applications):

- *Mpeg I Player* is a port of a free video Mpeg I stream player [9]. This application builds a graphic window, opens the video stream, and decodes the video stream until the end. We used an Mpeg I layer II video stream that generates 8 min of video in a 160 × 140 window.
- *Wave Player* is a wave riff file player. We built from scratch this application and we decomposed it into two parts: a sound player decodes the wave streams and generates the sound, and a video plug-in shows the frequency histogram of played sounds. We used a 16-bit stereo wave file that generates approximately 5 min of sound.
- *Gif89a Player* reads a gif89a stream and eternally loops until a user exit action. We built from scratch this application. Instead of decoding all the bitmap images included in a gif89a file once, we preferred to realize the decoding of each image on the fly (indexation and decompression) to introduce more computations. Moreover, this solution requires less memory than the first one (because we don't have to allocate an important number of bitmaps). We used a gif89 file that generates 78 successive 160 × 140 sized images, and we ran the application for 5 min.
- *Jmpg123 Player* is a Mpeg I sound player [11]. We adapted the original player to integrate it on the supported APIs of our Java Virtual machine. This player doesn't use graphics. For our results, we played an mp3 file at 128 Kb/s, for approximately 6 min.
- *Mine Sweeper* is a game using graphic APIs. This application characterized for us a general application using text and graphics management, and also user interactivity. We adapted the application from an existing one freely available [12]. We played for 30 min with this game.

We integrated these applications on our Jvm (see Section 8.3.2.5) and its supported APIs to be able to run them. Then we extended our Jvm to count for each Java opcode, the number of times it is executed by each application in order to have statistics on the opcode distribution. We used the categorization introduced in the previous section to present the results. Moreover, to understand the application behavior better, we will present and analyze in detail a sub-distribution for each category.

To precisely evaluate which opcodes influence the energy, we need a real estimation of energy consumption for each opcode to compare the several categories.

8.2.2.1 Global Distribution

Table 8.1 presents the results for the global opcode distribution. As we can see, load-store takes the major part of the execution (63.93% for all applications). It is important to note that this number is high for all the applications. Due to the complexity of decoding, Jmpg123 and Mpeg I players have the most important use of arithmetic opcodes. We also can see, for these two applications, that the use of stack opcodes is important due to the arithmetic algorithm used for the decoding. Wave and Gif89a players and Mine Sweeper, have the highest proportion of object opcodes, due to their fine-grained class decomposition. At last, Mine Sweeper which is a very user interacted application implies the most important use of PC opcodes. That is due to the graphic use which generates a huge number of conditional jumps.

Table 8.1   Global distribution of opcodes

	Arithmetic (%)	PC (%)	Stack (%)	Load-store (%)	Objects (%)
Mpeg I Player	19.22	7.03	8.04	64.82	0.89
Wave Player	14.89	9.37	5.22	68.29	2.23
Gif89a Player	13.65	9.35	5.60	69.41	1.99
Jmpg123 Player	24.48	3.50	11.86	59.58	0.58
Mine Sweeper	17.51	17.45	4.55	57.57	2.92

8.2.2.2 Sub-Distribution Analysis

Arithmetic
Table 8.2 presents a detailed distribution of arithmetic opcodes. First, only Jmpg123 uses intensively floating arithmetic double precision. It is important because floating point arithmetic needs more energy than integer or logic arithmetic. After analyzing the source code of Jmpg123, the decoder algorithm uses a lot of floating points. We think that it is possible to avoid the use of floating points for the decoding using a long representation in order to have the required precision, but the application's code has to be transformed. Gif89, Wave and Mpeg I Player use a lot of logic opcodes. In fact, these applications use lots of 'or' and 'and' operations to manage the screen bitmaps.

Table 8.2   Distribution of arithmetic opcodes

	Integer (%)	Long (%)	Float (%)	Double (%)	Logic (%)
Mpeg I Player	83.96	3.05	–	–	12.99
Wave Player	88.19	0.24	–	–	11.57
Gif89a Player	80.78	0.01	–	0.61	18.60
Jmpg123 Player	46.12	–	–	45.26	8.62
Mine Sweeper	91.90	0.27	–	–	7.83

PC
Table 8.3 presents the distribution of PC opcodes. Numbers show that the conditional jump opcodes are the most executed. We can also see that no subroutines are used. So, for this category, the energetic optimization of conditional jump is important.

Stack
Table 8.4 presents the distribution of stack based opcodes. We can remark that the *dup* is the most numbered except for Jmpg123 Player where the use of double precision floats gives a huge number of *dup2* (used to duplicate one double precision value on the stack). Moreover, no *pop2*, *swap*, *tableswitch* or *lookupswitch* are done (column 5). At last, Mine Sweeper makes a lot of pop opcodes. This is due to the unused return parameters of graphic API methods. As these parameters are on the stack, pop opcodes are generated to release them.

Table 8.3 Distribution of PC manipulation opcodes

	Conditional jump (%)	Goto (%)	Subroutine (%)	Switch (%)
Mpeg I Player	89.666	9.333	–	1.011
Wave Player	96.981	3.017	–	0.002
Gif89a Player	95.502	4.494	–	0.004
Jmpg123 Player	88.763	11.231	–	0.006
Mine Sweeper	94.158	5.782	–	0.070

Table 8.4 Distribution of stack opcodes

	Pop (%)	Dup (%)	Dup2 (%)	Other (%)
Mpeg I Player	0.00	99.34	0.65	–
Wave Player	0.04	99.30	0.66	–
Gif89a Player	0.06	99.94	–	–
Jmpg123 Player	–	21.63	78.37	–
Mine Sweeper	38.46	61.54	–	–

Load-Store

As we said previously, this category represent opcodes making memory transfers between the method's stack and the method's local variables, between the method's stack and the object heap, and between the constant pool and the stack. Table 8.5 presents the global distribution for these three sub-categories. Results indicate that Mpeg layers and Mine Sweeper use more local variable data than object data. For the Wave and Gif89a players, as we already mentioned previously, when we designed the applications, we decomposed them into many classes. This is why, according to the object decomposition, the number of accesses to the object heap is higher than the other applications. For Mpeg I Player, accesses to the constant pool corresponds to many constants.

Table 8.5 Distribution between stack-local and stack-heap

	Stack-local (%)	Stack-heap (%)	Stack-constant pool (%)
Mpeg I Player	59.22	38.06	2.72
Wave Player	42.28	56.63	1.09
Gif89a Player	37.91	60.64	1.45
Jmpg123 Player	60.95	37.39	1.66
Mine Sweeper	96.22	3.70	0.08

We calculated the distribution for each sub-category according to read (load) and write (store) opcodes. Moreover in Java, some opcode exists to push constants that are intensively used (for example, the opcode *iconst_1* puts the integer value 1 on the stack). These opcodes

Table 8.6 Detailed distribution of load-store between stack and local variables

	Integer			Long			Float	Double		
	Const	Load	Store	Const	Load	Store	–	Const	Load	Store
Mpeg I Player	12.27	65.71	19.80	0.75	0.73	0.73	–	0.00	–	–
Wave Player	6.42	81.45	12.00	0.01	0.90	0.02	–	–	–	–
Gif89a Player	5.29	83.21	11.49	0.00	0.00	0.00	–	–	–	–
Jmpg123 Player	10.69	31.89	3.86	–	–	–	–	0.78	29.93	22.85
Mine Sweeper	4.79	81.29	13.83	0.00	0.07	0.01	–	–	–	–

are simple and require less energy than other load-store opcodes. This is why we also distinguish them in the distribution.

Table 8.6 presents the results for the load-store between the stack and the local variables of methods. Loads are more important that store opcodes.

Results of the memory transfers between the stack and the heap are shown in Table 8.7. We can see, except for Mine Sweeper, that applications significantly use arrays. Moreover, as shown in Table 8.6, load operations are the most important. So, the load opcodes are going to drive the overall energetic consumption. For Jmpg123, field opcodes are important and are used to access data constants.

Table 8.7 Detailed distribution of load-store between stack and object heap

	Array		Field	
	Load (%)	Store (%)	Load (%)	Store (%)
Mpeg I Player	60.35	10.27	26.78	2.60
Wave Player	56.76	7.31	33.36	2.57
Gif89a Player	54.26	7.67	33.34	4.73
Jmpg123 Player	69.48	9.06	18.35	3.11
Mine Sweeper	51.49	1.49	45.44	1.58

Objects

Table 8.8 presents evaluation done for object opcodes. Results show that the invocation opcode should influence significantly (97.75% for all applications) the energy consumption. Table 8.8 shows that no *athrow* opcode occur (exception mechanism) and few new opcodes occur. It is clear that depending on the way the applications are coded, the proportion of these opcodes could be more important, but the invocation will be that of the majority executed opcode.

8.2.3 Analysis

We have shown that load-store opcodes, are the most executed opcodes, which is counter intuitive. Arithmetic opcodes are next.

Table 8.8 Distribution of objects opcodes

	New (%)	Athrow (%)	Invoke (%)	Other (%)
Mpeg I Player	0.01	–	99.97	0.02
Wave Player	0.21	–	92.75	7.04
Gif89a Player	0.14	–	99.78	0.08
Jmpg123 Player	2.14	–	97.83	0.03
Mine Sweeper	1.29	–	98.44	0.27

A real evaluation of the energy consumption for each opcode is needed to totally understand the problem. We group into three categories the approaches that could bring minimization of energy consumption in the context of Java:

8.2.3.1 Java Compilers

Existing compilation techniques could be adapted to Java compilers to minimize the energy consumption (by the use of less greedy energetic opcodes). An extension to these techniques could also permit the optimization of the number of load-store opcodes, or minimize the number of invocations. Moreover, the transformation of a floating-point based Java code into an integer based Java code will minimize the energy consumption. At last, the use of the switch jump opcode has to be preferred to a sequence of several conditional jumps opcode.

8.2.3.2 Opcode Realization

The most important work has to be done on Jvm opcode realization. An important Jvm flexibility is required in order to explore and quantify the energy consumption gain according to a specific energetic optimization. For example, the use of compilation techniques [1,2] that have energetic optimization features could improve the overall energy consumption of the Jvm. A careful management of memory areas [3] will also permit the optimization of the energy consumption of load-store opcodes. For example, the use of locked cache line techniques or private memories could bring a significant improvement of energy consumption by minimizing the number of electric signals needed to access the data. Moreover, some processor features need to be exploited. For example, for a DSP processor [10], the Jvm could create for a method the local variable memory area and the stack inside the local RAM of the DSP. In this way, accesses are going to be less greedy in energy consumption, but also more efficient due to the DSP possibility to access simultaneously different memory areas. Just-in-time compilers could generate during the execution a binary code requiring less energy. For example, on object opcodes, the amount of invocation opcodes can be reduced by in-lining of methods [16]. Nevertheless, at this moment these compilers generate an important memory volume making their use inappropriate in the WPDA context. Another way to improve performance is to accelerate type inclusion tests [8]. And last, but not least, hardware Java co-processors [5,6,7] could also permit the reduction of the energy for arithmetic opcodes. Moreover, due to the number of load-store opcodes, the memory hierarchy architecture should also influence the energy consumption.

8.2.3.3 Operating System

As the Jvm relies on an operating system (thread management, memory allocation and network), an energetic optimization of the operating system is complementary to these previous works. A power analysis of a real-time operating systems, as presented in [4] will permit the system to adapt the way the Jvm uses the operating system to decrease the energy consumption.

8.3 A Modular Java Virtual Machine

A flexible Java environment is required in order to work on the trade-off between memory, energy and performance. In order to be flexible, we designed a modular Java environment named Scratchy. Using modularity, it is possible to specialize some parts of the Jvm for one specific processor (for example to exploit low power features of the DSP). Moreover, as WPDA architecture orientation is to be shared memory heterogeneous multiprocessor based [25] support for managing the heterogeneity of data allocation schemes has been introduced.

In this section we first describe the several Java environment implantation possibilities; we then introduce our modular methodology and development environment; and then we present Scratchy, our modular Jvm.

8.3.1 Java Implantation Possibilities

8.3.1.1 Hardware Dependent Implantation

A Jvm permits the total abstraction of the embedded system from the application programmer's point of view. Nevertheless, for the Jvm developer, the good trade-off is critical because performance relies principally on the Jvm. For the developer, an embedded system is composed of, among other things, one or several core processors, memory architecture and some energy awareness features. This section characterizes these three categories and gives one simple example of the implementation of a Jvm part.

Core Processor
Each processor defines its own data representation capabilities from 8-bit up to 128-bit. To be efficient, the Jvm must realize a bytecode manipulation adapted to the data representation. Note that Ref. [13] reports that 5 billion out of 8 billion manufactured in 2000 were 8-bit microprocessors. The availability of a floating-point support (32- or 64-bit) in a processor could also be used by the Jvm to treat the float or double Java types. Regarding the available registers, the use of a subset of those can be exploited to optimize Java stack performance. One register can be used to represent the Java stack pointer. Memory alignment (constant, proportional to size with or without threshold, etc.) and memory access cost have to be taken into account to efficiently arrange object fields. In a multiprocessor case, homogeneous or heterogeneous data representation (little/big endian), data size and alignment have to be managed to correctly share Java objects or internal Jvm data structures between processors.

Memory
The diversity of memories (RAM, DRAM, flash, local RAM, etc.) has to be considered depending on their performance and their sizes. For example, local variable sets can be stored in a local RAM, and class files in flash. Finally, on a shared memory multiprocessor, the Jvm

must manage homogeneous or heterogeneous address space to correctly share Java objects or Jvm internal structures. Moreover, regarding the cache architecture (level one or two, type of flush, etc.), it must be used to implement Java synchronized and volatile attributes of an object field.

Energy

The use by the Jvm of energetic aware instruction set for the realization of the bytecode to minimize system energy consumption is a possibility. As well, considering the number of memory transfers is a big deal for the Jvm, as mentioned in the first part of this chapter.

8.3.1.2 Example

To illustrate the several possibilities of implementation, we describe in this section some experiments done with the interpreter engine of our Jvm on two different processors, Intel Pentium II and TI TMS320C55x DSP. To understand the following experiments we briefly describe (i) hardware features and tool chain characteristics of the TMS320C55x, (ii) the interpreter engine role, and (iii) two different possible implementations.

TI DSP TMS320C55x

The TMS320C55x family of Texas Instruments is a low power DSP. Its data space is only word – addressable. Its tool chain has an ANSI C compiler, an assembler and a linker. The C compiler has uncommon data types like char of 16-bit, and function pointer of 24-bit. The reader should refer to Ref. [10] for more details.

The Interpreter Engine

The interpreter engine of a Jvm is in charge of decoding the bytecode to call the appropriate function to perform the opcode operation. To implement the engine, the first solution is to use a classical loop to fetch one opcode, to decode it with a C switch statement (or similar statements in other languages) and to branch to the right piece of code. The second solution, called threaded code [17], is to translate, before execution, the original bytecode by a sequence of addresses which point directly to opcode implementations. At the end of each opcode implementation, a branch to the next address is performed and so on. This solution avoids the decoding phase of the switch solution but causes an expansion of code.

Experiments

We carried out experiments on these two solutions. We first coded a switched interpreter with ANSI C, and a threaded interpreter with gnu C (this implementation requires the gnu C's label as a values feature). We experimented with both versions on the 32-bit Intel Pentium II. We then compiled the switched interpreter on the TI DSP TMS320C55x. Due to the absence of gnu C features on the TI tool chain, we added assembler statements to enable the threaded interpreter to thus become a TMS320C55x specific one.

Table 8.9 shows the number of cycles used to execute one loop of a Java class-file to compute Fibonacci numbers and the number of memory bytes to store the bytecode (translated or not).

On the Intel Pentium II, the threaded interpreter saves 19% of cycles, whereas on the TI DSP TMS320C55x, it saves 62% of cycles. The difference of speed-up due to the same optimization is very important: more than three times on the TI DSP TMS320C55x.

Table 8.9 Example of different interpreter implementation

	Pentium II				TI DSP TMS320C55x		
	Lang	Cycle	Mem		Lang	Cycle	Mem
Switch	ANSI C	162	20	Switch	ANSI C	294	40
Threaded	gnu C	132	80	Threaded	C/asm	183	80
Ratio		19%	400%	Ratio		62%	200%
				Switch	ANSI C	294	40
				Switch2	ANSI C	622	20
				Ratio		211%	50%

On the other hand, on the Pentium II, there is a memory overhead of 400% between the threaded motor and the switched interpreter. This is because of the representation of one opcode by 1 byte in the first case and by one pointer in the last case. On the TI DSP TMS320C55x, the memory overhead is only 200% because of the representation of one opcode by 2 bytes (due to the 16-bit char type). Thus, we added a modified switched interpreter specific to the TMS320C55x ("switch 2") without memory overhead. It takes the high or low byte of one word depending of the program counter parity. This interpreter takes more than double the cycles of a classic switched one.

In conclusion, for this simple example, we obtained four interpreter implementations, two processor specifics, one C compiler specific and one generic. All have different performance and memory expansion behavior. This illustrates the difficulties involved in reaching the desired trade-off especially with respect to complex software like Jvm.

8.3.2 Approach: a Modular Java Environment

8.3.2.1 Objective: Resource Management Tradeoff Study

We believe that for one WPDA, a specific resource management trade-off has to be found to deal with the user choice. We consider that the trade-off can be abstracted through three criteria: *Mem* for memory allocation size, *Cpu* for efficiency of treatments and *Energy* for energy consumption. The less these criteria are, more optimum is the associated resource management. For example, a small *Mem* means that the memory is compacted. If *Cpu* is high that means that the Java environment is inefficient and if *Energy* is small that the Java environment takes care of energy consumption.

It is clear that, depending on the architecture of a WPDA, this trade-off has to be changed depending on the user needs. For example, if the user prefers to have an efficient Java environment, the trade-off has to minimize *Cpu* criterion, without taking care of *Energy* and *Mem* resources. As all these criteria depend on the architecture features, we believe that to one dedicated WPDA architecture is connected several strategies of trade-off that can be proposed to the user.

But focusing on this trade-off on a monolithic Java programmed environment is difficult and introduces an expensive cost. Due to the huge compactness of source code, it's very difficult to change at many levels the way to manage the hardware resources without rewriting the Jvm from scratch for each platform.

Our approach to reach the possibility to adapt our resource management for a WPDAs Java environment consists of splitting up the source code of the Java environment into software independent modules. In this way, we want to break with problems introduced by a monolithic design. Like this, it is possible to work on a particular module to obtain one trade-off between performance, memory and energy for that module by designing new resource management strategies. At the end, it is easier to reach the trade-off for all modules (so for the complete Jvm).

In this way, the usage of a resource can be redesign with a minimal cost of time. Moreover, the resource management strategies can be more open than a monolithic approach because the resource management in one module is isolated, but contributes to the global trade-off. We also believe that the experimentation of strategies is necessary to design new ones, compare with existing ones and choose the best one. Modularity is a way to realize these choices.

8.3.2.2 Modularity Related Works

Object-based solutions like *corba* [20] or *com* [18] are not suitable because the linking is too dynamic and introduces a high overhead for an embedded software. To reach the desired trade-off, it is necessary to introduce the minimum overhead in terms of Cpu, memory and energy.

Introduction of modularity using compilers like a generic package in *Modula-3* [21] or *Ada95* [19] is not a practical solution because it is uncommon to have compilers other than C or native assembler with embedded hardware. It is also uncommon to have source code of the compiler to add features.

The approach taken by *Knit* [24] for use with the OSKit is very interesting and quite close to our solution. It is a tool, with its own language to describe linking requirements. But it is too limited for our goals because it only addresses problems at the linking level, so Knit manages only function calls, and not data types. Management of data type is useful to optimize memory consumption and data access performance especially on shared memory multiprocessor hardware.

8.3.2.3 Scratchy Development Environment Principle

We designed the "Scratchy Development Environment" (SDE) to achieve modularity for a Java environment. This tool is designed to introduce no time or memory overhead on module merging and to be the most language and compiler independent as possible. SDE takes four inputs:

- A global specification file describes services (functions) and data types by using an Interface Definition Language (IDL);
- A set of modules implements services and data types in one of the supported language mappings;
- An implementation file inside each module describes the link between specification and implementation;
- Alignment descriptions indicate alignment constrains with their respective access cost for each processor in the targeted hardware.

SDE chooses a subset of modules, generates stubs for services, sets structures of data types, generates functions to dynamically allocate and access data types, etc. SDE works at source level, so it is the responsibility of compilers or preprocessors (through in-lining for example) to optimize a source which has potentially no overhead.

8.3.2.4 Modularity Perspectives

A modular environment classically allows replacing, completing, composing and intercepting functions. On top of that, for a specific hardware, SDE could be used to compensate the lack of features, to optimize an implementation or to exploit a hardware support:

- There is no direct use of compiler basic types. It is possible to easily implement missing types (e.g. 64-bit integer support);
- Structures are not managed by the compiler but by SDE. Therefore, it is possible to generate well-aligned structures compatible at the same time with several core processors as well as rearrange structure organization to minimize memory consumption. It is also possible to manage the trade-off between CPU, memory and energy. For example, due to the high frequency of object structure access, the access cost to objects is very important to optimize;
- SDE provides developers with mechanisms to access data. Therefore, a module in SDE could intercept data access to convert data between heterogeneous processors, and heterogeneous address space for example.

8.3.2.5 Scratchy Modular Java Virtual Machine

Scratchy is our Jvm that has been designed using the modular approach. To achieve the modular design, we wrote Scratchy totally "from scratch". It is composed of 23 independent modules.

From the application's point of view, we fully support CLDC [15] specification. However, Scratchy is closer to CDC than CLDC because we also support floating-point operations and we want to authorize classloader execution to permit the downloading of applications through the network.

Most modules are written in ANSI C compiler, few with the gnu C features and others with TI DSP TMS320C55x assembler statements. We have also designed some module implementations with ARM/DSP assembler language and with DSP optimized C compiler (to increase the efficiency of bytecode execution).

Considering the operating system, we designed a module named "middleware" to make the link with an operating system. This module supports a Posix general operating system (Linux, Windows Professional/98/2000) but also a real-time operating system with Posix compatibility (VxWorks).

8.3.3 Comparison with Existing Java Environments

In this section, we compare our approach to two different Jvms designed for embedded systems: JWorks from WindRiver Systems, and Kvm, the Jvm of J2m from Sun microsystems.

- JWorks is a port of Personal Java Jvm distribution on the VxWorks real-time operating system. Because VxWorks is designed to integrate a large range of hardware platforms and because JWorks is (from the operating system's point of view) an application respecting VxWorks APIs, JWorks could be executed on a large range of hardware platforms. Nevertheless, the integration of JWorks on a new embedded system is limited to VxWorks porting, without any reconsideration of the Jvm. A Jvm must take care of many different aspects. That way, JWorks cannot achieve the desired trade-off explained previously whereas Scratchy does, thanks to modularity.
- J2me is a Sun Java platform for small embedded devices. Kvm is one possible Jvm of J2me. It supports 16- and 32-bit cisc and risc processors, engenders a small memory footprint and keeps the code in an area of 128 KB. It is written for an ANSI C Compiler, with the size of basic types well defined (e.g. char on 8-bit, long on 32-bit). Regarding data alignment, an optional alignment can only be obtained for 64-bit data. The C compiler handles other alignments. Moreover, there is no possibility of managing a heterogeneous multiprocessor without rewriting C structures (due to data representation conversion). In fact, Kvm is one of six Jvms written by Sun more or less from scratch: HotSpot, Standard Jvm, Kvm, Cvm, ExactVm and Cardvm (it is a Java Card Vm but it indubitably has common mechanisms with a Jvm). That proves that it is impossible to tune a trade-off between CPU, memory and energy without rewriting all the parts of the Jvm.

8.4 Ongoing Work on Scratchy

Our main ongoing work is to design a Distributed Jvm (DJVM) in order to integrate an Omap shared memory heterogeneous multiprocessor [25]. Scratchy Jvm (presented in Section 8.3.2.5) is the base of our DJVM software implementation. Figure 8.1 presents an overview of our DJVM. The main objective of this approach is first to permit the use of the DSP as another processor to execute Java code and secondly to exploit the DSP low

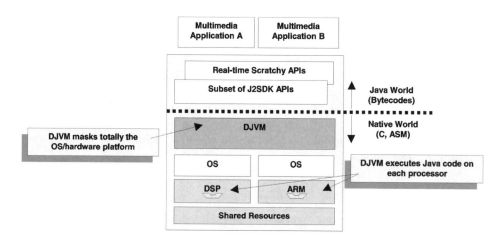

Figure 8.1 Distributed JVM overview

power features to minimize the overall energy consumption for Java applications. Using our DJVM it is the way to open the use of the DSP that is at this moment used for very specific jobs.

We briefly describe ongoing works we are carrying out on the DJVM in the context of multimedia applications:

8.4.1 Multi-Application Management

As a WPDA has a small memory, we introduce in the DJVM a support to execute several applications at the same time. In this way, only one Jvm process is necessary and the memory needed for the execution of Java applications can be reduced.

8.4.2 Managing the Processor's Heterogeneity and Architecture

We want each object to be shared by all the processors through the shared memory and accessed through the *putField* and *getField* opcodes. This is why we are designing modules to realize on the fly data representation conversion to take into account the plurality of data representations. Note that the heterogeneous allocation scheme of one object according to the specific object size of each processor is solved using our SDE development environment (see Section 8.3.2.3). At last, multiprocessor architectures dispose of multiple ways to allocate memory and to communicate between the processors. It is also an ongoing work to choose the best strategy that will result in a good performance.

8.4.3 Distribution of Tasks and Management of Soft Real-Time Constraints

Regarding the execution, our basic approach is to design a distribution rule that will transparently distribute the tasks among the multiprocessor. We consider a multimedia application as a soft real-time application (application deadlines are not guarantied but are just indicators for the scheduler). Thus, the distribution rules will take into account the efficiency of each processor to distribute the treatments in the best way. Moreover, a Java environment needs to provide a garbage collection process that will identify and free the unused objects. Due to soft real-time constraints of applications, the garbage collector strategy has to be designed in such a way that it does not disturb the application's execution [22].

8.4.4 Energy Management

Many works have been done on energy consumption reduction. We introduce [23] a new way to manage the energy in the context of a heterogeneous shared memory multiprocessor. The basic idea is to distribute treatments according to their respective energy consumption. In the case of the OMAP^{TM} platform, the use of the DSP results in a decrease in the energy consumption for all DSP-based treatments.

8.5 Conclusion

In this chapter we first presented the energy problem connected with Java in the context of WPDAs. We showed by experimentation made on a representative set of applications, that load-store opcodes and then arithmetic opcodes have the most important proportion (81.88% for all these opcodes). We secondly described our methodology to design and implement a Jvm on WPDA architectures. We presented our Jvm named Scratchy, which follows our modular methodology. This methodology is based on modularity specification and implementation. The flexibility introduced by modularity opens the possibilities of design, with regard to one module without reconsidering the implementation of all the other ones. It permits also the exploitation of specific features of processors such as the low-power instruction set of a DSP. Lastly, we presented our ongoing works on the distributed Jvm and presented some of our future extensions that are going to be integrated soon in Scratchy Distributed Java Virtual Machine.

To conclude, we think that achieving energy, memory and performance trade-off is a real challenge for Java on WPDA architectures. With a strong cooperation between hardware and software through a flexible Jvm, this challenge could be solved.

References

[1] Simunic, T., Benini, L. and De Micheli, G., 'Energy-Efficient Design of Battery-Powered Embedded Systems', In: *Proceedings of International Symposium on Low Power Electronics and Design*, 1999.

[2] Simunic, T., Benini, L. and De Micheli, G., 'Cycle-Accurate Emulation of Energy Consumption in Embedded Systems', In: *Proceedings of the Design Automation Conference*, 1999, pp. 867–872.

[3] Da Silva, J.L., Catthoor, F., Verkest, D. and De Man, H., 'Power Exploration for Dynamic Data Types Through Virtual Memory Management Refinement', In: *Proceedings of International Symposium on Low Power Electronics and Design*, 1998.

[4] Dick, R.P., Lakshminarayana, G., Raghunathan, A. and Hja, N.K., 'Power Analysis of Embedded Operating Systems', In: *Proceedings of Design Automation Conference*, 2000.

[5] Shiffman, H., 'JSTAR: Practical Java Acceleration For Information Appliances', *Technical White Paper*, JEDI Technologies, October 2000.

[6] Ajile, 'Overview of Ajile Java Processors', *Technical White Paper*, Ajile (http://www.ajile.com), 2000.

[7] Imsys, 'Overview of Cjip Processor', *Technical White Paper*, Imsys (http://www.imsys.se), 2001.

[8] Vitek, J., Horspool, R. and Krall, A., 'Efficient Type Inclusion Test'. In: *Proceedings of ACM Conference on Object Oriented Programming Systems, Languages and Applications (OOPSLA)*, 1997.

[9] Anders, J., Mpeg I Java Video Decoder. TU-Chemnitz, http://www.rnvs.informatik.tu/ chemnitz.de/~jan/MPEG/MPEG_Play.html, 2000.

[10] Texas Instruments, TMS320C5x User's Guide. http://www-s.ti.com/sc/psheets/spru056d/ spru056d.pdf, 2000.

[11] Hipp, M., Jmpg123: a Java Mpeg I Layer III. OODesign, http://www.mpg123.org/, 2000.

[12] Bialach, R., *A Mine Sweeper Java Game*, Knowledge Grove Inc., http://www.microjava.com/, 2000.

[13] Kopetz, H., 'Fundamental R&D Issues in Real-Time Distributed Computing', In: *Proceedings of the 3rd IEEE International Symposium on Object-oriented Real-time Distributed Computing (ISORC)*, 2000.

[14] Sun Microsystems, *The Java Virtual Machine Specification*, Second Edition, 1999.

[15] Sun Microsystems, *CLDC and the K Virtual Machine (KVM)*, http://java.sun.com/ products/cldc/, 2000.

[16] Sun Microsystems, 'The Java Hotspot Performance Engine Architecture', *White Paper*, 1999.

[17] Bell, J.R., 'Threaded Code', *Communication of the ACM* 16(6), 1973, pp. 370–372.

[18] Microsoft Corporation and Digital Equipment Corporation, *Component Object Model Specification*, October 1995.

[19] International Organization for Standardization, *The Language*, Ada 95 Reference Manual, January 1995.

[20] Object Management Group, *The Common Object Request Broker: Architecture and Specification*, Revision 2.3, June 1999.

[21] Harbison, S.P., *Modula-3*, Prentice Hall, Englewood Cliffs, NJ, 1991.

[22] Higuera, T., Issarny, V., Cabillic, G., Parain, F., Lesot, J.P. and Banâtre, M., '-based Memory Management for Real-time Java', In: *Proceedings of the 4th IEEE International Symposium on Object-oriented Real-time Distributed Computing (ISORC)*, 2001.

[23] Parain, F., Cabillic, G., Lesot, J.P., Higuera, T., Issarny, V., Banâtre, M., 'Increasing Appliance Autonomy using Energy-Aware Scheduling of Java Multimedia Applications', In: *Proceedings of the 9th ACM SIGOPS European Workshop – Beyond the PC: New Challenges*, 2000.

[24] Reid, A., Flatt, M., Soller, L., Lepreau, J. and Eide, E., 'Knit: Component Composition for System Software'. In: *Proceedings of the 4th Symposium on Operating Systems Design and Implementation (OSDI)*, San Diego, CA, October 2000, pp. 347–360.

[25] http://www.ti.com/sc/docs/apps/omap/overview.htm

9

Speech Coding Standards in Mobile Communications

Erdal Paksoy, Vishu Viswanathan, Alan McCree

9.1 Introduction

Speech coding is at the heart of digital wireless telephony. It consists of reducing the number of bits needed to represent the speech signal while maintaining acceptable quality. Digital cellular telephony began in the late 1980s at a time when speech coding had matured enough to make it possible. Speech coding has made digital telephony an attractive proposition by compressing the speech signal, thus allowing a capacity increase over analog systems.

Speech coding standards are necessary to allow equipment from different manufacturers to successfully interoperate, thereby providing a unified set of wireless services to as many customers as possible. Standards bodies specify all aspects of the entire communication system, including the air interface, modulation techniques, communication protocols, multiple access technologies, signaling, and speech coding and associated channel error control mechanisms. Despite the objective of achieving widespread interoperability, political and economic realities as well as technological factors have led to the formation of several regional standards bodies around the globe. As a result, we have witnessed the proliferation of numerous incompatible standards, sometimes even in the same geographic area.

There have been many changes since the emergence of digital telephony. Advances in speech coding have resulted in considerable improvements in the voice quality experienced by the end-user. Adaptive multirate (AMR) systems have made it possible to achieve optimal operating points for speech coders in varying channel conditions and to dynamically trade off capacity versus quality. The advent of low bit-rate wideband telephony is set to offer a significant leap in speech quality. Packet-based systems are becoming increasingly important as mobile devices move beyond a simple voice service. While the push to unify standards in the third generation universal systems has only been partially successful, different standards bodies are beginning to use the same or similar speech coders in different systems, making increased interoperability possible.

As speech coding algorithms involve extensive signal processing, they represent one of the main applications for digital signal processors (DSPs). In fact, DSPs are ideally suited for mobile handsets, and DSP architectures have evolved over time to accommodate the needs of speech coding algorithms. In this chapter, we provide the background necessary to understand modern speech coders, introduce the various speech coding standards, and discuss issues relating to their implementation on DSPs.

9.2 Speech Coder Attributes

Speech coding consists of minimizing redundancies present in the digitized speech signal, through the extraction of certain parameters, which are subsequently quantized and encoded. The resulting data compression is lossy, which means that the decoder output is not identical to the encoder input. The objective here is to achieve the best possible quality at a given bit-rate by minimizing the audible distortion resulting from the coding process. There are a number of attributes that are used to characterize the performance of a speech coder. The most important of these attributes are bit-rate, complexity, delay, and quality. We examine briefly each of these attributes in this section.

The bit-rate is simply the number of bits per second required to represent the speech signal. In the context of a mobile standard, the bit-rate at which the speech coder has to operate is usually set by the standards body, as a function of the characteristics of the communication channel and the desired capacity. Often, the total number of bits allocated for the speech service has to be split between speech coding and channel coding. Channel coding bits constitute the redundancy in the form of forward error correction coding designed to combat the adverse effects of bad channels. Telephone-bandwidth speech signals have a useful bandwidth of 300–3400 Hz and are normally sampled at 8000 Hz. At the input of a speech coder the speech samples are typically represented with 2 bytes (16 bits), leading to a raw bit-rate of 128 kilobits/second (kb/s). Modern speech coders targeted for commercial telephony services aim at maintaining high quality at only 4–16 kb/s, corresponding to compression ratios in the range of 8–32. In the case of secure telephony applications (government or satellite communication standards), the bit rates are usually at or below 4.8 kb/s and can be as low as 2.4 kb/s or even under 1 kb/s in some cases. In general, an increase in bit-rate results in an improvement in speech quality.

Complexity is another important factor affecting the design of speech coders. It is often possible to increase the complexity of a speech coder and thus improve speech quality. However, for practical reasons, it is desirable to keep the complexity within reasonable limits. In the early days when DSPs were not as powerful, it was important to lower the complexity so that the coder would simply be implementable on a single DSP. Even with the advent of faster DSPs, it is still important to keep the complexity low both to reduce cost and to increase battery life by reducing power consumption. Complexity has two principal components: storage and computational complexity. The storage component consists of the RAM and the ROM required to implement the speech coding algorithm. The computational complexity is the number of operations per second that the speech coder performs in encoding and decoding the speech signal. Both forms of complexity contribute to the cost of the DSP chip.

Another important factor that characterizes the performance of a speech coder is delay. Speech coders often operate on vectors consisting of consecutive speech samples over a time interval called a frame. The delay of a speech coder, as perceived by a user, is a function of the frame size and any lookahead capability used by the algorithm, as well as other factors related to the communication system in which it is used. Generally speaking, it is possible to increase the framing delay and/or lookahead delay and hence reduce coding distortion. However, in a real-time communication scenario, the increase of the delay beyond a certain point can cause a significant drop in communication quality. There are two reasons for the quality loss. First, the users will simply notice this delay, which tends to interfere with the flow of the conversation. Second, the problem of echoes present in communication systems is aggravated by the long coder delays.

Speech quality is invariably the ultimate determining factor of acceptability of a speech coding algorithm for a particular application. As we have already noted, speech quality is a function of bit-rate, delay, and complexity, all of which can be traded off for quality. The quality of a speech coder needs to be robust to several factors, which include the presence of non-speech signals, such as environmental noise (car, office, speech of other talkers) or music, multiple encodings (also known as tandeming), input signal level variations, multiple talkers, multiple languages, and channel errors. Speech quality is generally evaluated using subjective listening tests, where a group of subjects are asked to listen to and grade the quality of speech samples that were processed with different coders. These tests are called Mean Opinion Score (MOS) tests, and the grading is done on a five-point scale, where 5 denotes the highest quality [1]. The quality in high-grade wireline telephony is referred to as "toll quality", and it roughly corresponds to 4.0 on the MOS scale. There are several forms of speech quality tests. For instance, in a slightly different variation called the Degradation Mean Opinion Score (DMOS) test, a seven-point scale (-3 to $+3$) is used [1]. All speech coding standards are selected only after conducting rigorous listening tests in a variety of conditions. It must be added that since these listening tests involve human listeners and fairly elaborate equipment, they are usually expensive and time-consuming.

There are also objective methods for evaluating speech quality. These methods do not require the use of human listeners. Instead, they compare the coded speech signal with the uncoded original and compute an objective measure that correlates well with subjective listening test results. Although research on objective speech quality evaluation started several decades ago, accurate methods based on the principles of the human auditory system have become available only in the recent years. The International Telecommunications Union (ITU) recently adopted a Recommendation, denoted as P.862, for objective speech quality evaluation. P.862 uses an algorithm called Perceptual Evaluation of Speech Quality (PESQ) [2]. From published results and from our own experience, PESQ seems to provide a reasonably accurate prediction of speech quality as measured by MOS tests. However, at least for now, MOS tests continue to be the means used by nearly all industry standards bodies for evaluating the quality of speech coders.

9.3 Speech Coding Basics

Speech coding uses an understanding of the speech production mechanism, the mathematical analysis of the speech waveforms, and the knowledge of the human auditory apparatus, to minimize redundancy present in the speech signal. The speech coder consists of an encoder and a decoder. The encoder takes the speech signal as input and produces an output bitstream. This bitstream is fed into the decoder, which produces output speech that is an approximation of the input speech.

We discuss below three types of speech coders: waveform coders, parametric coders, and linear prediction based analysis-by-synthesis coders. Waveform coders strive to match the signal at the decoder output to the signal at the encoder input as closely as possible, using an error criterion such as the mean-squared error. Parametric coders exploit the properties of the speech signal to produce an output signal that is not necessarily closely similar to the input signal but still sounds as close to it as possible. Linear prediction based analysis-by-synthesis coders use a combination of waveform coding and parametric coding.

Figure 9.1 shows examples of typical speech waveforms and spectra for *voiced* and *unvoiced* segments. The waveforms corresponding to *voiced speech*, such as vowels, exhibit a quasi-periodic behavior, as can be seen in Figure 9.1a. The period of this waveform is called the pitch period, and the corresponding frequency is called the fundamental frequency. The corresponding voiced speech spectrum is shown in Figure 9.1b. The overall shape of the spectrum is called the spectral envelope and exhibits peaks (also known as formants) and valleys. The fine spectral structure consists of evenly spaced spectral harmonics, which corresponds to multiples of the fundamental frequency. *Unvoiced speech*, such as /s/, /t/, and /k/, does not have a clearly identifiable period and the waveform has a random character, as shown in Figure 9.1c. The corresponding unvoiced speech spectrum, shown in Figure 9.1d, does not have a pitch or harmonic structure and the spectral envelope is essentially flatter than in the voiced spectrum.

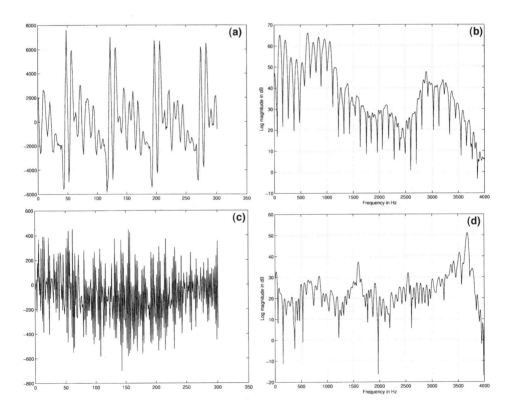

Figure 9.1 Example speech waveforms and spectra. (a) Voiced speech waveform (amplitude versus time in samples), (b) voiced speech spectrum, (c) unvoiced speech waveform (amplitude versus time in samples), (d) unvoiced speech spectrum

The spectral envelope of both voiced and unvoiced speech over each frame duration may be modeled and thus represented using a relatively small number of parameters, usually called spectral parameters. The quasi-periodic property of voiced speech is exploited to reduce redundancy using the so-called pitch prediction, where in its simple form, a pitch period of a waveform is approximated by a scaled version of the waveform from the immediately preceding pitch period. Speech coding algorithms reduce redundancy using both spectral modeling (*short-term redundancy*) and pitch prediction (*long-term redundancy*).

9.3.1 Waveform Coders

Early speech coders were waveform coders, based on sample-by-sample processing and quantization of the speech signal. These coders do not explicitly exploit the properties of the speech signal. As a result, they do not achieve very high compression ratios, but perform well also on non-speech signals such as modem and fax signaling tones. Waveform coders are, therefore, most useful in applications such as the public switched telephone network, which require successful transmission of both speech and non-speech signals. The simplest waveform coder is pulse code modulation (PCM), where the amplitude of each input sample is quantized directly. Linear (or uniform) PCM employs a constant (or uniform) step size across all signal amplitudes. Non-linear (or non-uniform) PCM employs a non-uniform step size, with smaller step sizes assigned to smaller amplitudes and larger ones assigned to larger amplitudes. μ-law PCM and A-law PCM are commonly used non-linear PCM coders using logarithmic, non-uniform quantizers. 16-bit uniform PCM (bit-rate = 128 kb/s) and 8-bit μ-law PCM or A-law PCM (64 kb/s) are commonly used in applications.

Improved coding efficiency can be obtained by coding the difference between consecutive samples, using a method called differential PCM (DPCM). In predictive coding, the difference that is coded is between the current sample and its predicted value, based on one or more previous samples. This method can be made adaptive by adapting either the step size of the quantizer used to code the prediction error or the prediction coefficients or both. The first variation leads to a technique called continuously variable slope delta modulation (CVSD), which uses an adaptive step size to quantize the difference signal at one bit per sample. For producing acceptable speech quality, the CVSD coder upsamples the input speech to 16–64 kHz. A version of CVSD is a US Department of Defense standard. CVSD at 64 kb/s is specified as a coder choice for Bluetooth wireless applications. Predictive DPCM with both adaptive quantization and adaptive prediction is referred to as adaptive differential PCM (ADPCM). As discussed below, ADPCM is an ITU standard at bit-rates 16–40 kb/s.

9.3.2 Parametric Coders

Parametric coders operate on blocks of samples called frames, with typical frame sizes being 10–40 ms. These coders employ parametric models attempting to characterize the human speech production mechanism. Most modern parametric coders use linear predictive coding (LPC) based parametric models. We thus limit our discussion to LPC-based parametric coders.

In the linear prediction approach, the current speech sample $s(n)$ is predicted as a linear combination of a number of immediately preceding samples:

$$\tilde{s}(n) = \sum_{k=1}^{p} a(k)s(n-k),$$

where $\tilde{s}(n)$ is the predicted value of $s(n)$, $a(k)$, $1 \leq k \leq p$ are the predictor coefficients, and p is the order of the predictor. The residual $e(n)$ is the error between the actual value $s(n)$ and the predicted value $\tilde{s}(n)$. The residual $e(n)$ is obtained by passing the speech signal through an inverse filter $A(z)$:

$$A(z) = 1 + \sum_{k=1}^{p} a(k)z^{-k}.$$

The predictor coefficients are obtained by minimizing the mean-squared value of the residual signal with respect to $a(k)$ over the current analysis frame. Computing $a(k)$ involves calculating the autocorrelations of the input speech and using an efficient matrix inversion procedure called Levinson–Durbin recursion, all of which are signal processing operations well suited to a DSP implementation [3].

At the encoder, several parameters representing the LPC residual signal are extracted, quantized, and transmitted along with the quantized LPC parameters. At the decoder, the LPC coefficients are decoded and used to form the LPC synthesis filter $1/A(z)$, which is an all-pole filter. The remaining indices in the bitstream are decoded and used to generate an excitation vector, which is an approximation to the residual signal. The excitation signal is passed through the LPC synthesis filter to obtain the output speech. Different types of LPC-based speech coders are mainly distinguished by the way in which the excitation signal is modeled.

The simplest type is the LPC vocoder, where vocoder stands for voice coder. The LPC vocoder models the excitation signal with a simple binary pulse/noise model: periodic sequence of pulses (separated by the pitch period) for voiced sounds such as vowels and random noise sequence for unvoiced sounds such as /s/. The binary model for a given frame is specified by its voiced/unvoiced status (voicing flag) and by the pitch period if the frame is voiced. The synthesized speech signal is obtained by creating the appropriate unit-gain excitation signal, scaling it by the gain of the frame, and passing it through the all-pole LPC synthesis filter, $1/A(z)$.

Other types of LPC-based and related parametric vocoders include Mixed Excitation Linear Prediction (MELP) [4], Sinusoidal Transform Coder (STC) [5], Multi-Band Excitation (MBE) [6], and Prototype Waveform Interpolation (PWI) [7]. For complete descriptions of these coders, the reader is referred to the cited references. As a quick summary, we note that MELP models the excitation as a mixture of pulse and noise sequences, with the mixture, called the voicing strength, set independently over five frequency bands. A 2.4 kb/s MELP coder was chosen as the new US Department of Defense standard [8]. STC models the excitation as a sum of sinusoids. MBE also uses a mixed excitation, with the voicing strength independently controlled over frequency bands representing pitch harmonics. PWI models the excitation signal for voiced sounds with one representative pitch period of the residual signal, with other pitch periods generated through interpolation. Parametric models other than LPC have been used in STC and MBE.

9.3.3 Linear Predictive Analysis-by-Synthesis

The concept of analysis-by-synthesis is at the center of modern speech coders used in mobile telephony standards [9]. Analysis-by-synthesis coders can be seen as a hybrid between parametric and waveform coders. They take advantage of blockwise linear prediction, while aiming to maintain a waveform match with the input signal. The basic principle of analysis-by-synthesis coding is that the LPC excitation vector is determined in a closed-loop fashion. The encoder contains a copy of the decoder: the candidate excitation vectors are filtered through the synthesis filter and the error between each candidate synthesized speech and the input speech is computed and the candidate excitation vector that minimizes this error is selected.

The error function most often used is the perceptually-weighted squared error. The squared error between the original and the synthesized speech is passed through a perceptual weighting filter, which shapes the spectrum of the error or the quantization noise so that it is less audible. This filter attenuates the noise in spectral valleys of the signal spectrum, where the speech energy is low, at the expense of amplifying it under the formants, where the relatively large speech energy masks the noise. The perceptual weighting filter is usually implemented as a pole-zero filter derived from the LPC inverse filter $A(z)$.

In analysis-by-synthesis coders, complexity is always an important issue. Certain simplifying assumptions made in the excitation search algorithms and specific excitation codebook structures developed for the purpose of complexity reduction make analysis-by-synthesis coders implementable in real-time. Most linear prediction based analysis-by-synthesis coders fall under the broad category of code-excited linear prediction (CELP) [10]. In the majority of CELP coders, the excitation vector is obtained by summing two components coming from the adaptive and fixed codebooks. The adaptive codebook is used to model the quasi-periodic pitch component of the speech signal. The fixed codebook is used to represent the part of the excitation signal that cannot be modeled with the adaptive codebook alone. This is illustrated in the CELP decoder block diagram in Figure 9.2.

The CELP encoder contains a copy of the decoder, as can be seen in the block diagram of Figure 9.3. The adaptive and fixed excitation search are often the most computationally complex part of analysis-by-synthesis coders because of the filtering operation and the correlations needed to compute the error function. Ideally, the adaptive and fixed codebooks

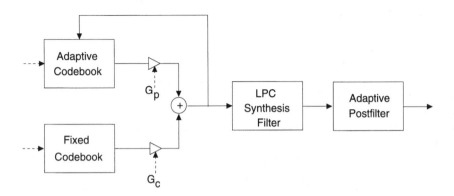

Figure 9.2 Basic CELP decoder

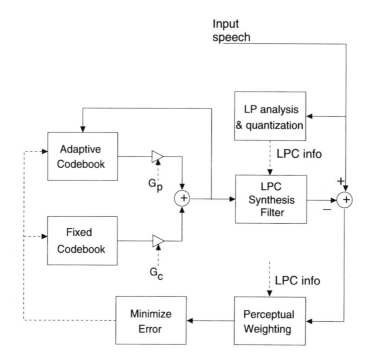

Figure 9.3   Basic CELP encoder block diagram

should be jointly searched to find the best excitation vector. However, since such an operation would result in excessive complexity, the search is performed in a sequential fashion, the adaptive codebook search first, followed by the fixed codebook search.

The adaptive codebook is updated several times per frame (once per *subframe*) and populated with past excitation vectors. The individual candidate vectors are identified by the pitch period (also called the pitch lag), which covers the range of values appropriate for human speech. The pitch lag can have a fractional value, in which case the candidate codevectors are obtained by interpolation. Typically, this pitch lag value does not change very rapidly in strongly voiced speech such as steady-state vowels, and is, therefore, often encoded differentially within a frame in state-of-the-art CELP coders. This helps reduce both the bit-rate, since only the pitch increments need to be transmitted, and the complexity, since the pitch search is limited to the neighborhood of a previously computed pitch lag.

Several varieties of analysis-by-synthesis coders are differentiated from each other mainly through the manner in which the fixed excitation vectors are generated. For example, in stochastic codebooks, the candidate excitation vectors can consist of random numbers or trained codevectors (trained over real speech data). Figure 9.4a shows an example stochastic codevector. Passing each candidate stochastic codevector through the LPC synthesis filter and computing the error function is computationally expensive. Several codebook structures can be used to reduce this search complexity.

For example, in Vector Sum Excited Linear Prediction (VSELP) [11], each codevector in the codebook is constructed as a linear combination of basis vectors. Only the basis vectors need to be filtered, and the error function computation can be greatly simplified by combining the contribution of the individual basis vectors. Sparse codevectors containing only a small number of non-zero elements can also be used to reduce complexity.

In multipulse LPC [12], a small number of non-zero pulses, each having its own individual gain, are combined to form the candidate codevectors (Figure 9.4b). Multipulse LPC (MP-LPC) is an analysis-by-synthesis coder, which is a predecessor of CELP. Its main drawback is that it requires the quantization and transmission of a separate gain for each fixed excitation pulse, which results in a relatively high bit-rate.

Algebraic codebooks also have sparse codevectors but here the pulses all have the same gain, resulting in a lower bit rate than MP-LPC coders. Algebraic CELP (ACELP) [13] allows an efficient joint search of the pulse locations and is widely used in state-of-the-art speech coders, including several important standards. Figure 9.4c shows an example algebraic CELP codevector.

At medium to high bit-rates (6–16 kb/s), analysis-by-synthesis coders typically have better performance than parametric coders, and are generally more robust to operational conditions, such as background noise.

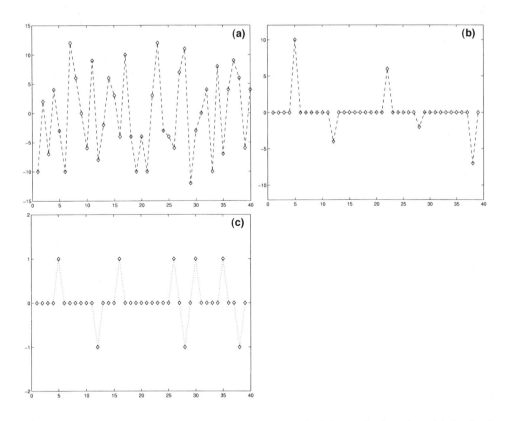

Figure 9.4 Examples of codevectors used in various analysis-by-synthesis coders. (a) Stochastic codevector, (b) multipulse LPC codevector, (c) algebraic CELP codevector

9.3.4 Postfiltering

The output of speech coders generally contains some amount of audible quantization noise. This can be removed or minimized with the use of an adaptive postfilter, designed to further attenuate the noise in the spectral valleys. Generally speaking, the adaptive postfilter consists of two components: the long-term (pitch) postfilter, designed to reduce the noise between the pitch harmonics and the short-term (LPC) postfilter, which attenuates the noise in the valleys of the spectral envelope. The combined postfilter may also be accompanied by a spectral tilt filter, designed to compensate for the low-pass effect generally caused by postfiltering, and by an adaptive gain control mechanism, which limits undesirable amplitude fluctuations.

9.3.5 VAD/DTX

During a typical telephone conversation, either one of the parties is usually silent for about 50% of the duration of the call. During these pauses, only the background noise is present in that direction. Encoding the background noise at the rate designed for the speech signal is not necessary. Using a lower coding rate for non-speech frames can have several advantages, such as capacity increase, interference reduction, or savings in mobile battery life, depending on the design of the overall communication system. This is most often achieved by the use of a voice activity detection (VAD) and discontinuous transmission (DTX) scheme. The VAD is a front-end algorithm, which classifies the input frames into speech and non-speech frames. The operation of the DTX algorithm is based on the information from the VAD. During non-speech frames, the DTX periodically computes and updates parameters describing the background noise signal. These are transmitted intermittently at a very low rate to the decoder, which uses them to generate an approximation to the background noise, called *comfort noise*. Most wireless standards include some form of VAD/DTX.

9.3.6 Channel Coding

In most communication scenarios and especially in mobile applications, the transmission medium is not ideal. To combat the effects of degraded channels, it is often necessary to apply channel coding via forward error correction to the bitstream. The transmitted bits are thus split between source (speech) and channel coding. The relative proportion of source and channel coding bits depends on the particular application and the expected operating conditions. Channel coding is generally done with a combination of rate-compatible punctured convolutional codes (RCPC) and cyclic redundancy check (CRC) based parity checking [14].

Generally speaking, all the bits in a transmitted bitstream are not of equal perceptual importance. For this reason, the bits are classified according to their relative importance. In a typical application, there are three or four such classes. The most important bits are usually called Class 0 bits and are most heavily protected. CRC parity bits are first computed over these bits. Then the Class 0 bits and the parity bits are combined and RCPC-encoded at the highest available rate. The remaining classes are RCPC-encoded at progressively lower rates. The function of RCPC is to correct channel errors. The CRC is used to detect any residual errors in the most important bits. If these bits are in error, the quality of the decoded frame is likely to be poor. Therefore, the received speech frame is considered corrupted and is often discarded. Instead, the frame data is replaced at the decoder with appropriately extrapolated (and possibly muted or otherwise modified) values from the past history. This process is generally known as error concealment.

9.4 Speech Coding Standards

As mentioned in the introduction, speech coding standards are created by the many regional and global standards bodies such as the Association of Radio Industries and Businesses (ARIB) in Japan, the Telecommunications Industry Association (TIA) in North America, the European Telecommunications Standards Institute (ETSI) in Europe, and the Telecommunication Standardization Sector of the International Telecommunication Union (ITU-T), which is a worldwide organization. These organizations have been responsible for the adoption of a number of digital speech coding standards in the past decades. These standards have emerged over time to satisfy changing needs.

9.4.1 ITU-T Standards

The ITU-T is responsible for creating speech coding standards for network telephony, including wireline and wireless telephony. The speech coders specified by the ITU-T are not created for specific systems, and therefore do not contain channel coding specifications. The ITU-T standards are sometimes used as the basis for developing standards appropriate for wireless and other applications. They are also used in the emerging field of voice over packet telephony. Table 9.1 lists five ITU-T speech coding standards discussed below.

Table 9.1   ITU-T speech coding standards

ITU standards		
Coder	Rate (kb/s)	Approach
G.711	64	Mu/A-law
G.726	16–40	ADPCM
G.728	16	LD-CELP
G.729	8	CS-ACELP
G.723.1	5.3/6.3	MP-LPC/ACELP

The first ITU-T coders pre-date the cellular standards and are waveform coders. G.711 is a 64 kb/s PCM coder, and G.726 is an ADPCM coder operating at 16–40 kb/s. The wideband coders, the G.722 split-band ADPCM algorithm and the more recent G.722.1, are discussed in Section 9.4.3. There are several more ITU-T standards, which we describe in this section.

G.728 is a 16 kb/s CELP coder with a very low delay [15]. The low delay is due to the extremely small frame size of five samples (0.625 ms). This small frame size presents some challenges, primarily since the traditional derivation of the LPC coefficients requires the buffering of a large frame of speech samples. Low-delay CELP coders must instead rely on backward-adaptive prediction, where the prediction coefficients are derived from previously quantized speech. G.728 does not use any pitch prediction. Instead, a very high-order backward-adaptive linear prediction filter with 50 taps is used. This coder provides toll quality for both clean speech and speech corrupted by background noise.

G.729 was also initially targeted to be a very low delay coder [16]. But the delay requirement was later relaxed, and the conjugate-structure (CS) ACELP coder with a 10 ms frame was selected. Several annexes to this coder were also standardized. G.729A is a reduced complexity version of the algorithm, G.729B is the silence compression (VAD/DTX) scheme, and G.729C is the floating point implementation. G.729 has also been extended down to 6.4 kb/s (G.729D) and up to 11.8 kb/s (G.729E). As mentioned in Section 9.4.2.2, G.729 has been adopted in Japan as an enhanced full-rate coder.

G.723.1 is a dual-mode coder, initially standardized for low bit-rate video-telephony. The 5.3 kb/s mode uses an algebraic codebook for the fixed excitation, and the 6.3 kb/s mode uses a variation of a multipulse coder. The quality is close to toll quality for clean speech, but the robustness to background noise and tandeming are not as good as some of the higher rate ITU-T coders.

9.4.2 Digital Cellular Standards

The first wave of digital cellular standards was motivated by the capacity increase offered by digital telephony over analog systems such as TACS (Total Access Communication Sytem) and NMT (Nordic Mobile Telephone) in Europe, JTACS (Japanese TACS) in Japan, and the Advanced Mobile Phone Service (AMPS) in North America. The continuing demand for capacity drove standards bodies to adopt half-rate (HR) standards, even before the first full-rate (FR) standards were widely deployed. After the first years of digital telephony, it became apparent that the speech quality offered by the full-rate standards was for the large part less than satisfactory. This motivated the next wave of coders, called enhanced full-rate (EFR) standards, which offered higher quality at the same bit-rates, thanks to advances in the field of speech coding.

The next generation of speech coders consists of multirate vocoders designed to more finely optimize the quality/capacity tradeoffs inherent in mobile communication systems. The ETSI/3GPP (Third Generation Partnership Project) adaptive multirate (AMR) standard is designed to provide the capability to dynamically adapt the ratio between source and channel coding bits depending on the instantaneous channel conditions. The Selectable Mode Vocoder is a variable rate speech coder that can function at several operating points providing multiple options in terms of the quality of service.

Recently, there has been an increasing interest in wideband standards for mobile telephony. The ETSI/3GPP wideband speech coder standardized in 2001 is also in the process of being adapted as an ITU-T wideband standard.

Despite the proliferation of all these incompatible standards, the various standards organizations are trying to work together to adopt approaches that facilitate interoperability. The third generation standards bodies (3GPP, 3GPP2) include representatives from the various regional bodies. Enhanced full-rate, AMR, and wideband coders are currently finding their ways into the 2.5 and 3G standards, and are offering the potential of increased interoperability. Figure 9.5 summarizes the evolution of various standardization bodies from the first generation analog systems to the emerging 3G systems.

This section describes the various speech coding standards for mobile telephony. Table 9.2 summarizes the various standards and is intended to serve as a quick reference. The channel

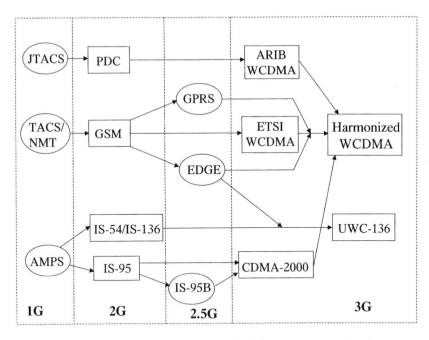

Figure 9.5 The evolution of cellular standards organizations

rate column is blank for the speech coding standards used in CDMA systems, since they do
not specify forward error correction.

9.4.2.1 First FR and HR Standards

The first cellular standards based on digital technology were deployed worldwide in the 1990s
and began to replace the existing analog standards in Japan, Europe, and the US.

The first standard to be deployed was GSM FR (GSM 06.10) [17]. It employs a method
called Regular Pulse Excitation-Long-Term Prediction (RPE-LTP), operating at a gross bit-
rate of 22.8 kb/s. In GSM FR, the LPC residual is first computed. The long-term predictor is
used to estimate the LPC residual from the past values of the reconstructed excitation signal.
The long-term prediction residual is downsampled and encoded with a block-adaptive PCM
algorithm. The speech coder rate is 13 kb/s, and the complexity is very low.

In North America, the full-rate standard for TDMA (Time Division Multiple Access)
cellular is TIA's IS-54 Vector Sum Excited Linear Prediction (VSELP) coder [11]. It has
a bit-rate of 7.95 kb/s, and the total channel rate is 13 kb/s. The main feature that provides
lower complexity and a degree of improved channel error resilience is the fixed excitation
codebook that consists of basis vectors, which are optimized off-line with a training procedure. The coder also contains an adaptive codebook. After the adaptive codebook search, the
basis vectors are orthogonalized with respect to the selected adaptive codevector. The candidate fixed codevectors are obtained as a linear combination of the basis vectors, where the
linear combination coefficients are either $+1$ or -1. With this structure, only the basis vectors
need to be passed through the synthesis filter. The filtered candidate vectors can then be

Table 9.2 Summary of speech coding standards for mobile telecommunications

Digital cellular standards

Coder	Rate (kb/s)	Channel rate	Approach	Date
GSM FR	13	22.8	RPE-LTP	1987
GSM HR	5.6	11.4	VSELP	1994
GSM EFR	12.2	22.8	ACELP	1995
GSM AMR	4.75–12.2	11.4–22.8	ACELP	1998
TIA IS54	7.95	13	VSELP	1989
TIA IS95	0.8–8.55		QCELP	1993
TIA Q13	0.8–13.3		QCELP	1995
TIA IS641	7.4	13	ACELP	1996
TIA EVRC	0.8–8.55		R-ACELP	1996
TIA SMV	0.8-8.5		R-ACELP	2001
PDC FR	6.7	11.2	VSELP	1990
PDC HR	3.45	5.6	PSI-CELP	1993
PDC EFR	8	11.2	ACELP	1999
PDC EFR	6.7	11.2	ACELP	2000

obtained by simple addition/subtraction of pre-filtered basis vectors, resulting in a complexity reduction over generic CELP coders. VSELP has a lower bit-rate compared to GSM FR. The quality is also good, but less than toll quality.

The Japanese Personal Digital Cellular Full-Rate (PDC FR) standard, operating at a lower total bit-rate of 11.2 kb/s, is also a VSELP coder. The speech bit-rate is 6.7 kb/s. Good quality at this reduced bit-rate is achieved using some enhancements to the VSELP algorithm such as the use of fractional pitch search and the differential encoding of pitch lags.

The North American TIA IS-95 Code Division Multiple Access (CDMA) system uses a CELP coder known as QCELP8 [18]. This is a variable rate speech coder, with four available bit-rates (8.55, 4.0, 2.0, 0.8 kb/s). It is a source-controlled variable rate coder, since the mode selection is achieved by analyzing the properties of the input signal. Despite the availability of four coding rates, the mode selection operates mostly in a manner similar to a VAD/DTX scheme. For this reason, the 2 and 4 kb/s modes are rarely used and QCELP8 mostly operates at 8.55 kb/s for active speech and 0.8 kb/s for background noise frames. It achieves a fairly low average bit rate but fails to deliver toll quality.

A higher rate, higher quality version of the QCELP algorithm known as QCELP13 (TIA/ EIA-722) has also been standardized for the North American TIA system. The four bit-rates used are 13.3, 6.3, 2.7, and 0.8 kb/s. It delivers toll quality at the expense of a substantial increase in bandwidth compared to QCELP8. The average bit-rate of QCELP13 has been measured to be around 5.6 kb/s as opposed to 3.6 kb/s for QCELP8.

Shortly after the adoption of the first full-rate standards, the standards bodies proceeded to standardize half-rate standards. In Europe, a 5.6 kb/s VSELP coder bearing many similarities to the North American and Japanese full-rate standards was selected as the GSM half-rate standard [19].

The Japanese half-rate standard (PDC HR) speech coder is called pitch-synchronous innovation CELP (PSI-CELP) and uses only 3.45 kb/s, with a total channel rate of 5.6 kb/s [20]. To achieve this low bit-rate, the frame size is increased to 40 ms, twice that of the other full-rate and half-rate standards. One of the main problems when the bit-rate of CELP coders is reduced to such low values is that it becomes difficult to preserve the periodicity in the input signal, especially when the pitch frequency is high (small pitch periods). To rectify this problem, the fixed excitation vectors are slightly modified: if the pitch value L is smaller than the subframe size, the fixed excitation vector is made periodic by repeating its first L samples until the end of the subframe. The resulting coder has a large delay and a high complexity. The quality is also not as good as that of PDC FR.

The TIA TDMA half-rate standardization activity initiated in 1991 was not completed because the candidates failed to meet the desired quality requirements.

9.4.2.2 Enhanced Full-Rate Standards

While the HR standards were being worked on, the FR systems were being deployed. Based on the experience in real life environments, it became apparent that the quality of the full-rate standards was not high enough and that there was a need for enhanced full-rate coders.

The technology that provided this needed improvement in quality, along with a reduction in complexity, is algebraic CELP (ACELP). This generation of coders also benefits from improvements in other areas of speech coding such as vector quantization of LPC parameters and fractional lag adaptive codebooks. ACELP is a family of excellent coders, which constitute the state of the art in current standards.

The GSM EFR coder operates at a speech coding rate of 12.2 kb/s. Additional CRC bits bring the rate up to 13 kb/s. The same channel coder used in GSM FR is also used in GSM EFR for backward compatibility reasons [21].

IS-641 is the TIA EFR coder. It is very similar to GSM EFR, except that fewer bits are used for the encoded parameters, resulting in a speech coding bit rate of 7.4 kb/s [22].

Japan has two enhanced full-rate (PDC EFR) coders. The first one is based on the ITU-T G.729 conjugate-structure ACELP (CS-ACELP) coder, operating at 8 kb/s. The second one has a speech bit-rate of 6.7 kb/s, and is part of the ETSI/3GPP AMR coder, which is discussed in Section 9.4.2.3. The channel bit-rate for both coders is still 11.2 kb/s.

The North American CDMA enhanced full-rate coder is called Enhanced Variable Rate Coder (EVRC) and is designated as TIA IS-127 [23]. The algorithm operates at the same rates as QCELP8. It is once again based on ACELP but also uses a new method called Relaxed CELP (RCELP), which is an example of generalized analysis-by-synthesis coding [24].

9.4.2.3 Adaptive Multirate Systems

In digital cellular communication systems, one of the major challenges is that of designing a coder that is able to provide high quality speech under a variety of channel conditions. Ideally, a good solution must provide the highest possible quality in the clean channel conditions while maintaining good quality in heavily disturbed channels. Traditionally, digital cellular applications use a fixed source/channel bit allocation that provides a compromise solution between clean and degraded channel performance. Clearly, a solution that is well suited for clean channels would use most of the available bits for source coding with only minimal error protection, while a solution designed for poor channels would use a lower rate speech coder protected with a large amount of forward error correction (FEC).

One way to obtain good performance across a wide range of conditions is to allow the network to monitor the state of the communication channel and direct the coders to adjust the allocation of bits between source and channel coding accordingly. This can be implemented via an adaptation algorithm whereby the network selects one of a number of available speech coders, called codec modes, each with a predetermined source/channel bit allocation. This concept is called AMR coding and is a form of network-controlled multimodal coding of speech [25]. The AMR concept is the centerpiece of the ETSI AMR standard, which specifies a new European cellular communication system designed to support an AMR mechanism in both the half-rate (11.4 kb/s) and the full-rate (22.8 kb/s) channels.

The AMR coder consists of 8 modes at 4.75, 5.15, 5.9, 6.7, 7.4, 7.95, 10.2, and 12.2 kb/s [26]. These modes are all based on the ACELP algorithm and include the three EFR coders, from Europe (GSM EFR at 12.2 kb/s), North America (TDMA-641 at 7.95 kb/s), and Japan (PDC EFR at 6.7 kb/s). The AMR coder has also been standardized for EDGE (Enhanced Data Rates for GSM Evolution) channels by ETSI and as the mandatory speech coder for 3G telephony and videoconferencing by 3GPP.

Another coder that has the capability of trading quality and capacity dynamically is being standardized by 3GPP2 and TIA and is scheduled to be finalized during 2001. The Selectable Mode Vocoder (SMV) has three modes of operation and aims to provide better quality at the same bit-rate as EVRC and the same quality as EVRC at a lower bit-rate. The speech coder is designed to be compatible with operation on TIA/EIA IS-95 CDMA systems and on the family of TIA/EIA IS-2000 (CDMA2000) systems.

As with QCELP8 and EVRC, SMV is a source-controlled variable bit-rate coder, consisting of four coder bit-rates (8.5, 4.0, 2.0, 0.8 kb/s) [27]. As opposed to its predecessors, SMV has a more sophisticated rate-selection mechanism. First of all, it is no longer a simple VAD + DTX scheme. The incoming frame of speech is analyzed and important perceptual features are detected, allowing the use of the intermediate bit-rates for a substantial portion of the time. Second, the rate selection mechanism itself has three modes of operation, which can be selected at the request of the network, depending on congestion or quality of service considerations. These three operating modes have specific quality and average bit-rate targets. Mode 0 (the Premium mode) has a target average rate not exceeding that of EVRC (around 3.6 kb/s), with better than EVRC quality. Mode 1 (the Standard mode) targets the same quality as EVRC at a much lower average rate (2.5 kb/s). Mode 2 (the Economy mode) aims for "near-EVRC" quality at an even lower average rate (1.9 kb/s).

9.4.3 Wideband Standards

Wideband speech coding, using the bandwidth from 0 to 7 kHz, offers the potential for a significant improvement in speech quality over traditional narrowband coding (0.3–3.4 kHz) at comparable bit-rates (8–32 kb/s). Wideband speech, sampled at 16 kHz, benefits from the crispness and higher intelligibility provided by the additional high-band and from the increase in perceived quality and "fullness" due to the extended low frequency content.

The first wideband speech coding standard adopted is the ITU-T's G.722 coder, which splits the input signal into two bands that are encoded each with an ADPCM algorithm. The low band has the option of using 4, 5, or 6 bits/sample, whereas the high-band always uses 2 bits, resulting in three bit-rate options at 48, 56, and 64 kb/s. Because of its high bit-rates, the G.722 coder has limited use, especially in wireless applications; however, it serves as an excellent quality benchmark for other wideband standards.

ITU-T's G.722.1 is a transform coder operating at 24 and 32 kb/s with good quality. It is recommended for hands-free applications such as conferencing where the input speech is somewhat noisy. It is robust to background noises and non-speech signals such as music, but it produces audible distortion for clean speech inputs, especially at 24 kb/s.

The first wideband speech coder targeted specifically for mobile applications is the AMR Wideband (AMR WB) coder [28] standardized by ETSI and 3GPP for a variety of mobile applications, such as GSM FR channels, GSM EDGE channels, and 3GPP's Wideband CDMA (WCDMA) systems. The coder is based on ACELP and supports nine bit-rates: 6.6, 8.85, 12.65, 14.25, 15.85, 18.25, 19.85, 23.05, and 23.85 kb/s. The 16 kHz input signal is downsampled to 12.8 kHz and the baseband is encoded with an ACELP algorithm, which is very similar to the narrowband AMR coder. In the highest rate mode of AMR WB, the high band information is encoded and transmitted, while in the remaining modes it is simply extrapolated from the baseband. The AMR WB coder appears set to become a very important part of future mobile communications systems. It is also in the process of being adopted as ITU-T wideband standard at 16 kb/s (Table 9.3).

Table 9.3 Wideband speech coding standards

Wideband standards		
Coder	(Rate (kb/s)	Approach
G.722	48, 56, 64	SB-ADPCM
G.722.1	24, 32	Transform
ITU WB	16, 24	ACELP
AMR WB	6.60–23.85	ACELP

9.5 Speech Coder Implementation

Once specified by a standards body, a speech coding standard is then implemented by multiple vendors. Speech coding algorithms are particularly well suited for DSP implementation, since they rely on intensive signal processing operations but are too complex for dedicated-hardware solutions. To guarantee that various implementations of the standard are interoperable and provide the appropriate speech quality, the standard must be specified in adequate detail, including conformance testing.

9.5.1 Specification and Conformance Testing

Speech coding standards are always described in detail in the form of a technical specification, typically accompanied by a reference version of the coder in a high-level language such as C. In some cases, such as the US Federal Standard LPC-10 vocoder (FS1015), compliance to the standard is specified only in terms of the bitstream of quantized parameters. This leaves manufacturers with the option of providing custom enhancements to their implementation, as long as they comply with this bitstream format. On the one hand, this has the advantage of providing room for improvements and product differentiation. On the other hand, it has the potential drawback of allowing inferior implementations that could undermine user acceptance of the standard. Conformance to the standard should then also include formal verification of speech quality, as measured by subjective listening tests. Unfortunately, this can be a time-consuming and expensive process.

To avoid this problem, the majority of recent speech coding standards specify compliance by the use of input and output test vectors. For each input test vector, the implementation must produce the corresponding output test vector. In floating-point or non-bit-exact standards, a small numerical deviation is allowed in the output vectors, but the implementation can still be numerically verified to be equivalent to the reference. In bit-exact fixed-point standards, the output test vectors from the implementation must exactly match those of the reference.

Even when the standard is specified in fixed-point, some features may be specified as just example solutions. As a result, coder functionalities such as channel decoding, frame erasure concealment, noise suppression, and link adaptation in the AMR standards can vary from one manufacturer to another, allowing some room for differentiation.

9.5.2 ETSI/ITU Fixed-Point C

For bit-exact standards, special libraries of arithmetic operators that simulate the behavior of fixed-point DSPs have been developed by the ITU-T and ETSI. These libraries allow C-language simulation of a generic 16/32 bit DSP on a PC or workstation. In addition to specifying bit-exact performance, these libraries include run-time measurement of DSP complexity, by counting basic operations and assigning weights to each. The resulting measurements of weighted millions of operations per second (WMOPS) can be used to predict the complexity of a DSP implementation of the algorithm. In addition, this fixed-point C code acts as a pseudo-assembly code to facilitate DSP development. A drawback of this approach, however, is that it may not result in a fully-efficient DSP implementation, since re-ordering operations or using extended-precision registers can slightly change numerical behavior.

The operations and associated weights are selected to model basic DSP operations. For example, the following operations have a complexity weight of 1 each:

- Add(var1,var2) and sub(var1,var2) perform 16-bit addition and subtraction with saturation.
- Abs_s(var1) provides a 16-bit absolute value.
- Shl(var1,var2) and shr(var1,var2) perform shifts by var2.
- Extract_h(L_var1) and extract_l(L_var1) extract the high and low 16 bits from a 32-bit word.
- L_Mac(L_var3,var1,var2) and L_msu(L_var3,var1,var2) are 32-bit multiply/addition and multiply/subtraction of 16-bit inputs.

Long operations typically have a weight of 2:

- L_add(L_var1,L_var2) and L_sub(L_var1,L_var2) perform 32-bit addition and subtraction.
- L_shl(L_var1,var2) and L_shr(L_var1,var2) perform shifts of a 32-bit value.

More complex operations can have much higher weights:

- Norm_s(var1) and norm_l(var1) compute the number of bits to shift 16 and 32-bit numbers for normalization, with weights of 15 and 30, respectively.
- Div_s(var1,var2) implements fractional integer division, with a weight of 18.

9.5.3 DSP Implementation

In mobile communication handsets, minimizing cost and maximizing battery life are important goals. Therefore, low-power fixed-point DSPs, such as the Texas Instruments TMS320C54x family, are widely used in handset applications. A good speech coder implementation on these DSPs can achieve a ratio of WMOPS to MIPS of 1:1, so that the DSP MIPS required is accurately predicted by the fixed-point WMOPS. Since current speech coders do not require all of the 40–100 available DSP MIPS, additional functions such as channel coding, echo cancellation, and noise suppression can also be performed simultaneously by the DSP. Earlier generations of DSPs could not implement all basic operators in single cycles, so that WMOPS to MIPS ratios were 1:1.5 or even 1:2. By contrast, the newer TMS320C55x family uses two multiply-accumulators to allow even more efficient implementations, for example with WMOPS to MIPS ratios of 1.5:1.

Base stations, on the other hand, need to handle many voice channels while keeping equipment size small. The most powerful DSPs, such as the TMS320C6000 family, are well suited for this application. The TMS320C6000 uses a highly parallel system architecture, emphasizing software-based flexibility, to achieve up to 4800 MIPS, easily allowing implementation of a large number of speech coders in parallel and thus leading to a high channel density per DSP. Base station speech coder implementations may use either floating or fixed-point DSPs.

9.6 Conclusion

Speech coding standards are an important part of the rapidly changing global landscape of mobile communications. New, higher rate channels, and packet based data services such as the General Packet Radio Service are driving the emerging 2.5G and 3G networks. The emergence of new versatile speech coders such as AMR and AMR WB and the increasing collaboration among the various standards bodies are providing some hope for harmonization of worldwide speech coding standards. Packet networks, where speech is simply treated as another data application, may also facilitate interoperability of future equipment. Finally, harmonized wideband speech coding standards can potentially provide a substantial leap in speech quality.

Acknowledgements

The authors would like to thank Servanne Dauphin, Juan Carlos De Martin, Mike McMahan, Toshiaki Ohura, Raj Pawate, and Masahiro Serizawa for providing information used in this chapter.

References

[1] 'Methods for Subjective Determination of Transmission Speech Quality', *ITU-T Recommendation P.800*, August 1996.

[2] Rix, A.W., Beerends, J.G., Hollier, M.P. and Hekstra, A.P. 'Perceptual Evaluation of Speech Quality (PESQ) – a New Method for Speech Quality Assessment of Telephone Networks and Codecs', *Proceedings of the IEEE International Conference on Acoustics, Speech, and Signal Processing*, 2001.

[3] Rabiner, L.R. and Schafer, R.W., *Digital Processing of Speech Signals*, Prentice-Hall, Englewood Cliffs, NJ, 1978.

[4] McCree, A.V. and Barnwell III, T.P., 'A Mixed Excitation LPC Vocoder Model for Low Bit Rate Speech Coding', *IEEE Transactions on Speech and Audio Processing*, Vol. 3, No. 4, July 1995, pp. 242–250.

[5] McAulay, R.J. and Quatieri, T.F., 'Low-Rate Speech Coding Based on the Sinusoidal Model', In: Sondhi, M. and Furui, S., editors, *Advances in Acoustics and Speech Processing*, Marcel Deckker, New York, 1992, pp. 165–207.

[6] Griffin, D.W. and Lim, J.S., 'Multi-Band Excitation Vocoder', *IEEE Transactions on Acoustics, Speech and Signal Processing*, Vol. 36, No.8, August 1988, pp. 1223–1235.

[7] Kleijn, W.B., 'Encoding Speech Using Prototype Waveforms', *IEEE Transactions on Acoustics, Speech, and Signal Processing*, Vol. 1, No.4, October 1993, pp. 386–399.

[8] McCree, A. Truong, K., George, E.B., Barnwell, T.P., and V. Viswanathan, 'A 2.4 Kbit/s MELP Coder Candidate for the New U.S. Federal Standard', *Proceedings of the IEEE International Conference on Acoustics, Speech, and Signal Processing*, May 1996, pp. 200–203.

[9] *Speech Coding and Synthesis*, Kleijn, W.B. and Paliwal, K.K., editors, Elsevier, Amsterdam, 1995.

[10] M.R. Schroeder, B.S. Atal, 'Code-Excited Linear Prediction (CELP): High-Quality Speech at Very Low Bit Rates', *Proceedings of the IEEE International Conference on Acoustics, Speech, and Signal Processing* (1985).

[11] Gerson, I.A. and Jasiuk, M.A., 'Vector Sum Excited Linear Prediction (VSELP) Speech at 8 kbps', *Proceedings of the IEEE International Conference on Acoustics, Speech, and Signal Processing*, Vol. 1, 1990, pp. 461–464.

[12] Atal, B.S. and Remde, J.R., 'A New Model of LPC Excitation for Reproducing Natural-Sounding Speech at Low Bit Rates', *Proceedings of the IEEE International Conference on Acoustics, Speech, and Signal Processing*, Vol. 1, 1989, pp. 614–617.

[13] Adoul, J.-P., Mabilleau, P., Delprat, M. and Morisette, S., 'Fast CELP Coding Based on Algebraic Codes', *Proceedings of the IEEE International Conference on Acoustics, Speech, and Signal Processing*, 1987, pp. 1957–1960.

[14] Lee, L.H., 'New Rate-Compatible Punctured Convolutional Codes for Viterbi Decoding', *IEEE Transactions on Communications*, Vol. 42, No. 12, Dec. 1994.

[15] Chen, J.-H., Cox, R.V., Lin, Y.-C., Jayant, N. and Melchner, M.J., 'A Low-Delay CELP Coder for the CCITT 16 kb/s Speech Coding Standard', *IEEE Journal on Selected Areas in Communications*, Vol. 10, No. 5, 1992, pp. 830–849.

[16] Salami, R., Laflamme, C., Adoul, J.-P., Kataoka, A., Hayashi, S., Moriya, T., Lamblin, C., Massalous, D., Proust, S., Kroon, P. and Shoham, Y., 'Design and Description of CS-ACELP: a Toll Quality 8 kb/s Speech Coder', *IEEE Transactions on Speech and Audio Processing*, Vol. 6, No.2, 1998, pp. 116–130.

[17] Vary, P., Hellwig, K., Hofman, R., Sluyter, R.J., Galand, C. and Rosso, M., 'Speech Codec for the European Mobile Radio System', *Proceedings of the IEEE International Conference on Acoustics, Speech, and Signal Processing*, Vol. 1, 1988, pp. 227–230.

[18] DeJaco, A., Gardner, W., Jacobs, P. and Lee, C., 'QCELP: The North American CDMA Digital Cellular Variable Rate Speech Coding Standard', *Proceedings of the IEEE Workshop on Speech Coding for Telecommnucations*, 1993, pp. 5–6.

[19] Gerson, I.A. and Jasiuk, M.A., 'A 5600-bps VSELP Speech Coder Candidate for HR GSM', *Proceedings of the IEEE Workshop on Speech Coding for Telecommnucations*, 1993, pp. 43–44.

[20] Ohya, T., Suda, H. and Miki, T., '5.6 kbits/s PSI-CELP of the HR PDC Speech Coding Standard', *IEEE Vehicular Technology Conference*, Vol. 3, 1994, pp. 1680–1684.

[21] Jarvinen, K., Vainio, J., Kapanen, P., Honkanen, T., Salami, R., Laflamme, C. and Adoul, J.-P., 'GSM Enhanced Full Rate Speech Codec', *Proceedings of the IEEE International Conference on Acoustics, Speech, and Signal Processing*, Vol. 2, 1997, pp. 771–774.

[22] Honkanen, T., Vainio, J., Jarvinen, K., Haavisto, P., Salami, R., Laflamme, C. and Adoul, J.-P., 'Enhanced Full Rate Speech Codec for IS-136 Digital Cellular System', *Proceedings of the IEEE International Conference on Acoustics, Speech, and Signal Processing*, Vol. 2, 1997, pp. 731–734.

[23] 'Enhanced Variable Rate Codec, Speech Service Option 3 for Wideband Spread Spectrum Digital Systems', *IS-127*, September 9, 1996.

[24] Kleijn, W.B., Kroon, P. and Nahumi, D., 'The RCELP Speech-Coding Algorithm', *European Transactions on Telecommunications*, Vol. 5, No. 5, 1994, pp. 573–582.

[25] A. Gersho, E. Paksoy, 'Variable rate speech coding', *Seventh European Signal Processing Conference*, Edinburgh, Scotland, September, 1994, pp. 13–16.

[26] Ekudden, E., Hagen, R., Johansson, I. and Svedberg, J., 'The Adaptive Multi-Rate Speech Coder', *Proceedings of the IEEE Workshop on Speech Coding*, 1999, pp. 117–119.

[27] Gao, Y., Benyassine, A., Thyssen, J., Su, H., Murgia, C., and Shlomot, E., 'The SMV Algorithm Selected by TIA and 3GPP2 for CDMA Applications', *Proceedings of the IEEE International Conference on Acoustics, Speech, and Signal Processing*, 2001.

[28] 3rd Generation Partnership Project; Technical Specification Group Services and Systems Aspects; Speech Codec speech processing functions; AMR Wideband speech codec; Transcoding functions (Release 4), 3GPP TS 26.190, v2.0, 2001.

10

Speech Recognition Solutions for Wireless Devices

Yeshwant Muthusamy, Yu-Hung Kao and Yifan Gong

10.1 Introduction

Access to wireless data services such as e-mail, news, stock quotes, flight schedules, weather forecasts, etc. is already a reality for cellular phone and pager users. However, the user interface of these services leaves much to be desired. Users still have to navigate menus with scroll buttons or "type in" information using a small keypad. Further, users have to put up with small, hard-to-read phone/pager displays to get the results of their information access. Not only is this inconvenient, but also can be downright hazardous if one has to take their eyes off the road while driving. As far as input goes, speaking the information (e.g. menu choices, company names or flight numbers) is a hands-free and eyes-free operation and would be much more convenient, especially if the user is driving. Similarly, listening to the information (spoken back) is a much better option than having to read it. In other words, speech is a much safer and natural input/output modality for interacting with wireless phones or other handheld devices.

For the past few years, Texas Instruments has been focusing on the development of DSP based speech recognition solutions designed for the wireless platform. In this chapter, we describe our DSP based speech recognition technology and highlight the important features of some of our speech-enabled system prototypes, developed specifically for wireless phones and other handheld devices.

10.2 DSP Based Speech Recognition Technology

Continuous speech recognition is a resource-intensive algorithm. For example, commercial dictation software requires more than 100 MB of disk space for installation and 32 MB for execution. A typical embedded system, however, has constraints of low power, small memory size and little to no disk storage. Therefore, speech recognition algorithms designed for embedded systems (such as wireless phones and other handheld devices) need to minimize resource usage (memory, CPU, battery life) while providing acceptable recognition performance.

10.2.1 Problem: Handling Dynamic Vocabulary

DSPs, by design, are well suited for intensive numerical computations that are characteristic of signal processing algorithms (e.g. FFT, log-likelihood computation). This fact, coupled with their low-power consumption, makes them ideal candidates for running embedded speech recognition systems. For an application where the number of recognition contexts is limited and vocabulary is known in advance, different sets of models can be pre-compiled and stored in inexpensive flash memory or ROM. The recognizer can then load different models as needed. In this scenario, a recognizer running just on the DSP is sufficient. It is even possible to use the recognizer to support several applications with known vocabularies by simply pre-compiling and storing their respective models, and swapping them as the application changes. However, if the vocabulary is unknown or there are too many recognition contexts, pre-compiling and storing models might not be efficient or even feasible. For example, there are an increasing number of handheld devices that support web browsing. In order to facilitate voice-activated web browsing, the speech recognition system must dynamically create recognition models from the text extracted from each web page. Even though the vocabulary for each page might be small enough for a DSP based speech recognizer, the number of recognition contexts is potentially unlimited. Another example is speech-enabled stock quote retrieval. Dynamic portfolio updates require new recognition models to be generated on the fly. Although speaker-dependent enrollment (where the person trains the system with a few exemplars of each new word) can be used to add and delete models when necessary, it is a tedious process and a turn-off for most users. It would be more efficient (and user-friendly) if the speech recognizer could automatically create models for new words. Such dynamic vocabulary changes require an online pronunciation dictionary and the entire database of phonetic model acoustic vectors for a language. For English, a typical dictionary contains tens of thousands of entries, and thousands of acoustic vectors are needed to achieve adequate recognition accuracy. Since a 16-bit DSP does not provide such a large amount of storage, a 32-bit General-Purpose Processor (GPP) is required. The grammar algorithms, dictionary look-up, and acoustic model construction are handled by the GPP, while the DSP concentrates on the signal processing and recognition search.

10.2.2 Solution: DSP-GPP Split

Our target platform is a 16-bit fixed-point DSP (e.g. TI TMS320C54x or TMS320C55x DSPs) and a 32-bit GPP (e.g. ARM™). These two-chip architectures are very popular for 3G wireless and other handheld devices. Texas Instruments' OMAP™ platform is an excellent example [1]. To implement a dynamic vocabulary speech recognizer, the computation-intensive, small-footprint recognizer engine runs on the DSP; and the computation non-intensive, larger footprint grammar, dictionary, and acoustic model components reside on the GPP. The recognition models are prepared on the GPP and transferred to the DSP; the interaction among the application, model generation, and recognition modules is minimal. The result is a speech recognition server implemented in a DSP-GPP embedded system. The recognition server can dynamically create flexible vocabularies to suit different recognition contexts, giving the perception of an unlimited vocabulary system. This design breaks down the barrier between dynamic vocabulary speech recognition and a low cost platform.

10.3 Overview of Texas Instruments DSP Based Speech Recognizers

Before we launch into a description of our portfolio of speech recognizers, it is pertinent to outline the different recognition algorithms supported by them and to discuss, in some detail, the one key ingredient in the development of a good speech recognizer: speech training data.

10.3.1 Speech Recognition Algorithms Supported

Some of our recognizers can handle more than one recognition algorithm. The recognition algorithms covered include:

- *Speaker-Independent (SI) isolated digit recognition.* An SI speech recognizer does not need to be retrained on new speakers. Isolated digits imply that the speaker inserts pauses between the individual digits.
- *Speaker-Dependent (SD) name dialing.* An SD speech recognizer requires a new user to train it by providing samples of his/her voice. Once trained, the recognizer will work only on that person's voice. For an application like name dialing, where you do not need others to access a person's call list, an SD system is ideal. A new user goes through an enrollment process (training the SD recognizer) after which the recognizer works best only on that user's voice.
- *SI continuous speech recognition.* Continuous speech implies no forced pauses between words.
- *Speaker and noise adaptation* to improve SI recognition performance. Adapting SI models to individual speakers and to the background noise significantly improves recognition performance.
- *Speaker recognition* – useful for security purposes as well as improving speech recognition (if the system can identify the speaker automatically, it can use speech models specific to the speaker).

10.3.2 Speech Databases Used

The speech databases used to train a speech recognizer play a crucial role in its performance and applicability for a given task and operating environment. For example, a recognizer trained on clean speech in a quiet sound room will not perform well in noisy in-car conditions. Similarly, a recognizer trained on just one or a few (< 5) speakers will not generalize well to speech from new speakers, as it has not been exposed to enough speaker variability. Our speech recognizers were trained on speech from the *Wall Street Journal* [2], TIDIGITS [3] and TI-WAVES databases. The *Wall Street Journal* database was used only for training our clean speech models. The TIDIGITS and TI-WAVES corpora were collected and developed in-house and merit further description.

10.3.2.1 TIDIGITS

The TIDIGITS database is a publicly available, clean speech database of 17,323 utterances from 225 speakers (111 male, 114 female), collected by TI for research in digit recognition [3]. The utterances consist of 1–5- and 7-digit strings recorded in a sound room under quiet conditions. The training set consists of 8623 utterances from 112 speakers (55 male; 57 female), while the test set consists of 8700 utterances from a different set of 113 speakers (56 male; 57 female). The fact that the training and test set speakers do not overlap allows us to do speaker-independent recognition experiments. This database provides a good resource for testing digit recognition performance on clean speech.

10.3.2.2 TI-WAVES

The TI-WAVES database is an internal TI database consisting of digit-strings, commands and names from 20 speakers (ten male, ten female). The utterances were recorded under three different noise conditions in a mid-size American sedan, using both a handheld and a hands-free (visor-mounted, noise-canceling) microphone. Therefore, each utterance in the database is effectively recorded under six different conditions. The three noise conditions were (i) parked (ii) stop-and-go traffic, and (iii) highway traffic. For each condition, the windows of the car were all closed and there was no fan or radio noise. However, the highway traffic condition generated considerable road and wind noise, making it the most challenging portion of the database. Table 10.1 lists the Signal-To-Noise Ratio (SNR) of the utterances for the different conditions.

The digit utterances consisted of 4-, 7- and 10-digit strings, the commands were 40 call and list management commands (e.g. "return call", "cancel", "review directory") and the names were chosen from a set of 1325 first and last name pairs. Each speaker spoke 50 first and last names. Of these, ten name pairs were common across all speakers, while 40 name pairs were unique to each speaker. This database provides an excellent resource to train and test speech

Table 10.1   SNR (in dB) for the TI-WAVES speech database

Microphone type	Parked		Stop-and-go		Highway	
	Average	Range	Average	Range	Average	Range
Hand-held	32.4	18.8–43.7	15.6	5.2–33.2	13.7	4.5–25.9
Hands-free	26.5	9.2–39.9	13.8	3.4–34.4	7.3	2.4–21.2

recognition algorithms designed for real-world noise conditions. The reader is directed to Refs. [9] and [17] for details on recent recognition experiments with the TI-WAVES database.

10.3.3 Speech Recognition Portfolio

Texas Instruments has developed three DSP based recognizers. These recognizers were designed with different applications in mind and therefore incorporate different sets of cost-performance trade-offs. We present recognition results on several different tasks to compare and contrast the recognizers.

10.3.3.1 Min_HMM

Min_HMM (short for *MIN*imal *H*idden *M*arkov *M*odel) is the generic name for a family of simple speech recognizers that have been implemented on multiple DSP platforms. Min_HMM recognizers are isolated word recognizers, using low amounts of program and data memory space with modest CPU requirements on fixed-point DSPs.

Some of the ideas incorporated in Min_HMM to minimize resources include:

- No traceback capability, combined with efficient processing, so that scoring memory is fixed at just one 16-bit word for each state of each model.
- Fixed transitions and probabilities, incorporated in the algorithm instead of the data structures.
- Ten principal components of LPC based filter-bank values used for acoustic Euclidean distance.
- Memory can be further decreased, at the expense of some additional CPU cycles, by updating autocorrelation sums on a sample-by-sample basis rather than buffering a frame of samples.

Min_HMM was first implemented as a speaker-independent recognition algorithm on a DSP using a TI TMS320C5x EVM, limited to the C2xx dialect of the assembly language. It was later implemented in C54x assembly language by TI-France and ported to the TI GSM chipset. This version also has speaker-dependent enrollment and update for name dialing.

Table 10.2 shows the specifics of different versions of Min_HMM. Results are expressed in *% Word Error Rate* (WER), the percentage of words mis-recognized (each digit is treated as a word. Results on the TI-WAVES database are averaged over the three conditions (parked, stop-and-go and highway). Note that the number of MIPS increases dramatically with noisier speech on the same task (SD Name Dialing).

Table 10.2 Min_HMM on the C54x platform (ROM and RAM figures are in 16-bit words)

Task	Speech database	ROM	RAM	MIPS	Results (%WER)
SI isolated digits	TIDIGITS	4K program; 4K models	1.5K data	4	1.1
SD name dialing (50 names)	TI-WAVES handheld	4K program	25K models; 6K data	16	1.1
SD name dialing (50 names)	TI-WAVES hands-free	4K program	25K models; 6K data	61	3.4

10.3.3.2 IG

The Integrated Grammar (IG) recognizer differs from Min_HMM in that it supports continuous speech recognition and allows flexible vocabularies. Like Min_HMM, it is also implemented on a 16-bit fixed-point DSP with no more than 64K words of memory. It supports the following recognition algorithms:

- Continuous speech recognition on speaker-independent models, such as digits and commands.
- Speaker-dependent enrollment, such as name dialing.
- Adaptation (training) of speaker-independent models to improve performance.

IG has been implemented on the TI TMS320C541, TMS320C5410 and TMS320C5402 DSPs. Table 10.3 shows the resource requirements and recognition performance on the TIDIGITS and TI-WAVES (handheld) speech databases. Experiments with IG are described in greater detail in Refs. [4–6].

Table 10.3   IG on the TI C54x platform (ROM and RAM figures are in 16-bit words)

Task	Speech database	ROM	RAM	MIPS	Results (%WER)
SI continuous digits	TIDIGITS	8K program	8K search	40	1.8
SD name dialing (50 names)	TI-WAVES handheld	8K program	28K models; 5K search	20	0.9

10.3.3.3 TIESR

The Texas Instruments Embedded Speech Recognizer (TIESR) provides speaker-independent continuous speech recognition robust to noisy background, with optional speaker-adaptation for enhanced performance. TIESR has all of the features of IG, but is also designed for operation in adverse conditions such as in a vehicle on a highway with a hands-free microphone. The performance of most recognizers that work well in an office environment degrades under background noise, microphone differences and speaker accents. TIESR includes TI's recent advances in handling such situations, such as:

- On-line compensation for noisy background, for good recognition at low SNR.
- Noise-dependent rejection capability, for reliable out-of-vocabulary speech rejection.
- Speech signal periodicity-based utterance detection, to reduce false speech decision triggering.
- Speaker-adaptation using name-dialing enrollment data, for improved recognition without reading adaptation sentences.
- Speaker identification, for improved performance on groups of users.

TIESR has been implemented on the TI TMS320C55x DSP core-based OMAP1510 platform. The salient features of TIESR and its resource requirements will be discussed in greater detail in the next section. Table 10.4 shows the speaker-independent recognition results (with no adaptation) obtained with TIESR on the C55x DSP. The results on the TI-WAVES database include %WER on each of the three conditions (parked, stop-and-go, and highway). Note the perfect recognition (0% WER) on the SD Name Dialing task in the 'parked' condition. Also, the model size, RAM and MIPS increase on the noisier TI-WAVES digit data (not surprisingly), compared to the clean TIDIGITS data. The RAM and MIPS figures for the other TI-WAVES task are not yet available.

Table 10.4   TIESR on C55x DSP (RAM and ROM figures are in 16-bit words)

Task	Speech database	ROM	RAM	MIPS	Results (%WER)
SI continuous digits	TIDIGITS	6.7K program; 18K models;	4K	8	0.5
SI continuous digits	TI-WAVES hands-free	6.7K program; 22K models	10K	21	0.6/2.0/8.6
SD name dialing (50 names)	TI-WAVES hands-free	6.7K program; 50K models	–	–	0.0/0.1/0.3
SI commands (40 commands)	TI-WAVES hands-free	6.7K program; 40K models	–	–	0.5/0.8/3.4

10.4 TIESR Details

In this section, we describe two distinctive features of TIESR in some detail, noise robustness and speaker adaptation. Also, we highlight the implementation details of the grammar parsing and model creation module (on the GPP) and discuss the issues involved in porting TIESR to the TI C55x DSP.

10.4.1 Distinctive Features

10.4.1.1 Noise Robustness

Channel distortion and background noise are the two of the main causes of recognition errors in any speech recognizer [11]. Channel distortion is caused by the different frequency responses of the microphone and A/D. It is also called convolutional noise because it manifests itself as an impulse response that "convolves" with the original signal. The net effect is a non-uniform frequency response multiplied with the signal's linear spectrum (i.e. additive in the log spectral domain). Cepstral Mean Normalization (CMN) is a very effective technique [12] to deal with it because the distortion is modeled as a constant additive component in the cepstral domain and can be removed by subtracting a running mean computed over a 2–5 second window.

Background noise can be any sound other than the *intended* speech, such as wind or engine noise in a car. This is called additive noise because it can be modeled as an additive component in the linear spectral domain. Two methods can be used to combat this problem: spectral subtraction [14] and Parallel Model Combination (PMC) [13]. Both algorithms estimate a running noise energy profile, and then subtract it from the input signal's spectrum or add it to the spectrum of all the models. Spectral subtraction requires less computation because it needs to modify only one spectrum of the speech input. PMC requires a lot more computation because it needs to modify the spectra of all the models; the larger the model, the more computation required. However, we find that PMC is more effective than spectral subtraction.

CMN and PMC cannot be easily combined in tandem because they operate in different domains, the log and linear spectra, respectively. Therefore, we use a novel joint compensation algorithm, called Joint Additive and Convolutional (JAC) noise compensation, that can compensate both the linear domain correction and log domain correction simultaneously [15]. This JAC algorithm achieves large error rate reduction across various channel and noise conditions.

10.4.1.2 Speaker Adaptation

To achieve good speaker-independent performance, we need large models to model different accents and speaking styles. However, embedded systems cannot accommodate large models, due to storage resource constraints. Adaptation thus becomes very important. Mobile phones and PDAs are "personal" devices and can therefore be adapted for the user's voice. Most embedded recognizers do not allow adaptation of models (other than enrollment) because training software is usually too large to fit into an embedded system. TIESR, on the other hand, incorporates training capability into the recognizer itself. It supports supervision alignment and trace output (where each input speech frame is mapped to a model). This capability enables us to do Maximum Likelihood Linear Regression (MLLR) phonetic class adaptation [16,17,19]. After adaptation, the recognition accuracy usually improves significantly, because the models effectively take channel distortion and speaker characteristics into account.

10.4.2 Grammar Parsing and Model Creation

As described in Section 10.2, in order to support flexible recognition context switching, a speech recognizer needs to create grammar and models on demand. This requires two major information components: an online pronunciation dictionary and decision tree acoustics. Because of the large sizes of these components, a 32-bit GPP is a natural choice.

10.4.2.1 Pronunciation Dictionary

The size and complexity of the pronunciation dictionary varies widely for different languages. For a language with more *regular* pronunciation, such as Spanish, a few hundred rules are enough to convert text to phone accurately. On the other hand, for a language with more *irregular* pronunciation, such as English, a comprehensive online pronunciation dictionary is required. We used a typical English pronunciation dictionary (COMLEX) with 70,955 entries; it required 1,826,302 bytes of storage in ASCII form. We used an efficient way to represent this dictionary using only 367,599 bytes, a 5:1 compression. Our compression technique was such that there was no need to decompress the dictionary to do a look-up, and there was no extra data structure required for the look-up either; it was directly computable in low-cost ROM. We also used a rule-based word-to-phone algorithm to generate a phonetic decomposition for any word not found in the dictionary. Details of our dictionary compression algorithm are given in Ref. [8].

10.4.2.2 Decision Tree Acoustics

A decision tree algorithm is an important component in a medium or large vocabulary speech recognition system [7,18]. It is used to generate context-dependent phonetic acoustics to build recognition models. A typical decision tree system consists of hundreds of classification trees, used to classify a phone based on its left and right contexts. It is very expensive to store these trees on disk and create searchable trees in memory (due to their large sizes). We devised a mechanism to store the tree in binary form and create one tree at a time during search. The tree file was reduced from 788 KB in ASCII form to 32 KB in binary form (ROM), a 25:1 reduction. The searchable tree was created and destroyed one at a time, bringing the memory usage down to only 2.5 KB (RAM). The decision tree serves as an index mechanism for acoustic vectors. A typical 10K-vector set requires 300 KB to store in ROM. A larger vector set will provide better performance. It can be easily scaled depending on the available ROM size. Details of our decision tree acoustics compression are given in Ref. [8].

10.4.2.3 Resource Requirements

Table 10.5 shows the resource requirements for the grammar parsing and model creation module running on the ARM9 core. The MIPS numbers represent averages over several utterances for the digit grammars specified.

Table 10.5 Resource requirements on the ARM9 core for grammar creation and model generation

Item	Resource	Comments
Program size	57 KB (ROM)	
Data (breakdown below)	773 KB (ROM)	
Dictionary	418.2 KB	Online pronunciation dictionary
Acoustic vectors	314.1 KB	Spectral vectors
Decision tree	27.9 KB	
Monophone HMM	6.5 KB	HMM temporal modeling
Decision tree table	3.0 KB	
Decision tree questions	1.2 KB	
Question table	0.9 KB	
Phone list	0.2 KB	ASCII list of English monophones
CPU	23.0 MIPS	Four or seven continuous digits grammar
	22.7 MIPS	One digit grammar

10.4.3 Fixed-Point Implementation Issues

In addition to making the system small (low memory) and efficient (low MIPS), we need to deal with fixed-point issues. In a floating-point processor, all numbers are normalized into a format with sign bit, exponent, and mantissa. For example, the IEEE standard for *float* has one sign bit, an 8–bit exponent, and a 23-bit mantissa. The exponent provides a large dynamic range: $2^{128} \sim = 10^{38}$. The mantissa provides a fixed level of precision. Because every *float* number is *individually* normalized into this format, it always maintains a 23-bit precision as long as it is within the 10^{38} dynamic range. Such good precision covering a large dynamic range frees the algorithm designer from worrying about scaling problems. However, it comes at the cost of more power, larger silicon, and higher cost. In a 16-bit fixed-point processor, on the other hand, the only format is a 16-bit integer, ranging from 0 to 65535 (unsigned) or

-32768 to $+32767$ (signed). The numerical behavior of the algorithm has to be carefully normalized to be within the dynamic range of a 16-bit integer at every stage of the computation.

In addition to the data format limitation, another issue is that some operations can be done efficiently, while others cannot. A fixed-point DSP processor usually incorporates a hardware multiplier so that addition and multiplication can be completed in one CPU cycle. However, there is no hardware for division and it takes more than 20 cycles to do it by a routine. To avoid division, we want to pre-compute the inverted data. For example, we can pre-compute and store $1/\sigma^2$ instead of σ^2 for the Gaussian probability computation. Other than the *explicit* divisions, there are also *implicit* divisions hidden in other operations. For example, *pointer arithmetic* is used heavily in the memory management in the search algorithm. Pointer subtraction actually incurs a division. Division can be approximated by multiplication and shift. However, pointer arithmetic cannot tolerate any errors. Algorithm design has to take this into consideration and make sure it is accurate under all possible running conditions.

We found that 16-bit resolution was not a problem for our speech recognition algorithms [10]. With careful scaling, we were able to convert computations such as Mel-Frequency Cepstral Coefficients (MFCC) used in our speech front-end and Parallel Model Combination (PMC) used in our noise compensation, to fixed-point precision with no performance degradation.

10.4.4 Software Design Issues

In an embedded system, resources are scarce and their usage needs to be optimized. Many seemingly innocent function calls actually use a lot of resources. For example, string operation and memory allocation are both very expensive. Calling one string function will cause the entire string library to be included, and malloc() is not efficient in allocating memory. We did the following optimizations to our code:

* Replace all string operations with efficient integer operations.
* Remove all malloc() and free(). Design algorithms to do memory management and garbage collection. The algorithms are tailored for efficient utilization of memory.
* Local variables consume stack size. We examine the allocation of local and global variables to balance memory efficiency and program modularity. This is especially important for recursive routines.
* Streamline data structures so that all model data are stored efficiently and designed for computability, as opposed to using one format for disk storage and another for computation.

10.5 Speech-Enabled Wireless Application Prototypes

Figure 10.1 shows the schematic block diagram of a speech-enabled application designed for a dual-processor wireless architecture (like the OMAP1510). The application runs on the GPP, while the entire speech recognizer and portions of the Text-To-Speech (TTS) system run on the DSP. The application interacts with the speech recognizer and TTS via a speech API that encapsulates the DSP-GPP communication details. In addition, the grammar parsing

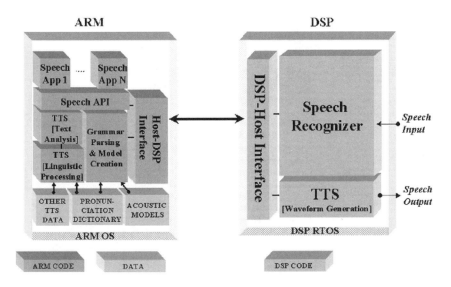

Figure 10.1   Speech-enabled application schematic using the DSP-GPP architecture

and model creation software runs on the GPP and interacts with the DSP recognizer, as described in Section 10.2.2.

The TTS system shown in Figure 10.1 is assumed to be a concatenative TTS system (similar to the one described in Ref. [20]). The text analysis and linguistic processing modules of the TTS system are resource-intensive and require large databases. As such, they are best suited to run on the GPP. The waveform generation component of the TTS system runs on the DSP. Note that the TTS system and the Grammar Parsing modules are shown sharing a common pronunciation dictionary. While this may not be true of some TTS systems in existence today, it is indeed possible and is the preferred scenario, in order to conserve storage on the ARM™. The "Other TTS Data" box refers to the point-of-speech lexicon, trigram language models, letter-to-sound rules and binary trees used by the Text Analysis and Linguistic Processing modules.

10.5.1 Hierarchical Organization of APIs

The Speech API module in Figure 10.1 merits further description. The application on the ARM™ interacts with the speech recognizer and TTS system using a hierarchy of progressively finer-grained APIs. This is shown in Figure 10.2. The application talks to the Speech API layer, which could be either the Java Speech API (JSAPI) [26] or a variant of Microsoft's SAPI [27]. JSAPI and SAPI are two of the most commonly used standard speech APIs. This API layer is implemented in terms of a set of basic speech functions, called Speech Primitives (SP). The SP layer contains functions to start and stop a recognizer, pause and resume audio, load grammars, return results, start and stop the TTS, set the TTS speech rate, select the TTS 'speaker', etc. in a format dictated by the speech recognizer and the TTS system. The SP layer in turn is implemented in terms of the DSP/BIOS Bridge API. The DSP/BIOS Bridge API takes care of the low-level ARM-DSP communication and the transfer of data between the

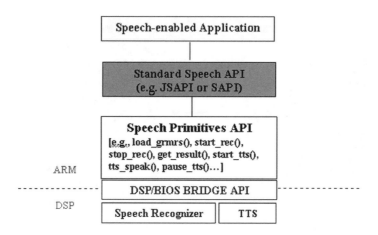

Figure 10.2   Hierarchical organization of APIs

application and the recognizer and TTS system. This hierarchical API architecture has the following advantages:

- The application and the standard API layers (JSAPI and SAPI) are totally independent of the implementation details of the lower-level APIs. This encapsulation makes it easier to incorporate changes into the lower-level APIs, without having to rework the higher-level APIs.
- The existence of the SP layer reduces the amount of development needed to implement the mutually incompatible JSAPI and SAPI standards, as they are implemented in terms of a common set of functions/methods in the SP layer.

With this architecture in mind, we have developed several application prototypes that are specifically designed for the wireless domain. In the following sections, we describe four system prototypes for:

- Internet information retrieval (InfoPhone);
- Voice e-mail;
- Voice navigation; and
- Voice-enabled web browsing.

The first three systems, designed primarily for hands-busy, eyes-busy conditions, use speaker-independent speech recognizers, and can be used with a restricted display or no display at all. These systems use a client–server architecture with the client designed to be resident on a GPP-DSP combination on the phone or other handheld device. The fourth prototype for web browsing, called VoiceVoyager, was originally developed for desktop browsers (Netscape and Microsoft IE). We are currently in the process of modifying it for a client–server, wireless platform with a wireless microbrowser. Of the four systems, the InfoPhone prototype is the first one to be ported to a GPP-DSP platform; we have versions using both IG (on TI C541) and TIESR (TI C55x DSP; on OMAP1510). Work is underway to port the other three applications to a DSP-GPP platform as well.

10.5.2 InfoPhone

InfoPhone is a speech-enabled Java application that is best described as a 3G wireless data service prototype. It allows speech-enabled retrieval of useful information such as stock quotes, flight schedules and weather forecasts. Users can choose one of flights, stocks and weather from a top-level menu and interact with each "service" by speech commands. "Keypad" (non-speech) input is also available as a back-up. The application incorporates separate grammars for company names (for stocks), flight numbers and city names (for weather). We have developed versions of this demo using both IG (on C541) and TIESR (C5510; OMAP™). In this section, we will be describing the OMAP-enabled version that runs on the pre-OMAP™ EVM (ARM925 and TI C5510 DSP).

The application runs on the ARM925 under Symbian OS (previously known as EPOC) Release 5 [21] and communicates with the TIESR recognizer running on the C5510 under OSE [22], via the TI JSP Speech API. The JSP API is a Java API, developed in-house, that handles all of the GPP-DSP communication and allows any Java application to be speech-enabled with a DSP based speech recognizer. Figure 10.3 shows the block diagram of the system architecture.

Speech input to the application is processed by TIESR and the information request is sent to an InfoServer application running on a remote server that accesses the appropriate website, retrieves the HTML page, extracts just the essential information and transmits it to the application. The results of the information retrieval are displayed on the 320×240 LCD screen that is part of the pre-OMAP™ EVM. We are in the process of incorporating a TTS system on the ARM™ and the C5510. Once this is done, then the information retrieved will be played back by the TTS system, resulting in true eyes-free, hands-free operation.

Users can switch between the three "services" dynamically. Keypad input is always active and users can switch back and forth between voice and tactile input at any time.

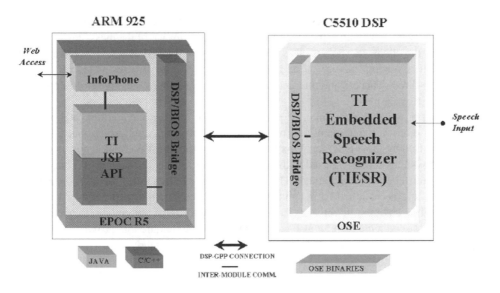

Figure 10.3   OMAP™-enabled InfoPhone architecture.

10.5.3 Voice E-mail

Over the past several years, the cellular telephone has become an important mobile communication tool. The use of voice-mail has also increased over the same time period. It would be convenient if a mobile user could use a single device (such as a cellular phone) to access both his e-mail and voice-mail. This eliminates the hassle of dealing with multiple devices and also allows multimodal messaging; a user can call up the sender of an e-mail message to respond verbally to his e-mail, or send e-mail in response to a voice-mail message. TI has developed a Voice E-mail (VE) system prototype that addresses these issues.

10.5.3.1 System Overview

The VE system has a client–server architecture and is completely voice-driven. Users talk to the system and listen to messages and prompts played back by the speech synthesizer. The system has a minimal display (for status messages) and is designed to operate primarily in a "displayless" mode, where the user can effectively interact with the system without looking at a display. The current system is an extension of previous collaborative work with MIT [23] and handles reading, filtering, categorization and navigation of e-mail messages. It also has the capability to "compose" and send e-mail using speech-based form filling. Work is underway to incorporate voice-mail send and receive (using caller ID information).

An important aspect of the displayless user interface is that the user should, at all times, know exactly what to do, or should be able to find out easily. To this end, we have incorporated an elaborate context-dependent help feature. If the user gets lost, he also has the ability to reset all changes and start over from the beginning. An optional display can be incorporated into the VE system to provide visual feedback as well.

10.5.3.2 Client–Server Architecture

The VE server handles all of the e-mail/voice-mail functions. It accesses the e-mail and voice-mail servers and handles the receiving, sending and storage of the messages. It communicates with the client via sockets. The VE server is implemented as a Java application. We use Microsoft Exchange Server as the mail server. The VE server uses MAPI (Microsoft's Mail API) to directly access and interact with mail objects such as message stores, messages, sender lists, etc.

The client provides the user interface and handles the reading, navigation, categorization, and filtering of e-mail and voice-mail messages. It is completely agnostic about the type of mail server used by the VE server. This feature ensures that the VE client is not specific to a single mail system and can be used with any mail server as long as the interface between the VE server and client is maintained. The VE client has both speech recognition and TTS capabilities, and is designed to not maintain constant connection to the server (to reduce connection time charges). It connects to the server only to initiate or end a session, check for new mail or to send a message. It also has an extensive help feature that provides guidance to beginners of the system and on request. The client is implemented as a Java applet.

10.5.3.3 User Interface

The user can speak to the system in a natural, continuous speaking style. Several alternates to each phrase are allowed (for example, "any messages from John Smith?" and "is there a message from John Smith?"). There is also a rejection feature that handles incorrect speech input; it prompts the user for more information if the recognition score falls below an empirically determined threshold. To minimize fatigue, the error prompts in case of a rejection are randomized. Further, if more than three consecutive rejections occur, the system initiates context-dependent help to guide the user. The TTS system operates in e-mail mode; that is, it can correctly speak out the e-mail headers.

10.5.4 Voice Navigation

Car navigation systems have been available for some time, but they have received only limited use. We can partly attribute this to the user interface available for such systems: often unnatural, sometimes clumsy, and potentially unsafe. Some systems use a touch screen while others use a rotating knob to enter destination addresses one alpha-numeric character at a time. We have developed a system to obtain maps and/or directions for different places in a city as naturally as possible, by voice I/O only. It could be incorporated into either a built-in device in a car or a cellular phone. This navigation system is primarily aimed at hands-busy, eyes-busy conditions such as automobile driving. An optional display is provided for situations where the user may safely look at the screen, when the car is parked. All textual information is played back to the user via a TTS system. A dialog manager is used to handle all interactions with the user.

10.5.4.1 Client–Server Architecture

The car navigation device acts as a client. The user interacts with the client which in turn communicates with a remote server to process user utterances. A Global Positioning System (GPS) connected to the client tracks the location of the user at any point in time. A web-based map service on the server provides maps and directions. We currently use the MapQuest™ website as our map server (www.mapquest.com). Further, a yellow pages server is used to find businesses near the user's current location. We use the GTE SuperPages™ website as our yellow pages server (www.superpages.com).[1] Our speech recognizer processes the user's utterances and passes the result to the dialog manager, which then interprets these utterances in context. If the appropriate information needed to issue a query has been given, the dialog manager will query the appropriate server to get a response. Otherwise, it may interact further with the user. For example, if the user says "Where is the DoubleTree Hotel?" and the system has knowledge of multiple hotels of the same name, it will first interact with the user to resolve this ambiguity before querying the map server.

The navigation application has been designed so that the user may query the system using natural speech. The speech interface provides a natural way for users to specify the destination, while the presence of a dialog manager ensures that users can have their queries satisfied

---

[1] Our prototypes access these websites programmatically and post-process the complete HTML page(s) retrieved from them to extract just the information needed. This information is passed onto the TTS system and the client display.

even in the presence of missing, ambiguous, inconsistent, or erroneous information. The dialog manager also assists in constraining the grammars for the speech recognizer and in providing context-sensitive help. This dialog manager has been described in greater detail in Ref. [24]. In case of any errors on the part of the user or the system, the user may say "Go back" at any time to undo the effect of the previous utterance. It also supports a rejection feature that requests the user to repeat something if the system does not have enough confidence in its recognition.

10.5.4.2 Navigation Scenarios

This application covers different scenarios in which a user may need directions to some place. In some cases, the user may know the exact address or cross streets of the destination and might query the system for directions to these locations (for example, "How do I get to 8330 LBJ Freeway in Dallas?"). In addition, the system has knowledge of a list of common points of interest for the current city. These may include hotels, hospitals, airports, malls, universities, sports arenas, etc. and the user can get directions to any of these by referring to them by name (for example, "I need to go to the Dallas Museum of Art"). Finally, there are often instances where a user is interested in locating some business near his/her current location. For example, the user may just say "Find a movie theater around here". In such situations, the system needs to access the yellow pages server to find the list of movie theaters, interact with the user to identify the one of interest, and then query the map server for maps and/or directions. The phone number of the identified business can also be provided to the user on demand.

10.5.5 Voice-Enabled Web Browsing

We have developed an interface to the Web, called VoiceVoyager that allows convenient voice access to information [25]. VoiceVoyager uses a speaker-independent, continuous speech, arbitrary vocabulary recognition system that has the following specific features for interacting with the Web:

- Customizable speakable commands for simple browser control;
- Speakable bookmarks to retrieve pages by random access using customized phrases;
- Speakable links to select any hypertext link by simply speaking it; and
- Smart pages for natural spoken queries specific to pages.

To support these features, VoiceVoyager has the ability to incorporate new grammars and vocabularies "on the fly". The ability to handle a flexible vocabulary, coupled with the ability to dynamically modify and create new grammars in the recognizer (as described in Section 10.2.1), is crucial to VoiceVoyager's ability to speech-enable arbitrary web pages, including those that the user has never visited before.

Since VoiceVoyager was originally developed as a PC desktop tool, the following discussion uses terms such as Hypertext Mark-up Language (HTML) and "pages", concepts that are somewhat specific to desktop-based web browsers. For a WAP [28] microbrowser, the corresponding analogues would be Wireless Markup Language (WML) and "cards" (or "decks"), respectively. A deck is a set of WML cards (or pages). It is to be noted that the features of VoiceVoyager described below are not specific to the PC domain, they apply just as well to wireless microbrowsers, be they WAP or i-Mode or any other type.

10.5.5.1 Speakable Commands

To control the browser, VoiceVoyager provides spoken commands to display help pages, scroll up or down, go back or forward, display the speakable commands and bookmarks, add a page to the speakable bookmarks, and edit phrases for the speakable bookmarks. Voice-Voyager has default phrases for these commands, but the user may change them, if desired, to more convenient ones.

10.5.5.2 Speakable Bookmarks

To reach frequently accessed pages, users may add pages to their speakable bookmarks. When adding a page currently displayed in the browser, VoiceVoyager uses the title of the page to construct a grammar for subsequent access by voice. The initial grammar includes likely alternatives to allow, for example, either "NIST's" or "N.I.S.T's" in a page entitled "NIST's Home Page". The user may then add additional phrases to make access to the information more convenient or easier to remember. The speakable bookmarks remain active at all times giving users instant access to important information.

10.5.5.3 Speakable Links

Every time VoiceVoyager encounters a page on the Web, it parses HTML content to determine the links and the Uniform Resource Locators (URLs) associated with them. Voice-Voyager then transforms the string of words into a grammar that allows likely alternatives as mentioned above. It checks several phonetic dictionaries for pronunciations and uses a text-to-phone mapping if these fail. We currently use a proper name dictionary, an abbreviation/ acronym dictionary, and a 250,000 entry general English dictionary. The text-to-phone mapping proves necessary in many cases, including, for example, pages that include invented words (for example, "Yahooligans" on the Yahoo page).

10.5.5.4 Smart Pages

On some occasions, the point-and-click paradigm associated with links falls short. For a more flexible voice-input paradigm, we developed a mechanism called smart pages. Smart pages are simply web pages that contain a link to a grammar appropriate for a page or set of pages. When a smart page is downloaded onto a client browser, the grammar(s) associated with that page are also downloaded and dynamically incorporated into the speech recognizer. Using standard web conventions, web page authors may specify what users can say and interpret the recognized words appropriately for the context.

10.6 Summary and Conclusions

Unlike desktop speech recognition systems, embedded speech recognizers have to contend with constraints of limited memory, low power and little to no disk storage. Combining its expertise in speech recognition technology and its leadership in DSP platforms, TI has developed several speech recognizers for the C54x and C55x platforms. Despite conforming to low-cost, low-memory constraints of DSPs, these recognizers handle a variety of useful

Table 10.6   Summary of Texas Instruments' DSP based speech recognizers

Recognizer	Recognition Tasks	TI Platforms	Current use/deployment
Min_HMM	SD name dialing	C5x	TI GSM Chipset
	SI isolated digits	C54x	
IG	SD name dialing	C541	Body-worn PCs
	SI continuous speech	C5410	
	Speaker adaptation	C5402 DSK	
TIESR	SD name dialing	C5510	OMAP1510
	Robust SI recognition		
	Speaker adaptation		
	Speaker identification		

recognition tasks, including isolated and continuous digits, speaker-dependent name-dialing, speaker-independent continuous speech recognition under adverse noise conditions (using both handheld and hands-free in-car microphones). Table 10.6 summarizes our portfolio of recognizers.

The four system prototypes (InfoPhone, Voice E-mail, Voice Navigation and VoiceVoyager) demonstrate the speech capabilities of a DSP-GPP platform. They are a significant step towards providing GPP-DSP-based speech recognition solutions for 3G wireless platforms.

References

[1] Chaoui, J., Cyr, K., Giacalone, J.-P., de Gregorio, S., Masse, Y., Muthusamy, Y., Spits, T., Budagavi, M. and Webb, J., 'Open Multimedia Application Platform: Enabling Multimedia Applications in Third Generation Wireless Terminals", *Texas Instruments Technical Journal*, October–December 2000.

[2] Paul, D.B. and Baker, J.M., 'The Design for the *Wall Street Journal* Based CSR Corpus", *Proceedings of ICSLP 1992*.

[3] Leonard, R.G., 'A Database for Speaker-Independent Digit Recognition", *Proceedings of ICASSP 1984*.

[4] Kao, Y.H. 'A Multi-Lingual, Speaker-Independent, Continuous Speech Recognizer on TMS320C5x Fixed-Point DSP", *Proceedings of ICSPAT 1997*.

[5] Kao, Y.H., 'Minimization of Search Network in Speech Recognition", *Proceedings of ICSPAT 1998*.

[6] Kao, Y.H., 'N-Best Search Algorithm for Continuous Speech Recognition", *Proceedings of ICSPAT 1998*.

[7] Kao, Y.H., 'Building Phonetic Models for Low Cost Implementation Using Acoustic Decision Tree Algorithm", *Proceedings of ICSPAT 1999*.

[8] Kao, Y.H. and Rajasekaran, P.K., 'Designing a Low Cost Dynamic Vocabulary Speech Recognizer on a GPP-DSP System", *Proceedings of ICASSP 2000*.

[9] Ramalingam, C.S., Gong, Y., Netsch, L.P., Anderson, W.W., Godfrey, J.J. and Kao, Y.H., 'Speaker-Dependent Name-Dialing in a Car Environment with Out-of-Vocabulary Rejection', *Proceedings of ICASSP 1999*.

[10] Kao, Y.H., Gong, Y., 'Implementing a High Accuracy Continuous Speech Recognizer on a Fixed-Point DSP', *Proceedings of ICASSP 2000*.

[11] Gong, Y., 'Speech Recognition in Noisy Environments: A Survey', *Speech Communications*, Vol. 16, No. 3, 1995, pp. 261–291.

[12] Atal, B., 'Effectiveness of Linear Prediction Characteristics of the Speech Wave for Automatic Speaker Identification and Verification', *Journal of the Acoustical Society of America*, Vol. 55, 1974, pp. 1304–1312.

[13] Gale, M.J.F. and Young, S., 'An Improved Approach to the Hidden Markov Model Decomposition of Speech and Noise', *Proceedings of ICASSP 1992*.

[14] Boll, S.F., 'Suppression of Acoustic Noise in Speech Using Spectral Subtraction', *Acoustics, Speech and Signal Processing (ASSP) Journal*, Vol. ASSP-27, No. 2, 1979, pp. 113–120.

[15] Gong, Y., 'A Robust Continuous Speech Recognition System for Mobile Information Devices', *Proceedings of Workshop on Hands-Free Speech Communication*, Kyoto, Japan, April 2001.

[16] Leggetter, C.J. and Woodland, P.C., 'Maximum Likelihood Linear Regression for Speaker Adaptation of Continuous Density HMMs', *Computer Speech and Language*, Vol. 9, No. 2, 1995, pp. 171–185.

[17] Gong, Y. and Godfrey, J.J., 'Transforming HMMs for Speaker-Independent Hands-Free Speech Recognition in the Car', *Proceedings of ICASSP 1999*.

[18] Bahl, L.R., de Souza, P.V., Gopalakrishnan, P. and Picheny, M., 'Decision Trees for Phonological Rules in Continuous Speech', *Proceedings of ICASSP 1991*.

[19] Gong, Y., 'Source Normalization Training for HMM Applied to Noisy Telephone Speech Recognition', *Proceedings of Eurospeech 1997*, Rhodes, Greece, September 1997, pp. 1555–1558.

[20] Black, A.W. and Taylor, P., 'Festival Speech Synthesis System: System Documentation Edition 1.1 (for Festival Version 1.1.1)'. Human Communication Research Centre Technical Report HCRC/TR-83, University of Edinburgh, 1997.

[21] The Symbian Platform. http://www.epocworld.com.

[22] Enea OSE Systems. http://www.enea.com.

[23] Marx, M., 'Towards Effective Conversational Messaging', MS Thesis, MIT, June 1995.

[24] Agarwal, R., 'Towards a PURE Spoken Dialogue System for Information Access', *Proceedings of the ACL/EACL Workshop on Interactive Spoken Dialog Systems*, 1997.

[25] Hemphill, C.T. and Thrift, P.R., 'Surfing the Web by Voice', *Proceedings of Multimedia 1995*, San Francisco, CA, November 1995.

[26] Java Speech Application Programming Interface. http://java.sun.com/products/java-media/speech.

[27] Microsoft Speech Application Programming Interface version 5.0. http://www.microsoft.com/speech/technical/SAPIOverview.asp.

[28] The Wireless Application Protocol Forum. http://www.wapforum.org.

11

Video and Audio Coding for Mobile Applications

Jennifer Webb and Chuck Lueck

11.1 Introduction

Increased bandwidth for Third Generation (3G) communication not only expands the capacity to support more users, but also makes it possible for network providers to offer new services with higher bit rates for multimedia applications. With increased bit rates and programmable DSPs, several new types of applications are possible for mobile devices, that include audio and video content. No longer will there be a limited 8–13 kbps, suitable only for compressed speech. At higher bit rates, the same phone's speakers and DSP with different software can be used as a digital radio. 3G cellular standards will support bit rates up to 384 kbps outdoors and up to 2 Mbps indoors. Other new higher-rate indoor wireless technologies, such as Bluetooth (802.15), WLAN (802.11), and ultra wideband, will also require low-power solutions. With low-power DSPs available to execute 100s of MIPS, it will be possible to decode compressed video, as well as graphics and images, along with audio or speech. In addition to being used for spoken communication, mobile devices may become multifunctional multimedia terminals.

Even the higher 3G bit rates would not be sufficient for video and audio, without efficient compression technology. For instance, raw 24-bit color video at 30 fps and (640 × 480) pixels per frame requires 221 Mbps. Stereo CD with two 16-bit samples at 44.1 kHz requires 1.41 Mbps [1]. State-of-the-art compression technology makes it feasible to have mobile access to multimedia content, probably at reduced resolution.

Another enabler to multimedia communication is the *standardization* of compression algorithms, to allow devices from different manufacturers to interoperate. Even so, at this time, multiple standards exist for different applications, depending on bandwidth and processing availability, as well as the type of content and desired quality. In addition, there are popular non-standard formats, or de facto standards. Having multiple standards practically requires use of a programmable processor, for flexibility.

Compression and decompression require significant processing, and are just now becoming feasible for mobile applications with high-performance, low-power, low-cost DSPs. For video and audio, the processor must be fast enough to play out and/or encode in real-time, and power consumption must be low enough to avoid excessive battery drain. With the availability of affordable DSPs, there is the possibility of offering products with a greater variety of cost-quality-convenience combinations.

As the technological hurdles of multimedia communication are being solved, the acceptance of the technology also depends on the availability of content, and the availability of high-bandwidth service from network providers, which also depends on consumer demand and the cost of service at higher bit rates. This chicken-and-egg problem has similarities to the situation in the early days of VCRs and fax machines, with the demand for playback or receive capability interdependent on the availability of encoded material, and vice versa. For instance, what good is videophone capability, unless there are other people with videophones? How useful is audio decoding, until a wide selection of music is available to choose from? There may be some reluctance to offer commercial content until a dominant standard prevails, and until security/piracy issues have been resolved. The non-technical obstacles may be harder to overcome, but are certainly not insurmountable.

Motivations for adding audio and video capability include product differentiation, Internet compatibility, and the fact that lifestyles and expectations are changing. Little additional equipment is needed to add multimedia capability to a communications device with an embedded DSP, other than different software, which offers manufacturers a way to differentiate and add value to their products. Mobile devices are already capable of accessing simplified WAP Internet pages, yet at 3G bit rates, it is also feasible to add the richness of multimedia, and access some content already available via the Internet. Skeptics who are addicted to TV and CD audio may question if there will be a need or demand for wireless video and audio; to some degree, the popularity of mobile phones has shown increased demand for convenience, even if there is some increase in cost or degradation in quality. Although wireless communications devices may not be able to provide a living-room multimedia experience, they can certainly enrich the mobile lifestyle through added video and audio capability.

Some of the possible multimedia applications are listed in Table 11.1. The following sections give more detail, describing compression technology, standards, implementation on a DSP, and special considerations for mobile applications. Video is described, then audio, followed by an example illustrating requirements for implementation of a multimedia mobile application.

Table 11.1 Many new mobile applications will be possible with audio and video capability

	No sound	One-way speech	Two-way speech	Audio
No display		Answering machine	Phone	Digital radio
Image	E-postcard	News	Ordering tickets, fast food	Advertisement
One-way video	Surveillance	Sports coverage	Telemedicine	Movies; games; music videos
Two-way video	Sign language		Videophone	

11.2 Video

Possible mobile video applications include streaming video players, videophone, video e-postcards and messaging, surveillance, and telemedicine. If the video duration is fairly short and non-real-time, such as for video e-postcards or messaging, the data can be buffered, less compression is required, and better quality is achievable. For surveillance and telemedicine, a sequence of high-quality still images, or low frame-rate video, may be required. An application such as surveillance may use a stationary server for encoding, with decoding on a portable wireless device. In contrast, for telemedicine, paramedics may encode images on a wireless device, to be decoded at a fixed hospital computer. With streaming video, complex off-line encoding is feasible. Streaming decoding occurs as the bitstream is received, somewhat similar to television, but much more economically with reduced quality. For one-way decoding, some buffering and delay are acceptable. Videophone applications require simultaneous encoding and decoding, with small delay, resulting in further quality compromises. Of all the mobile video applications mentioned, two-way videophone is perhaps the first to come to mind, and the most difficult to implement.

Wireless video communication has long been a technological fantasy dating back before Dick Tracy and the Jetsons. One of the earliest science-fiction novels, *Ralph 124C 41 +* ("one to foresee"), by Hugo Gernsback, had a cover depicting a space-age courtship via videophone, shown in Figure 11.1 [2]. Gernsback himself designed and manufactured the first mass-produced two-way home radio, the Telimco Wireless, in 1905 [3]. The following year, Boris Rosing created the world's first television prototype in Russia [4], and transmitted

Figure 11.1   This Frank R. Paul illustration, circa 1911, depicts video communication in 2660

silhouettes of shapes in 1907. It must have seemed that video communication was right around the corner. Video is the logical next step beyond wireless speech, and much progress has been made, yet there are a number of differences that pose technological challenges.

Some differences in coding video, compared with speech, include the increased bandwidth and dynamic range, and higher dimensionality of the data, which have led to the use of variable bit rate, predictive, error-sensitive, lossy compression, and standards that are not bit-exact. For instance, each block in a frame, or picture, may be coded with respect to a non-unique block in the previous frame, or it may be coded independently of the previous frame; thus, different bitstreams may produce the same decoded result, yet some choices will result in better compression, lower memory requirements, or less computation. While variability in bit rate is key to achieving higher compression ratios, some target bit rate must be maintained to avoid buffer overflows, and to match channel capacity. Except for non-real-time video, e.g. e-postcards, either pixel precision or frame rate must be adjusted dynamically, causing quality to vary within a frame, as well as from frame to frame. Furthermore, a method that works well on one type of content, e.g. talking head, may not work as well on another type of content, such as sports. Usually good results can be achieved, but without hand tweaking, it is always possible to find "malicious" content, contrived or not, that will give poor encoding results for a particular real-time encoder. For video transmitted over an error-prone wireless channel, it is similarly always possible to find a particular error pattern that is not effectively concealed by the decoder, and propagates to subsequent frames. As difficult as it is to encode and transmit video robustly, it is exciting to think of the potential uses and convenience that it affords, for those conditions under which it performs sufficiently well.

11.2.1 Video Coding Overview

A general description of video compression will provide a better understanding of the processing complexity for DSPs, and the effect of errors from mobile channels. In general, compression removes redundancy through prediction and transform coding. For instance, after the first frame, motion vectors are used to predict a 16×16 macroblock of pixels using a similar block from the previous frame, to remove temporal redundancy. The Discrete Cosine Transform (DCT) can represent an 8×8 block of data in terms of a few significant non-zero coefficients, which are scaled down by a quantization parameter. Finally, Variable Length Coding (VLC) assigns the shortest codewords to the most common symbols. Based on the assumption that most values are zero, run-length coded symbols represent the number of zero values between non-zero values, rather than coding all of the zero values separately. The encoder reconstructs the frame as the decoder would, to be used for motion prediction for the next frame. A typical video encoder is depicted in Figure 11.2. Video compression requires significant processing and data transfers, as well as memory, and the variable length coding makes it difficult to detect and recover from bitstream errors.

Although all video standards use similar techniques to achieve compression, there is much latitude in the standards to allow implementers to trade off between quality, compression efficiency, and complexity. Unlike speech compression standards, video standards specify only the decoder processing, and simply require that the output of an encoder must be decodable. The resulting bitstream depends on the selected motion estimation, quantization, frame rate, frame size, and error resilience, and various implementation trade-offs are summarized in Table 11.2.

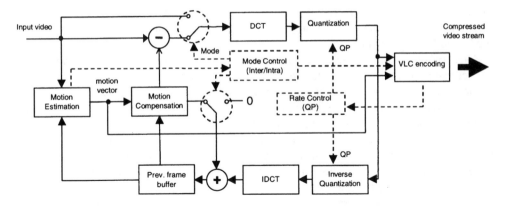

Figure 11.2 A typical video encoder with block motion compensation, discrete cosine transform and variable length coding achieves high compression, leaving little redundancy in the bitstream

Table 11.2 Various video codec implementation trade-offs are possible, depending on the available bit rate, processor capabilities, and the needs of the application. This table summarizes key design choices and their affect on resource requirements

Feature\impact	MIPS	Data transfers	Bit rate	Memory	Code size
Motion estimation	√	√	√ depending on motion	√ if no ME	
Quantization	√ decoder		√		
Frame rate	√	√	√		
Frame size	√	√	√	√	
Error resilience	√		√		√

For instance, implementers are free to use any motion estimation technique. In fact, motion compensation may or may not be used. The complexity of motion estimation can vary from an exhaustive search over all possible values, to searching over a smaller subset, or the search may be skipped entirely by assuming zero motion or using intracoding mode, i.e. without reference to the previous frame. A simpler motion estimation strategy dramatically decreases computational complexity and data transfers, yet the penalty in terms of quality or compression efficiency may (or may not) be small, depending on the application and the type of content.

Selection of the Quantization Parameter (QP) is particularly important for mobile applications, because it affects the quality, bit rate, buffering and delay. A large QP gives coarser quality, and results in smaller, and more zeroed, values; hence a lower bit rate. Because variable bit rate coding is used, the number of bits per frame can vary widely, depending on how similar a frame is to the previous frame. During motion or a scene change, it may be necessary to raise QP to avoid overflowing internal buffers. When many bits are required to code a frame, particularly the first frame, it takes longer to transmit that frame over a fixed-rate channel, and the encoder must skip some frames until there is room in its buffer (and the decoder's buffer), which adds to delay. It is difficult to predetermine the best coding strategy in real-time, because using more bits in a particular region or frame may force the rate control to degrade quality elsewhere, or may actually save bits, if that region provides a better prediction for subsequent frames.

Selection of the frame rate affects not only bit rate, but also data transfers, which can impact battery life on a mobile device. Because the reference frame and the reconstructed frame require a lot of memory, they are typically kept off-chip, and must be transferred into on-chip memory for processing. At higher frame rates, a decoder must update the display more often, and an encoder must read and preprocess more data from the camera. Additional data transfers and processing will increase the power consumption proportionally with the frame rate, which can be significant. The impact on quality can vary. For a given channel rate, a higher frame rate generally allows fewer bits per frame, but may also provide better motion prediction. For talking head sequences, there may be little degradation in quality at a higher frame rate, for a given bit rate. However, if there is more motion, QP must be raised to maintain the target bit rate at a higher frame rate, which degrades spatial quality. Generally, a target of 10–15 frames per second, or lower, is considered to be adequate and economical for mobile applications.

To extend the use of video beyond broadcast TV, video standards also support smaller frame sizes, to match the lower bit rates, smaller form factors, power and cost constraints of mobile devices. The Common Intermediate Format (CIF) is 352×288 pixels, so named because its size is convenient for conversion from either NTSC 640×480 or PAL 768×576 interlaced formats. Content in CIF format may be scaled down by a factor of two, vertically and horizontally, to obtain Quarter CIF (QCIF) with 176×144 pixels. Sub-QCIF (SQCIF) has about half as many pixels as QCIF, with 128×96 pixels. SQCIF can be formed from QCIF by scaling and/or cropping the image. In some cases, cropping only removes surrounding background pixels, and SQCIF is almost as useful as QCIF, but for sports or panning sequences, it is usually better to maintain the full field of view. Without cropping the QCIF images, there will be a slight, hardly noticeable, change in aspect ratio. A SQCIF display may be just the right size for a compact handheld communicator, but on a high-resolution display that is also used for displaying documents, SQCIF may seem too small; one option is to scale up the output. For mobile communication, smaller is generally better, resulting in better quality for a given bit rate, less processing and memory required (lower cost), less drain on the battery, and less noticeable coding artifacts.

Typical artifacts for wireless video include blocking, ringing, and distortion from channel errors. Because the DCT coefficients for 8×8 blocks are quantized, there may be a visible discontinuity at block boundaries. Ringing artifacts occur near object boundaries during motion. These artifacts are especially visible at lower bit rates, with larger formats, and when shown on a high-quality display. If the bitstream is corrupted from transmission over an error-prone channel, colors may be altered, and objects may actually appear to break up, due to errors in motion compensation. Because frames are coded with respect to the previous frame, errors may persist and propagate through motion, causing severe degradation. Because wireless devices are less likely to have large, high-quality displays, the blocking and ringing artifacts may be less of a concern, but error resilience is essential.

For wireless applications, one option is to use channel coding or retransmission to correct errors in the bitstream, but this may not always be affordable. For transmission over circuit-switched networks, errors may occur randomly or in bursts, during fading. Techniques such as interleaving are effective to break up bursts, but increase buffering requirements and add

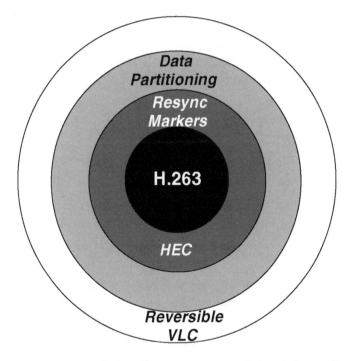

Figure 11.3   MPEG-4 simple profile includes error resilience tools for wireless applications. The core of MPEG-4 simple profile is baseline H.263 compression. In addition, the standard supports RMs to delineate video packets, HEC to provide redundant header information, data partitioning within video packets, and reversible VLC within a data partition

delay. Channel coding can reduce the effective bit error rate, but it is difficult to determine the best allocation between channel coding and source coding. Because channel coding uses part of the bit allocation, either users will have to pay more for better service, or the bit rate for source coding must be reduced. Over packet-switched networks, entire packets may be lost, and retransmission may create too much delay for real-time video decoding. Therefore, some measures must be taken as part of the source coding to enhance error resilience.

The encoder can be implemented to facilitate error recovery through adding redundancy and resynchronization markers to the bitstream. Resynchronization markers are inserted to subdivide the bitstream into *video packets*. The propagation of the errors in VLC codewords can be limited if the encoder creates smaller video packets. Also, the encoder implementation may reduce dependence on previous data and enhance error recovery through added header information, or by intracoding more blocks. Intracoding, resynchronization markers, and added header information can significantly improve error resilience, but compression efficiency is also reduced, which penalizes quality under error-free conditions.

The decoder can be implemented to improve performance under error conditions, through error detection and concealment. The decoder must check for any inconsistency in the data, such as an invalid codeword, to avoid processing and displaying garbage. With variable length codewords, an error may cause codewords to be misinterpreted. It is generally not possible to determine the exact location of an error, so the entire video packet must be discarded. What to display in the place of missing data is not standardized. Concealment methods may be very elaborate or very simple, such as copying data from the previous frame. After detecting an error, the decoder must find the next resynchronization marker to resume decoding of the next video packet. Error checking and concealment can significantly increase the computational complexity and code size for decoder software.

11.2.2 Video Compression Standards

The latest video standards provide increased compression efficiency for low bit rate applications, and include tools for improved error resilience. The H.263 standard [5] was originally released by ITU-T in 1995 for videophone communication over Public Switched Telephone Network (PSTN), targeting bit rates around 20 kbps, but with no need for error resilience. Many of the same experts helped develop the 1998 ISO MPEG-4 standard, and included compatibility with baseline H.263 plus added error-resilience tools [6,7], in its *simple* profile. The error-resilience tools for MPEG-4 are Resynchronization Markers (RMs), Header Extension Codes (HECs), Data Partitioning (DP), and Reversible Variable Length Codes (RVLCs). The RM tool divides the bitstream into video packets, to limit propagation of errors in VLC decoding, and to permit resynchronization when errors occur. The HEC tool allows the encoder to insert redundant header information, in case essential header data are lost. The DP tool subdivides each packet into partitions, putting the higher-priority codewords in a separate partition, to allow recovery of some information, even if another partition is corrupted. Use of RVLC allows a partition with errors in the middle to be decoded in both the forward and reverse direction, to attempt to salvage more information from both ends of the partition. These tools are described in greater detail in Ref. [8]. Figure 11.3 depicts schematically the relationship between *simple* profile MPEG-4 and baseline H.263.

H.263 version 2, also called H.263 + , includes several new annexes, and H.263 version 3, a.k.a. H.263++ and a few more, to improve quality or compression efficiency or error resilience. H.263+ Annex K supports a slice structure, similar to the MPEG-4 RM tool. Annex W includes a mechanism to repeat header data, similar to the MPEG-4 HEC tool. H.263+ Appendix I describes an error tracking method that may be used if a feedback channel is available for the decoder to report errors to the encoder. H.263+ Annex D specifies a RVLC for motion data. H.263++ Annex V specifies data partitioning and RVLCs for header data, contrasted with MPEG-4, which specifies a RVLC for the coefficient data. The large number of H.263+(+) Annexes allows a wide variety of implementations, which poses problems for testing and interoperability. To encourage interoperability, H.263++ Annex X specifies profiles and levels, including two *interactive and streaming wireless video* profiles.

Because there is not a single dominant video standard, two specifications for multimedia communication over 3G mobile networks are being developed by the Third Generation Partnership Project (3GPP) [9] and 3GPP2 [10,11]. 3GPP2 has not specified video codecs at the time of writing, but it is likely their video codec options will be similar to 3GPP's. 3GPP mandates support for baseline H.263, and allows simple profile MPEG-4 or H.263++ wireless Profile 3 as options.

Some mobile applications, such as audio players or security monitors, may not be bound by the 3GPP specifications. There will likely be demand for wireless gadgets to decode streaming video from www pages, some of which, e.g. RealVideo, are proprietary and not standardized. For applications not requiring a low bit rate, or that can tolerate delay and very low frame rates, another possible format is motion JPEG, a series of intracoded images. Without motion estimation, block-based intracoding significantly reduces cycles, code size, and memory requirements, and the bitstream is error-resilient, because there is no interdependence between frames. JPEG-2000 has added error resilience and scalability features, but is wavelet based, and much more complex than JPEG. Despite standardization efforts, there is no single dominant video standard, which makes a programmable DSP implementation even more attractive.

11.2.3 Video Coding on DSPs

Before the availability of low-power, high-performance DSPs, video on a DSP would have been unthinkable. Conveniently, video codecs operate on byte data with integer arithmetic, and few floating point operations are needed, so a low-cost, low-power, fixed-point DSP with 16-bit word length is sufficient. Division requires some finagling, but is only needed for quantization and rate control in the encoder, and for DC and AC (coefficient) prediction in the decoder, as well as for some more complex error concealment algorithms. Some effort must be taken to obtain the IDCT precision that is required for standard compliance, but several good algorithms have been developed [12]. H.263 requires that the IDCT meet the extended IEEE-1180 spec [13], but the MPEG-4 conformance requirements are actually less stringent. It is possible to run compiled C code in real-time on a DSP, but some restructuring may be necessary to fit in a DSP's program memory or data memory.

Processing video on a DSP, compared to a desktop computer, requires more attention to memory, data transfers, and localized memory access, because of the impact on cost, power consumption and performance. Fast on-chip memory is relatively expensive, so most of the data are kept in slower off-chip memory. This makes it very inefficient to directly access a frame buffer. Instead, blocks of data are transferred to an on-chip buffer for faster access. For video, a DSP with Direct Memory Access (DMA) is needed to transfer the data in background, without halting the processing. Because video coding is performed on a 16×16 macroblock basis, and because of the two-dimensional nature of the frame data, typically a multiple of 16 rows are transferred and stored in on-chip memory at a time for local access. To further increase efficiency, processing routines, such as quantization and inverse quantization, may be combined, to avoid moving data in and out of registers.

The amount of memory and data transfers required varies depending on the format, frame rate, and any preprocessing or postprocessing. Frame rate affects only data transfers, not the memory requirement. The consequences of frame size, in terms of memory and power consumption, must be carefully considered. For instance, a decoder must access the previous decoded frame as a reference frame, as well as the current reconstructed frame. A *single* frame in YUV 4:2:0 format (with chrominance data subsampled) requires 18, 38, and 152 kbytes for SQCIF, QCIF, and CIF, respectively. For two-way video communication, two frames of memory are needed for decoding, another two for encoding, and preprocessed or postprocessed frames for the camera or display may be in RGB format, which requires twice as much memory as 4:2:0 format! Some DSPs limit data memory to 64 kbytes, but platforms designed for multimedia, e.g. OMAP™ platform [14], provide expanded data memory.

The amount of processing required depends not only on format and frame rate, but also on content. Decoder complexity is highly variable with content, since some macroblocks may not be coded, depending on the amount of motion. Encoder complexity is less variable with content, because the motion estimation must be performed whether the macroblock is eventually coded or not. Efficient decoding consumes anywhere from 5 to 50 MIPS, while encoding can take an order of magnitude more, depending on the complexity of the motion estimation algorithm. Because most of the cycles are spent for motion estimation and the IDCT, coprocessors are often used to speed up these functions.

Besides compression and decompression, video processing may require significant additional processing concomitantly, to interface with a display or camera. Encoder preprocessing from camera output may involve format conversion from various formats, e.g. RGB to YUV or 4:2:2 YCrYCb to 4:2:0 YUV. If the camera processing is also integrated, that could include white balance, gamma correction, autofocus, and color filter array interpolation for the Bayer output from a CCD sensor. Decoder postprocessing could include format conversion for the display, and possibly deblocking and deringing filters, as suggested in Annex F of the MPEG-4 standard, although this may not be necessary for small, low-cost displays. The memory and processing requirements for postprocessing and preprocessing can be comparable to that of the compression itself, so it is important not to skimp on the peripherals!

More likely than not, hand-coded assembly will be necessary to obtain the efficiency required for video. As DSPs become faster, efficiency may seem less critical, yet it is still important to conserve battery life, and to allow other applications to run concurrently. For instance, to play a video clip with speech requires running video decode and speech decode, simultaneously. Both should fit in memory and run in real-time, and if there are cycles to spare, the DSP can enter an idle mode to conserve power. For this reason, it is still common practice to use hand-coded assembly, at least for critical routines. Good development tools and assembly libraries of commonly used routines help reduce time to market. The effort and expense to hand-code in assembly are needed to provide competitive performance and are justifiable for mass-produced products.

11.2.4 Considerations for Mobile Applications

Processing video on a DSP is challenging in itself, but transmitting video over a wireless network adds another set of challenges, including systems issues of how to packetize it for network transport, and how to treat network-induced delays and errors. Additional processing is needed for multimedia signaling, and to send or receive transport packets. Video packets transmitted over a packet-switched network require special headers, and the video decoder must be resilient to packet loss. A circuit-switched connection can be corrupted by both random and burst errors, and requires that video and speech be multiplexed together. Additional standards besides compression must be implemented to transmit video over a wireless network, and that processing may be performed on a separate processor.

There are several standards that support transmission of video over networks, including ITU-T standard H.324, for circuit-switched two-way communication, H.323 and IETF's Session Initiation Protocol (SIP), for packet-switched two-way communication, and Real Time Streaming Protocol (RTSP) for one-way video streaming over IP. Besides transmitting the compressed bitstream, it is necessary to send a sequence of control messages as a mechanism to establish the connection and signal the type and format for video. SIP and RTSP specify text-based protocols, similar to HTTP, whereas H.323 and H.324 use a common control standard H.245 for messaging. These standards must be implemented efficiently with a small footprint for mobile communicators. Control messaging and packetization are more suitable for a microcontroller than a DSP, so the systems code will typically run on the microcontroller (MCU) part of a DSP + MCU platform.

For transmission over packet-switched networks, control messages are usually transmitted reliably over Transmission Control Protocol (TCP), and the bitstreams via faster but unreliable User Datagram Protocol (UDP), as depicted in Figure 11.4. There are some exceptions, with RTSP and SIP allowing signaling over UDP for fast set-up. A bitstream sent over UDP will not pass through a firewall, so TCP is sometimes used for the media itself. In addition to UDP packetization, Real-time Transport Protocol (RTP) packet headers contain information such as payload type, a timestamp, sequence number, and a marker bit to indicate the last packet of a video frame [15], since packets may arrive out of order. The way the bitstream is packetized will affect performance and recovery from packet loss. To avoid too much overhead from packet headers, and systems calls to send and receive packets, it may be most efficient to send an entire frame in a packet, in which case, an entire video frame may be lost. For full recovery, the bitstream may contain Intracoded frames periodically, which are costly because of the associated higher bit rate and delay. 3GPP is currently supporting the use of SIP for two-way communication and RTSP for one-way streaming over packet-switched networks.

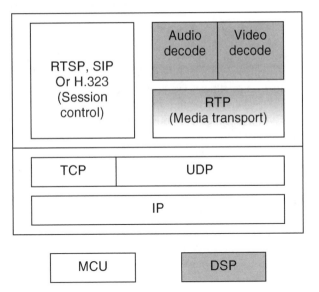

Figure 11.4   Typical protocol stack used to transport video over a packet-switched network [16]

For transmission over circuit-switched networks, multiple logical channels, e.g. video and audio, are multiplexed together into packets, which are transmitted via modem over a single physical channel. The ITU umbrella standard for circuit-switched multimedia communication is H.324, which cites the H.223 standard for multiplexing data. The H.324 protocol stack is depicted in Figure 11.5. H.223 includes the option of adding channel coding to the media in an adaptation layer, in addition to what is provided by the network. There is a mobile version of H.223, called H.223M, which includes annexes giving extra error protection to packet headers, and H.324M is the corresponding mobile version of H.324. 3GPP has specified its own variant of H.324M, called 3G-324M, which supports a subset of the modes and annexes.

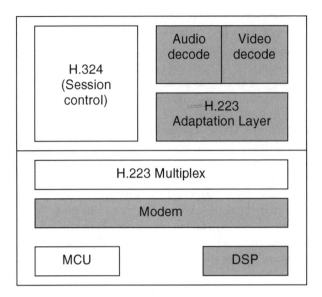

Figure 11.5   H.324 stack used to transport video over a circuit-switched network [16]

Besides the problem of transmitting video over a network, there are issues with power consumption and error-resilience for mobile devices. Memory and data transfers for video will drain a battery much faster than speech, making it more economical to use a smaller frame size and frame rate, when practical, rather than demanding the largest frame size and frame rate at all times. As mentioned previously, there are some methods supported by the standards that can be used to improve error resilience for video, but wireless video quality, like speech, will not be as good as its wire-line counterpart. Although users may have to adjust their expectations for video, many new applications are possible that put to good use faster, lower-power DSPs, and the increasing capacity offered by 3G standards.

Finally, some may question the need for video on mobile devices, since few people can watch a screen while walking or driving. There are potential business uses, such as for telemedicine or surveillance. Yet it is true that the living-room television paradigm doesn't fit. Mobile video applications are more likely to provide added value to mobile communicators, offering convenience for people who are spending more time away from home.

11.3 Audio

Fueled by the excitement of music distribution over the Internet, compressed audio formats such as MP3 have gained great popularity in recent years. Along with this phenomenon has come the introduction of a large number of portable audio products designed for playback of audio content by users who no longer want to be tied to their PC. These early players, the so-called "MP3 players", are typically flash memory based, using either internal storage or removable flash memory cards, and often require a link to a PC to download content. Digital audio jukeboxes for the car and home have also surfaced which allow the user to maintain and access large audio databases of compressed content.

Future applications of audio compression will soon expand to include an array of wireless devices, such as digital radios, which will allow the user to receive CD quality audio through a conventional FM radio band. For stereo material, high-quality audio playback is possible at rates of 96 kbps (essentially 15:1 compression compared to CD) or even lower. At these rates streaming audio over wireless connection will be possible using 3G technology. Increased bandwidths will lead to new applications such as audio-enabled cell phones and portable streaming audio players, which will allow the user access to content being streamed via wireless link or the Internet. The introduction of such applications presents new challenges to the DSP design engineer in terms of MIPS, memory, and power consumption

11.3.1 Audio Coding Overview

Fundamentally, the perceptual audio coder relies on the application of psychoacoustic principles to exploit the temporal and spectral masking properties of the human ear. Although audio compression can be achieved using strictly lossless coding techniques which take advantage of statistical redundancies in the input signal, significantly greater compression ratios can be achieved by using lossy compression in which the introduction of distortion is tightly controlled according to a psychoacoustically based distortion metric.

A high-level block diagram of a generic perceptual audio coder is shown in Figure 11.6. The major components of the audio encoder include the filterbank, the joint coding module, the quantizer module, the entropy coding module, and the psychoacoustic model. The audio decoder contains the counterparts to most of these components, and some audio algorithms may even employ an additional psychoacoustic model in the decoder as well. Normally, however, the encoder contains significant components that have no counterpart in the decoder, so most perceptual audio codecs are asymmetric in complexity. Most real-world systems will have some variation of this basic design, but the core blocks and their operation are essentially the same.

The input into the audio encoder typically consists of digitally-sampled audio, which has been segmented into blocks, or frames, of audio samples. To smooth transitions between

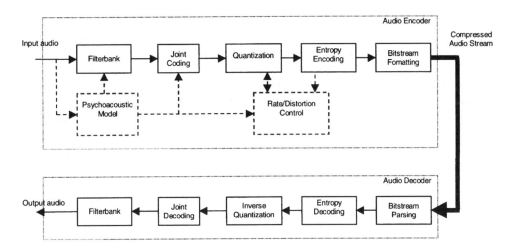

Figure 11.6 Block diagram of generic perceptual audio coder for general audio signals

consecutive input blocks, the input frames may overlap each other. An overlap of 50% is typical in modern algorithms. The filterbank performs a time-frequency transformation on each input block, resulting in a frequency-domain representation of the input signal which varies from block-to-block. Since many musical signals, such as string instruments, are composed of slowly varying sinusoidal components, only a few frequency components will have significant energy, effectively reducing the amount of data which needs to be coded. By modifying the block size, trade-offs in time and frequency resolution can be made. A larger block size will yield a higher resolution in the frequency-domain but a lower resolution in the time-domain. One of the most common filterbank transformations in use is the Modified Discrete Cosine Transform (MDCT), which projects the input block onto a set of basis functions consisting essentially of windowed cosine functions [17].

For transient audio signals, such as percussive instruments or speech, some compression algorithms provide the ability to dynamically adjust the block size to accommodate large variations in the temporal characteristics of the signal. This technique is sometimes referred to as block switching or window switching. Smaller window sizes centered around signal transients allow the encoder to localize, temporally, large coding errors around sections of the signal with large energies.

Because stereo and multichannel audio material typically have a high correlation among audio channels, significantly greater compression can be achieved by joint channel coding techniques. For stereo coding, a technique commonly applied is Middle/Side (MS) stereo processing. In this technique, the left channel is replaced with the sum of the left and right channels (middle signal), and the right channel is replaced with the difference of the left and right channel (side signal). For stereo material with high similarity in the left and right channels, most of the signal energy will now exist in the sum channel, and considerably less signal energy will reside in the difference channel. Since the difference signal can generally be coded using fewer bits than the original right channel, a reduction in bit rate can be achieved. At the decoder, the transmitted sum and difference signals are reconstructed, and an inverse sum and differencing procedure is used to obtain the decompressed right and left channels.

The quantizer converts the high-precision spectral coefficients from the filterbank into a set of reduced-precision integers, which are then typically coded using entropy codes. The quantizer used can be either uniform or non-uniform, but quite often a non-uniform quantizer, in which the step sizes are smaller for smaller amplitudes, is used to give a higher Signal-to-Noise Ratio (SNR) at low signal levels, and thus reduce audibly correlated distortion in the audio at low volumes. Dithering may also be used to achieve the same effect. The quantizer is designed so that the number of quantization levels, and hence the resulting SNR, can be adjusted across various frequency regions to match some predetermined psychoacoustic threshold.

One of the key elements of the encoder is the psychoacoustic model. The psychoacoustic model has the responsibility of determining and controlling the parameters of the encoding in such a way as to minimize the perceptual distortion in the reconstructed audio. Such parameters may include the desired filterbank block size, the desired number of bits to use to code the present frame, and the SNRs needed for various groupings of spectral data. The selection of SNR for a particular band will ultimately dictate the resolution of the quantizer and hence the number of bits used.

As with video encoding, perceptual audio encoding is inherently a variable rate process.

Dynamics in musical content lead to variability in required bit rate. Although the long-term average bit rate may be constrained to be constant, the instantaneous bit rate is typically allowed to vary. Variations in instantaneous bit rate require a buffer in the decoder to store data when the instantaneous bit rate becomes larger than the average and to retrieve data when the bit rate becomes lower than the average. Since changes in signal dynamics cannot be predicted, the allocation of bits on a instantaneous per-frame basis is not a trivial task in the encoder, and typically must be determined using parameters based on heuristic measurements taken from sample test data.

11.3.2 Audio Compression Standards

In recent years, public interest in compressed audio formats has skyrocketed due to the popularity of MP3 and the availability of downloadable content via the Internet. Although MP3 is perhaps the most commonly known compression standard in use today, there are currently many audio formats available, including the more recently-developed MPEG audio standards as well as a large number of privately-developed proprietary formats.

Now commonly referred to as MP3, ISO/IEC MPEG-1 Audio Layer 3 was standardized in 1992 by the Motion Pictures Expert Group (MPEG) as part of the MPEG-1 standard, a comprehensive standard for the coding of motion video and audio [18]. MPEG-1 Layer 3 is one of the three layers which make up the MPEG-1 audio specification. The three layers of MPEG-1 Audio were selected to give users the ability to select varying performance/complexity trade-offs, with Layer 3 having the most complexity but also the best sound quality. Contrary to what the labeling suggests, the three layers are not entirely supersets of each other. Layer 2 is much like a layer around Layer 1, with each using the same filterbank and a similar quantization and coding scheme. Layer 3, however, has a modified filterbank and an entirely different way of encoding spectral coefficients. Layer 3 achieves greater compression than either Layers 1 or 2, but at the cost of increased complexity.

In 1994, ISO/IEC MPEG-2 Audio was standardized. MPEG-2 audio provided two major extensions to MPEG-1. The first was multichannel audio, because MPEG-1 was limited to solely mono and stereo coding. This paved the way to new applications such as multi-track movie soundtracks, which typically are recorded in what is called 5.1 format, for five primary audio channels (left, right, center, left surround, right surround) and one low-frequency (subwoofer) channel. The second extension that MPEG-2 provided was the support for lower sampling frequencies and, consequently, lower transmission data rates.

Promising better compression without the burden of having to support backward compatibility, MPEG-2 Advanced Audio Coding (AAC) was developed as a non-backward compatible (with respect to MPEG-1) addition to MPEG-2 Audio. Standardized in April of 1997 [19], AAC became the first codec to achieve transparent quality audio (by ITU definition) at 64 kbps per audio channel [20]. AAC has also been adopted as the high-quality audio codec in the new MPEG-4 compression standard, and has also been selected for use in the Japanese HDTV standard. AAC at 96 kbps stereo has been shown to be slightly better, in terms of perceived audio quality, as MP3 at 128 kbps. Table 11.3 shows a rough quality comparison of several MPEG audio standards, where the diff grade represents a measurement of perceived audio degradation.

Recently standardized, MPEG-4 audio supports a wider range of data rates, from high-quality coding at 64–392 kbps all the way down to 2 kbps for speech. MPEG-4 audio is

Table 11.3 Perceptual comparison of MPEG audio standards

Algorithm – stereo bit rate (kbps)	Diff grade	Perceived degradation
AAC-128	−0.47	Perceptible but not annoying
AAC-96	−1.15	Perceptible and slightly annoying
MP2-192	−1.18	Perceptible and slightly annoying
MP3-128	−1.73	Perceptible and slightly annoying
MP2-160	−1.75	Perceptible and slightly annoying
MP2-128	−2.14	Perceptible and annoying

comprised of a number of different codecs, each specializing in a particular bit rate range and signal type. The high-quality data rates within MPEG-4 are supported by a slightly modified version of MPEG-2 AAC. An alternative codec, TWIN-VQ, which has a modified quantization and coding scheme, is also supported. At lower data rates, several additional codecs are included (see Figure 11.7), such as wide and narrow band CELP for speech and two parametric coders for speech or music, HILN and HVXC.

To enhance the performance of MPEG-2 AAC, a number of extensions have been added within the MPEG-4 framework, effectively producing an MPEG-4 AAC version. These AAC extensions include a improved prediction module, fine and course bit rate scalability, and extensions for error robustness in error-prone environments.

In addition to the various MPEG audio standards, which are co-developed by a number of contributing organizations, many proprietary audio coders have been privately developed and introduced into the marketplace. One of the most familiar and successful of these is Dolby Laboratories' AC-3, or Dolby Digital™. Developed in the early 1990s, AC-3 has most commonly been used for multi-track movie soundtracks. AC-3 is now commonly used in

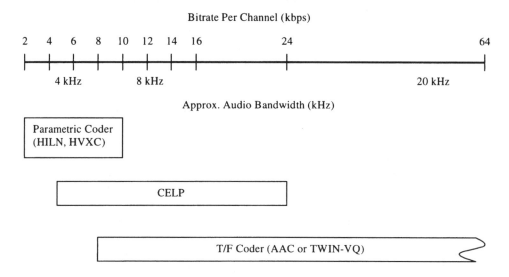

Figure 11.7 Codecs supported within MPEG-4 audio for general audio coding

the US for DVD and HDTV [17]. Other audio codecs include Lucent's EPAC, AT&T's PAC, RealAudio's G2, QDesign's QDMC, and NTT's TWIN-VQ. Recently, Microsoft has also entered the audio coding arena with their Window Media Audio (WMA) player.

11.3.3 Audio Coding on DSPs

Today's low-power DSPs provide the processing power and on-chip memory to enable audio applications which, until recently, would have been impossible. Compared to video decoding, the required MIPS, memory, and data transfer rates of an audio decoder are considerably lower. However, an efficient DSP implementation is not necessarily any less challenging. DSPs typically have limited on-chip memory, and developing real-time algorithms which fit into this memory requires special attention. In addition, precision requirements at the output of the audio decoder make numerical accuracy, particularly for fixed-point processors, a critical component of the DSP algorithm design.

One of the main advantages of the DSP is its programmability. This ability allows it to support multiple audio formats in the same platform and, in addition, provide upgradability to other current and future standards. The large number of audio coding formats in use today make development on the DSP extremely attractive. In flash memory-based audio players, the DSP program necessary to decode a particular format, such as MP3, can be stored in flash memory along with the media that is to be decoded. This is typically a small fraction of the media size, and an insignificant fraction of the total flash memory size. For wireless applications, upgrades for new audio formats may be downloaded through a wireless link.

The input data into the audio codec generally consists of 16-bit signed integer data samples. Typically, a number of different sampling rates between 8 and 48 kHz can be supported, but a 44.1-kHz sampling rate is quite common as it is the sampling rate used in the familiar CD format. Maintaining 16-bit precision at the output of the decoder, which is often required for high-quality audio applications, requires significant computation power, particularly for fixed-point applications. For floating-point processors, the development cycle can be shorter because 32-bit floating-point processing is adequate for most signal processing, and compiled C code can be used in many areas with reasonable MIPS and code size. Development on a fixed-point processor, however, requires greater development resources because algorithms must be carefully designed to maintain adequate precision, and often the resulting code must be hand-coded in assembly to maintain low MIPs while keeping code space small. For 16-bit processors, this will typically mean performing computations using double-precision (32-bit values). In addition, all memory buffers along the primary data path must maintain double-precision values to ensure adequate precision at each computational stage.

Algorithm MIPS are determined by a number of factors, including input sampling rate, required precision, and the efficiency of the compiler for C code. Sampling rate has a large effect on the overall computational requirement of the decoder, because this will ultimately determine the overall number of frames processed per second. Variations in bit rate may also lead to variations in MIPS, but these variations are mostly the result of the variable length decoding. An efficient Huffman decoder design can greatly reduce the variations due to bit rate and potentially increase the maximum bit rate supported by the decoder.

A number of factors affect the ROM and RAM requirements of an audio decoder. One of the main factors affecting memory size is the frame size (number of audio samples per

Table 11.4 Frame sizes for various audio coding standards. Frame size affects memory requirements for an audio decoder

Algorithm	Frame length (long/short)
MPEG-2 AAC	1024/8 × 128
MPEG-1 Layer 3	576/3 × 192
MPEG-1 Layer 1/2	576
Dolby AC-3	256/2 × 128

processing block) of the decoder, since this affects the sizes of internal storage buffers, the size of data tables such as sine tables, and the size of output buffers. See Table 11.4 for a listing of frame sizes for various MPEG standards. For those algorithms which can dynamically change the filterbank block sizes, the length of both long and short blocks are listed. The AAC decoder, for instance, is an algorithm with a frame size of 1024 audio samples. This can be either a single long block of 1024 samples, or a set of eight short blocks each with 128 samples in the case transient signals. Implementing this algorithm will require internal RAM buffers of 1024 coefficients per channel for data processing, provided that all computations can be performed in-place. A implementation in double-precision effectively doubles the sizes of the required memory buffers. In addition, sine tables proportional to frame size will be required for performing the filterbank IMDCT (actual memory will depend on the implementation), as well as possible buffers for storing the 1024 output samples per channel.

As with most DSP applications, designing audio coders for minimum memory usage typically results in an increase in computational requirements. For instance, performing operations in-place conserves memory space, but may increase MIPS due to data rearrangement or copying. Creative table construction, such as storing a 1/2 or 1/4 period of a sine wave instead of a full period, will significantly reduce table storage but will also increase MIPS.

In addition, particularly for streaming audio applications, an input buffer is needed to store data from the incoming bitstream. This will often reside in the on-chip memory of the DSP. In most compression standards, audio bitstreams are inherently variable rate. Even so-called "constant-rate" streams have short-term variability in bit rate but have a constant average rate. For flash memory based players, where data can be retrieved from flash memory as needed, the input buffer only needs to be large enough to absorb delays in retrieving data from flash memory. For streaming applications where data is being received at a constant rate, the buffer must be large enough to absorb variability in bit rate of the bitstream. Many standards specify the maximum input buffer size to keep the decoder input buffer from overflowing when the instantaneous bit rate rises or underflowing when the instantaneous bit rate falls. Within AAC, for example, the decoder input buffer must be at least as large as 6144 bits per channel. An AAC encoder must keep track of these changes in instantaneous bit rate and adjust its bit allocation accordingly.

11.3.4 Considerations for Mobile Applications

When audio is transmitted over an error-prone transmission channel, such as wireless link, special considerations must be made to protect against transmission errors, as well as to conceal uncorrectable errors that occur in the bitstream. Within MPEG-4 audio, for instance, two sets of error robustness tools have been added to improve performance in error-prone environments. The first is a set of codec-specific error resilience modifications which are designed to make the bitstream more robust against transmission errors. The second is a set of general-purpose error protection tools in the form of error correction and detection codes to retrieve corrupted data when possible.

To handle error detection and correction, MPEG-4 provides a common set of error protection codes, which can be applied to any of the codecs supported within MPEG-4 audio. These correction/detection codes have a wide range of performance and redundancy and can be applied unequally to various parts of the bitstream to provide increased error protection to critical parts. MPEG-4 Audio coding algorithms provide a classification of each bitstream field according to its error sensitivity. Based on this classification, the bitstream is divided into several sections, each of which can be separately protected, such that more error sensitive parts are protected more strongly. In addition, several techniques are employed to make the bitstream more resilient to errors. This includes use of reversible variable length codes, as in video, to help recover a larger portion of the variable-length coded data when a bit error occurs. This is illustrated in Figure 11.8 using AAC as an example.

Error concealment techniques can be used to reduce the perceived degradation in the decoded audio when detectable, but uncorrectable, bitstream errors occur. Simple concealment techniques involve muting or repeating a frame of data when an error occurs. More sophisticated techniques attempt to reconstruct the missing sections of signals using signal

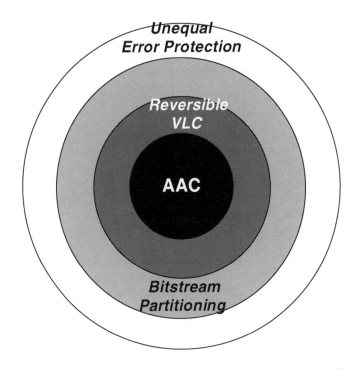

Figure 11.8   Error resilience tools for mobile applications available within ISO/IEC MPEG-4 Audio

modeling techniques. Fraunhofer IIS, for instance, has produced a proprietary error concealment technique for AAC which attempts to reconstruct missing parts of the signal to properly
match the adjacent, error-free signal parts [21].

Another consideration for mobile applications is coding delay, particularly for applications
which require two-way communication between users. While perceptual audio coders may
provide efficient coding of general audio signals at low rates, they can have algorithmic
delays of up to several hundred milliseconds. To address this issue, MPEG-4 has provided
a low-delay audio coder, based on AAC, which has an algorithmic delay of only 20 ms
compared to 110 ms of the standard AAC codec (at 24 kHz). This low-delay coder has a
reduced frame size, modified filterbank, and eliminates block switching to remove decision
delays in the encoder. Unlike most speech coders, it allows perceptual coding of most general
signal types, including music, while providing low-delay communication between users.

11.4 Audio and Video Decode on a DSP

As an example of the requirements for 3G multimedia applications, consider a music video
player. Table 11.5 represents three scenarios, to show a range of processing, power, and bit
rate requirements. Processing MIPS do not necessarily scale proportionally with bit rate or
format. The exact MIPS numbers depend on the platform, the implementation, coprocessors,
and the content, so actual DSP performance will vary. The assumed parameters give a general
flavor of the many possibilities for applications with video and audio.

The power dissipation depends on the processing MIPS, data transfers between the DSP
and external memory, and the LCD. For this example, we have not included MIPS for
baseband processing or protocol stack, e.g. RTSP, to stream the data over the network, but
assume the data is played from compact flash memory. Suppose the mobile device has a 3.7-
V battery with 650 mAh capacity. The DSP may have a core supply of 1.6 V and an I/O
supply of 3.3 V. Suppose the DSP consumes 0.05 mW/MIPS and 0.1 mA per Mword16 per
second (M16ps) DMA transfer at 1.5 V, and the LCD uses 25 mW. The mid-range application
example requiring 51 MIPS and 1 M16ps DMA will consume

$$51 \text{ MIPS} \times 0.05 \text{ mW/MIPS} = 2.6 \text{ mW}$$

for processing, and

$$1 \text{ M16ps DMA} \times 0.1 \text{ mA/M16ps} \times 1.5 \text{ V} = 0.15 \text{ mW}$$

for data transfers, plus 25 mW for the LCD. For a 3.7-V battery at 80% efficiency, and 650
mAh capacity, battery life would be about

$$650 \text{ mAh} \times (0.8 \times 3.7 \text{ V})/28 \text{ mW} = 68 \text{ h } 42 \text{ min}$$

Note that the LCD is the dominant factor for power consumption, with a low-power DSP.
For 95 MIPS and 4 M16ps DMA and a 100-mW LCD, the battery life is reduced to about 18
h. Low-power LCD technology is improving, and power consumption is an order of magnitude lower for still images. Also note that streaming the data over the wireless network,
running RTSP instead of reading from compact flash, would further increase processing
requirements.

One can appreciate that mobile multimedia applications are only now becoming practical,
comparing this example to what is available for 2G devices. The 2G mobile standards support

Table 11.5 Three scenarios for audio and video decode illustrate a range of requirements for 3G multimedia applications

	Low-end	Mid-range	Higher-rate
Video decoder			
Frame size	SQCIF	QCIF	CIF
	(128×96)	(176×144)	(352×288)
Frame rate (fps)	10	15	15
Bit rate (kbps)	64	256	592
Decode cycles (MIPS)	10	20	40
YUV to 16-b RGB (MIPS)	2	6	25
Data memory YUV (KB)	37	76	304
Data memory 16-b RGB display (KB)	25	51	203
DMA for decode	369 kbps	1.1 Mbps	4.6 Mbps
DMA for 16-b RGB output	246 kbps	0.76 Mbps	3.0 Mbps
LCD power (mW)	13	25	100
Audio decoder			
Bit rate (kbps)	64	96	128
Sample rate (kHz)	32	32	44.1
MIPS	20	25	30
Data memory (RAM and ROM) (KB)	30	30	30
DMA transfers, DMA output (kbps)	128	128	176
Total bit rate (kbps)	128	384	720
Total MIPS	32	51	95
Total DMA (M16ps)	~0.4	~1	~4
Total data memory (KB)	92	157	537
Power consumption for 0.05 mW/MIPS DSP + LCD (mW)	~15	~30	~100
Battery life assuming 3.7-V, 650 mAh (h)	120 + (>5 days)	60+	15+
Music video duration on 64 MB flash (min)	68	23	12

bit rates of only 8–13 kbps. TMS320C54x DSPs in 2G phones are capable of 40 MIPS, with 64 Kword16 of data memory and no DMA, and consume 0.32 mW/MIPS. The new 3G mobile standards and lower power DSPs extend the capability of mobile devices far beyond the realm of speech.

References

[1] Eyre, J. and Bier, J., 'DSPs Court the Consumer', *IEEE Spectrum*, March 1999, pp. 47–53.

[2] Gernsback, H., *Ralph 124C 41+, A Romance of the Year 2660*, 1911, Buccaneer Books, originally published by Gernsback.

[3] Kyle, D., *A Pictorial History of Science Fiction*, The Hamlyn Publishing Group Limited, Holland, 1976.

[4] *A Timeline of Television History*, http://www.ev-stift-gymn.guetersloh.de/massmedia/timeline/liste.html or search the Internet for more links.

[5] ITU-T Recommendation H.263, *Video Coding for Low Bit Rate Communication.*

[6] Talluri, R., 'Error-Resilient Video Coding in the ISO MPEG-4 Standard', *IEEE Communication Magazine*, June 1998.

[7] Budagavi, M. and Talluri, R., 'Wireless Video Communications,', In: Gibson, J., *Mobile Communications Handbook*, 2 ed., CRC Press, Boca Raton, FL, 1999.

[8] Budagavi, M., Heinzelman, W.R., Webb, J. and Talluri, R., 'Wireless MPEG-4 Video Communication on DSP Chips', *IEEE Signal Processing Magazine*, January 2000, pp. 36–53.

[9] 3GPP website, http://www.3gpp.org/

[10] Bi, Q., Zysman, I. and Menkes, H., 'Wireless Mobile Communications at the Start of the 21st Century', *IEEE Communications Magazine*, Janyary 2001, pp. 110–116.

[11] Dixit, S., Guo, Y., Antoniou, Z., 'Resource Management and Quality of Service in Third Generation Wireless Networks', *IEEE Communications Magazine*, February 2001, pp. 125–133.

[12] Chen, W.-H., Smith, C.H. and Fralick, S.C. 'A Fast Computational Algorithm for Discrete Cosine Transform', *IEEE Transactions on Communications*, Vol. 25, No. 9, September 1977, pp. 1004–1009.

[13] IEEE Std 1180–1990, *IEEE Standard Specification for the Implementation of 8×8 Inverse Discrete Cosine Transform.*

[14] Chaoui, J., Cyr, K., de Gregorio, S, Giacolone, J.-P., Webb, J., and Masse, Y. 'Open Multimedia Application Platform: Enabling Multimedia Applications in Third Generation Wireless Terminals Through a Combined RISC/DSP Architecture', *Proceedings of the IEEE International Conference on Acoustic Speech and Signal Processing*, 2001.

[15] RTP Payload Format for MPEG-4 Audio/Visual Streams, IETF RFC2026, http://www.isi.edu/

[16] Budagavi, M., Internal Communication, 2001.

[17] Painter, T. and Spanias, A., 'Perceptual Coding of Digital Audio', *Proceedings of IEEE*, April 2000.

[18] ISO/IEC JTC1/SC29/WG11 MPEG. International Standard IS 11172-3 Information Technology – Generic Coding of Moving Pictures and Associated Audio, Part 3: Audio, 1991.

[19] ISO/IEC JTC1/SC29/WG11 MPEG. International Standard IS 13818-7 Information Technology – Generic Coding of Moving Pictures and Associated Audio, Part 7: Advanced Audio Coding, 1997.

[20] Meares, D., Watanabe, K. and Schreirer, E., 'Report on the MPEG-2 AAC Stereo Verification Tests', ISO/IEC JTC1/SC29/WG11 MPEG document N2006, February 1998.

[21] Fraunhofer IIS website, http://www.iis.fhg.de/amm/techinf/mpeg4/error.html

12

Security Paradigm for Mobile Terminals

Edgar Auslander, Jerome Azema, Alain Chateau and Loic Hamon

As was highlighted in Chapter 1, the mobile phone has become a personal device that has strong potential to morph into a mobile wallet, or a mobile entertainment device. In both cases, merchants or service providers will want to bill for content or service provision. There lies an opportunity for fraud. When your phone only stores names of people you call, there is not much interest in hacking the device; when it contains your credit card information, however, a lot of energy will be spent trying to steal or hack your mobile phone. Telecommunication fraud has been documented already in the early days of the telegraph, though every attempt was then made to guard the secrecy of transmissions. There was clearly money to be made at the time by speculators if, for instance, they could find a way to transmit news from the stock market in, say, Paris, to other parts of the country faster than the daily papers. A curious case of fraud was discovered on the Paris to Bordeaux line. Two bankers, the brothers Francois and Joseph Blanc, had bribed the telegraph operators at a station just behind Tours to introduce a specific pattern of errors into the transmissions, to signal the direction in which the stock market was moving in Paris to an accomplice in Bordeaux. The telegraph operators near Tours received their instructions from Paris by ordinary (stage-coach) mail, in the form of packages wrapped in either white or gray paper, indicating whether the stock market in Paris had gone up or down. The fraud had been in operation for 2 years when it was discovered in August of 1836. With the advent of the Internet, many more spectacular frauds have been documented. A very good reference is *Secrets and Lies: Digital Security in a Network World*, by Bruce Schneier, Wiley Computer Publishing. Figure 12.1 illustrates that the mobile phone is likely to be the device that accesses Internet "the most", making security an even more important mater. Figure 12.1 also illustrates that a mobile phone is potentially the most attractive personal billboard for advertisers; of course, users will not enjoy being bombarded with spam advertising, but subtle value-added advertising might be appreciated. The resulting personalized advertising or coupon gained via affinity programs might result in impulse purchase, at the touch of a button or a finger print on your mobile phone. This prospect motivates even further the industry to develop secure schemes for mobile commerce, what we call "the security paradigm for mobile terminals". Mobile-commerce/security capabilities is a "must have" as an enabler of value added services. We will explore

Mobile Phones: Personal bilboards!

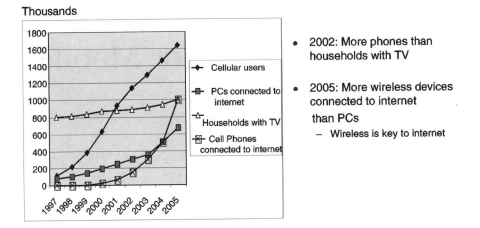

Thousands

- 2002: More phones than households with TV

- 2005: More wireless devices connected to internet

  than PCs
  - Wireless is key to internet

Figure 12.1   Mobile phones connected to the Internet

in this chapter some of the challenges presented by the wireless security issue as well as implementation choices.

12.1 Mobile Commerce General Environment

When the basic service delivered by mobile telephony was just voice, the value chain was relatively simple and people were billed per minute of usage, either via a subscription package or via a pre-paid scheme. With the possibility to deliver content, new players come into the picture: portal providers, content providers, application providers, mobile Internet service providers, payment processing providers and security providers. Carriers will have to decide the type of vertical integration role they want to play. We see evidence of several choices ranging from carriers outsourcing most roles (we even see the emergence of Mobile Virtual Network Providers (MVNOs), like Virgin Mobile, who lease a network and provide value-added services and brand-related innovations), to carriers who integrate content providers or packagers as illustrated by the recent acquisition of MP3.Com by Vivendi-Universal who also own the mobile operator SFR. The convergence of telecommunications, computing and entertainment will present new quality of service and intellectual property rights ownership and protection challenges.

The security issues that need to be addressed and solved in the mobile phone environment are:

- *Confidentiality*: ensure that only communication parties are able to understand the content of the transferred information.
- *Integrity*: ensure that information has not been altered during transmission: changes should not be possible without detection.

- *Authentication*: ensure that other communication party is who he/she claims to be.
- *Non-repudiation*: ensure that the sender cannot deny sending the message.
- *consumer protection*: pseudonym and anonymity.
- *Protection against clone*.

The security solutions generally examined are: encryption, SSL/WTLS, and Wireless Public Key Infrastructure (PKI). Encryption scrambles data according to an algorithm so that eavesdropping is made useless during transmission. Confidentiality is then achieved, but integrity, authentication and non-repudiation are not necessarily achieved. SSL/WTLS offers encoded connection and server authentication but non-repudiation is not achieved. Wireless PKI, combined with other techniques, provides digital signature over mobile networks and user/server authentication.

All the solutions, hardware and software, are based on asymmetric encryption techniques.

- The need for hardware based security solutions versus a software-only one is motivated by:
- a more difficult and expensive implementation to analyze;
- duplication is more challenging;
- tampering can be detected and the system can more efficiently react to a tampering attempt.
- End-to-end solutions (client and server).
- Stand-alone smart cards will not be able to support the full range of applications and in particular the one related to content protection. Because of the very limited processing and I/O speed of a smart card.

The three main types of applications have to be supported for products (i.e. three different industries and three different actors):

1. *Secure e-transaction*: make a financial transaction, i.e. pay a product and/or a service directly with a mobile phone. Being able to confirm identity is crucial in being able to bill the end user for the service. Public key technology combined with user identification and tamper proofing techniques are the backbone of wireless e-transactions.
2. *Secure e-content*: ensure the security of the platform for content and application software download (e.g. MP3 download, application software download, virus protection, copyrights and digital right management...). The main contents are:

   - *Application software*: currently provided by handset manufacturers.
   - *Music and video*: three solutions are today available on the market and have already been endorsed and used by the majors (Universal, Warner, EMI, BMG, Sony)

3. *Secure e-billing*: secure billing could be provided via a local solution running on the handset. This local solution will monitor and meter the usage of wireless data communications and will be fully controlled by the hardware based secure central billing system.

12.2 Secure Platform Definition

In order to provide the necessary level of security to the security layers of the chosen protocol stack (WAP or equivalent) and to the PKI system running on the mobile device, it would make sense to merge intimately the security elements within the existing platform as built-in features. A single fabric solution with a sharing of the computing resources seems effectively the best solution to offer a seamless integration of the security sensitive applications in the protocol stack with a minimum silicon cost penalty. Nevertheless, an integrated solution will lead to the introduction in the existing platform of specific software and hardware mechanisms to support the expected level of security.

12.2.1 Security Paradigm Alternatives

Building a secure framework is a complex task. Two approaches can be envisaged.

The first approach, which can be called "fine-grained security paradigm", implies that a complex custom software "security manager" controls:

- Installed application code and data spaces.
- Security functions code and data spaces.
- Security functions upgrades.
- Application life-cycle management.

This approach has several disadvantages, and is not preferred:

- It requires complex software for dynamic allocation of code and data.
- It requires complex hardware for functions and data boundaries monitoring.

The second approach, which can be called "coarse-grained security paradigm", implies that the fine-grained security is left to the Java based solution, and that the custom "security manager" implements coarse-grained security for security functions and security system.

This approach has several advantages:

- Hardware complexities are much reduced.
- It is not necessary to track functions for security, which removes much software complexity.
- Code downloads that do not go through the Java environment are restricted to security downloads and are easier to control.

This second approach is the one preferred. The secure platform software component is a Java based solution, and is responsible for the fine-grained security. The security manager of the hardware based component is responsible for the coarse-grained security.

12.2.2 Secure Platform Software Component

The secure platform software component is responsible for overall system dynamic software security:

- It provides access to a service controller for applications.
- It opens connections from the service controller to particular services (keyboard, LCD, cryptography).
- It manages the request between applications and services.
- It manages the service sharing or exclusivity.

It is also responsible for application life-cycle management:

- Download
- Initialization
- Installation
- Registration
- Inter application firewall

The software component role and advantages will be detailed in Section 12.3.

12.2.3 Secure Platform Hardware Component

The secure platform hardware component solution will be detailed in Section 12.4.

12.3 Software Based Security Component

12.3.1 Java and Security

Java technology offers crucial benefits for portability and safety:

- A well-defined semantics independent of the hosting processor/Operating System (OS): a Java program will always behave exactly in the same manner whatever the host device processor and OS.
- A completely typed intermediate code (byte-code) that can be verified: verified Java code cannot break complete typification and, as a consequence, cannot forge pointers and manipulate memory or go beyond limits of data structure and thus access freely the memory. Java can provide perfectly safe software firewalling, without any help from the OS or the hardware (no memory management unit or microprocessor needed).
- Data built by a Java application cannot be accessed by another application if stored in object fields, except exported static fields.
- The memory is automatically freed once used, with the included garbage collection.
- The Java virtual machine includes embedded byte-code verification mechanisms. Moreover, classes loaded through the network are stored in separated space from local classes.

For all these reasons, the security software component will use Java language.

12.3.2 Definition

The secure platform software component is a Java Application Environment (JAE). It provides a complete Java API to program applications. This JAE is built as a profile on top of a CLDC configuration (KVM), which is the smallest configuration defined in the family J2ME proposed by Sun for embedded devices.

This component offers a multi-application framework in which applications can be integrated via download. As a framework, it should define an exhaustive set of device services. It defines device services needed to perform secure transactions, such as payment, identification, cryptographic services, secure storage services, but also services needed for distant connections using Internet or WAP based protocols.

The security features built into the software component design are important not only for secure transactions but also to control other kinds of applications that could be installed on the device besides secure transaction ones, such as games, etc.

12.3.3 Features for Security

12.3.3.1 Access Manager

The central feature of the secure platform software component is the concept of "access manager". When launched, an application using the secure platform security API gets from the shell an `AccessManager` object. This object is personalized for the particular application that receives it. An application cannot access directly any device service or peripheral. The access is made in two steps:

- First, the application has to ask its personal access manager to provide a "service control" for a particular kind of device service. The access manager can reject this request if the application has not the rights associated to the use of this kind of services (e.g. an application may not be allowed to use cryptographic service).
- If the access manager provides a service control object, this is not yet an access to a device service. The service control object, which is an emanation of the access manager, has then to be connected to a particular device service of the correct type. The control can deny the access to a particular service if the application has not the right to use it (for instance, an application can have the right to access a customer smart card slot but not some other smart card slots internal to the device).

The advantage of this approach is that it allows an efficient and easy to implement control of applications.

A classical alternative is to have the services protecting themselves from their clients, checking for each of the methods they export the right of their caller, which has to be identified then by scanning the call stack, to use this method. This is what is done in full Java: each method of the API has to call the so-called "security manager" that retrieves on the call stack and checks the right of the caller. This approach has proven to be very difficult to achieve a flexible and complete security, and moreover requires too many dynamic checks.

The secure platform approach delegates this security control to a gatekeeper, the "access manager", personalized for each application, which allows a flexible fine tuning of the security control that has not to be implemented by the device services themselves, but by the shell. The alternative provided by the secure platform is more flexible, easy and also efficient as it requires no scanning of the call stack.

12.3.3.2 Indirect Access to Services

The secure platform software component is characterized by the indirect access to services. An application is never allowed to get a direct reference on a service implementation; it can only handle service controls that can only be connected themselves (i.e. have a communication channel) to a service implementation. With this approach, an application has no way of knowing the real implementation and API of the service it uses, it can only communicate with it through the fixed API provided by the control. This is important both for portability of applications and security of the hosting device. A second advantage is that the secure platform software component can manage through the service controls the sharing of device services. Most of these services are of exclusive use, i.e. cannot be used simultaneously by two concurrent applications. This can be translated by the fact, easy to verify by secure platform software component implementation, that these services cannot have a communication channel with two service controls simultaneously.

12.3.3.3 Asynchronous Communication

All the communication from the application to the device services, via service controls, is made exclusively through asynchronous requests, and the communication from device services to the application via events. The secure platform software component distinguishes two kinds of events: completion events indicating the completion (or abortion) of a task launched by an asynchronous request, and status events, used by the services to alert spontaneously their client of some change of situation. More generally, the choice of an event-driven style of programming allows to naturally take into account device services that are not local to the hosting device but can be remote and, as a consequence, with unpredictable time of answer (distributed platform: e.g. connected wireless). In addition, the use of status events allows the application to react immediately to unexpected events.

12.3.4 Dependency on OS

The dependencies between the secure platform software component and the hosting device OS are very weak. In fact, with the exception of the access to peripheral drivers almost no interaction with the OS is needed.

The secure platform software component should control its security with respect to downloaded applications and the security of the applications themselves entirely on its own and independently of the OS.

The only case where a secure platform software component platform relies on the OS for its security is for protection from the attacks of non-certified native applications that would have been downloaded. If there are such applications, the isolation of the software component in its own secure environment via secure storage should be required from the hardware based security component.

12.4 Hardware Based Security Component: Distributed Security

The distributed security is an optimized combination of selected hardware blocks together with a protected software execution environment. In this application, the platform processor will host the security application. A partitioned "secure mode" will be created so that it operates as a separate "virtual security processor" while it is executing security operations. A set of critical security blocks, such as crypto-processors, can be added to the peripheral bus of the processor with an access restricted to the sole secure mode operation.

The main assumptions, which guided the implementation of distributed security, are:

- All re-writable software in the system must be considered originally as un-trusted.
- The operating system is un-trusted

12.4.1 Secure Mode Description

A secure mode processing means creating an environment for protecting sensitive information from access or tampering by un-trusted software. It relies on the presence of special purpose hardware creating a boundary between services that trusted software might access, and those that are available to any code.

Typically, the services controlled by the secure mode hardware are sections of memory. Mainly:

- Program ROM
- Program RAM (optional)
- Secure storage RAM
- Root key store

However, by extension, that control can also be used to place I/O devices (such as a keypad, LCD, touch-screen, smart card) inside the secure boundary by protecting access to I/O or memory-mapped registers. In any case, the hardware resources must be linked together by proper control of the software itself. That is, there can exist no possible flows by which un-trusted code can either fool the hardware into allowing it secure mode privileges, or get trusted code to perform tasks it shouldn't.

If the boundary is properly created, there should be no way to utilize the normal operation of the processor to move information from inside the boundary to outside, except through controlled operations. For example, an operation may be provided to decrypt a message, and the operation may return the plain text result for use by un-trusted software, but there should be no way to view the keys used to do the decrypting.

Note that normal operation of the processor includes executing flawed "user-mode" software, which for example might corrupt the process call stack or overflow a variable boundary. Depending on the potential cost of compromise, additional mechanisms beyond the implementation of secure mode may be required to protect against abnormal operation of the processor. These mechanisms can be protective to prevent tampering or reactive to inform of a tampering attempt and generate an appropriate response.

12.4.1.1 Secure Mode Activation

The security manager is running in supervisor mode with an additional privilege level granted by the secure configuration bit, which can be viewed as an outer privilege level encompassing the processing unit sub-system.

The security manager and the standard secure routines from the secure library are resident and located in the embedded program ROM. Thus the corresponding code can be considered as trusted.

On user application call to a security function through a dedicated API of the OS, the OS will branch to *the single entry point* of the security manager. This single entry point allows control of the activation of the secure mode and the access to the secure resources. The OS using the normal procedure will handle the saving of the system context.

The first task of the security manager will be to mask all the interruptions. After executing the sequence of instructions that prevents any preemption of the secure mode privilege by un-trusted software. Hardware based state-machine spying of the fetched instruction address will set the secure bit to grant the access to security dedicated resources and to restrict concurrent access from other processors to shared resources.

Two strategies are offered:

1. Disabling the program and data cache, which has the advantage of easing the procedures of entry in and exit from secure mode with the drawback of lower processing performances.
2. Enabling program and data caches with the consequent improvement of processing performances at the cost of an increased complexity of the entry and exit procedures. One needs to insure that instructions fetched during entry procedures are effectively executed and that a proper cleaning is achieved on exit.

Note: the root key store must be located in a non-cacheable section.

12.4.1.2 Interrupts Management

If interruptions are allowed in secure mode, the security manager must take over the handling of all the interruptions. The management of interruptions in secure mode requires handling all interruptions over the secure mode. An indirection table for interrupt vectors is located in the secure ROM in order to control the application call.

On interrupt occurrence in secure mode, the security manager will save context in a private stack stored in the secure storage RAM. It will clean all the registers and the secure bit will be released before giving back the hand to the OS.

If interrupt occurs in non-secure mode, then the program will directly jump to the OS, which will manage the context saving with the normal procedure. The interruption of a secure application with another secure application will require a complex management of the secure stack stored in the secure storage RAM (i.e. use of session keys for temporary encryption of application related data).

12.4.1.3 Secure Mode Configuration

The secure mode configuration is set with the secure bit. This bit is controlled by a hardware state-machine, which has the following features:

- Spying the program address fetched.
- Control the integrity of the entry sequence starting from the single entry point.
- Set and release the secure configuration bit.

The logic must not be scanable to prevent any tampering attempt to set the secure bit out of the secure boundary of the security manager.

Access Control to Security Dedicated Resources
The processor will have access to the secure components when only in secure mode.

Secure components considered are:

- Secure program ROM. This ROM element can be partitioned to allow access to one area by the processor in secure mode only and to another area in regular mode. The access control will be only valid from the security manager entry point. The boot area will be free access.

- Secure program RAM (optional). Dedicated to the execution of non-resident secure applications downloaded from external FLASH device.
- Secure storage RAM
- Key root store

Access Control to Shared Resources
While in secure mode, the concurrent access to the shared peripherals must be prohibited. Components considered are:

- MMI peripherals (keyboard, LCD, FPR sensor)
- Smart card physical interface
- Crypto-processors (optional)

Additionally, the emulation capability or any test mode configurations must be prohibited thus requiring disabling of the embedded JTAG TAP controller. Note that in emulation mode, the secure mode cannot be entered.

12.4.1.4 Impact on OS

The OS manages the calls to the "security API". Collaboration between the OS kernel and the security manager is mandatory. The aim is to minimize the modifications to the OS and keep them as generic as possible. If the secure mode needs to be exited through interruption, it will impact the interruption management of the OS with the necessary use of indirection tables.

12.4.2 Key Management

Key management is the main feature to consider when building a secure system. The compromise of key material is the most dangerous fault in a security system since it can allow an adversary to attack a system silently. The key management combines the secure library functions with the proper hardware protection logic to safely generate keys, use these keys and store those keys outside of the security boundary without compromise. This allows a large number of keys to be used in the system, without requiring a large amount of protected key storage RAM in the device.

The main tasks of the key manager are:

- Key generation
- Key storage
- Key usage control
- Key archiving and recovery
- Key hierarchical ring

The key management scheme must allow system applications to build their own key management mechanism. A hierarchical key structure allows multiple applications to share a single cryptographic library, while keeping the keys separate. User identification data such as PIN or FPR are used as input data of the key management system for authentication purpose. The key management system should not be intrusive on the system and must not cause performance degradation on traffic encryption or other time-sensitive tasks.

12.4.2.1 Key Generation

The robustness of the keys generated is highly dependent of the randomness of the output data generated by the Random Number Generator (RNG). A truly entropy driven RNG must rely on a hardware based RNG device.

The key generation will be run in the secure mode under the control of the security manager. The secure library incorporates a full set of symmetric and asymmetric key generation commands. Asymmetric keys (public keys) are generated according to the existing standards like X9.42/X9.43 and PKCS

12.4.2.2 Key Encryption

The confidentiality of the keys and certificates stored off-chip relies on their encryption with a symmetrical algorithm using the device permanent root key. AES algorithm should be favored for its robustness and key/block sizes flexibility. Triple-DES algorithm is also possible.

12.4.3 Data Encryption and Hashing

The encrypted data baud rate expected in mobile equipment is not in the same order of magnitude as the one required for networking applications. Consequently, the choice of a hardware implementation versus a software implementation must not be considered as the prime criterion (Table 12.1).

Table 12.1 Recommended cryptographic algorithms

Cryptographic algorithms	Today	Tomorrow
Symmetric	DES/3DES	AES Rijndael
Asymmetric	RSA	ECC
Digital signature	RSA/DSA	ECDSA
Hashing	SHA-1	–
Bus encryption	DES (ECB)	Custom

A main criterion to consider for a hardware implementation of a crypto-algorithm is the best level of protection offered by a crypto-processor because the host processor does not have access to the keys.

However, the secure boundary offered by the proposed secure architecture with dedicated program and data memories to run the secure applications is tending to a software implementation of the crypto-algorithms as long as the processing time is acceptable at the application level.

The choice of hardware versus software implementation of crypto-algorithms is also dependent on the type of devices and these will be discussed further in the later sections of this chapter.

12.4.4 Distributed Security Architecture

12.4.4.1 Secure Components

Program ROM
The secure mode firmware is executed from an embedded ROM. Thus no authentication of the security software is needed. Additionally, ROM is offering a greater density than RAM and consequently a better silicon area versus capacity ratio. The ROM will contain the security manager and the secure library in addition to the processor boot session and indirection interruptions table.

The size of the ROM is expected to be in the range of 64–128 Kbytes depending mainly of the secure library specification.

Program RAM (Optional)
There may be a need to extend the security software for adding more algorithms or other custom secure applications, such as the software based security component described earlier. The corresponding program code will be stored off-chip in an encrypted form and downloaded in the program RAM. Before any execution, the security manager will be required to authenticate this extended software with its digital signature (HMAC). This extended software will execute equally in the secure mode.

Secure Storage RAM
Eight Kbytes of secure embedded storage RAM will be used for:

- Key material generation.
- Dynamic key storage.
- Data encryption or hashing, using symmetrical or asymmetrical keys.
- Manage secure data (private stack, scratchpad…).

This RAM will allow the security manager to create a secure run-time environment for building its own stack space, scratch RAM, and a heap for big-number mathematical operations to support asymmetrical algorithms. The security manager supports the procedure of "clean-up" of the stack and scratch areas of any key material to prevent any potential accidental key material leakage between different secure applications.

Non-Volatile Root Key Store
The non-volatile storage of the root key is based on electrically programmable fuses. The programming of these fuses is done subsequently to the fabrication process of the chip from a value generated from the embedded RNG.

Additional fuses can be used for a device personalization (custom application, test device…).

Die Identifier
A device identifier or die ID is provided as a seed for the generation of a bound key. As a result of the hashing of this die ID and of an application key, this bound key will be used to bind an application firmware to the physical platform through the encryption of the firmware and the creation of its HMAC.

The die identifier is a 64-bit word based on electrically programmable fuses. It is programmed at silicon manufacturer factory level.

Randomizer

The randomizer generates the seeds for hashing, encryption and key generation. It is based on a true RNG, which can be associated with additional logic to guarantee the generation of an unbiased output (i.e. to a von Neumann hardware corrector).

12.4.5 Tampering Protection

They are several techniques for attacking a security system:

- Physical modifications to the device or its external interface.
- Modification of programs running on the device.
- Environmental attacks on power, clock or emissions.

The device must provide a resistance to the static and dynamic observation of internal data and to the modification or alteration of these data. The level of tamper proofing must be balanced with the impact of the disclosure of the secret data to the attacker (implementation cost versus tamper consequence). The ultimate goal is to keep a device secure even if attackers carried out a successful tampering attack on another device from the same class and to make the attacked device out of service after the tampering attack (destructive tampering).

12.4.5.1 Tamper Resistance

The first level of tamper resistance must be integrated in the secure system architecture itself. The recommended rule is to bind the embedded secrets to the device itself. The acquaintance of the secrets does not allow tampering with another device. This level of system resistance relies on a hierarchical key ring structure with each key ring level secured by the level below (i.e. encryption of the private key of the key pair of one level with the public key of the underlying level). The root level is based on the "root" or device key.

When operating in secure mode, the JTAG access to the on-chip emulation logic of the processors must be disabled, thus preventing any external possibility to take the control on the OS execution. Any other modes of observation of internal data transfer on external ports of the device must be prohibited.

The use of a PLL to provide the system clock to the processors prevents any manipulation of the clocks. The configuration in bypass mode of the embedded PLL cell must be prohibited when operating in secure mode.

The resistance to crypto-analysis relies on the prevention of an analysis of the electronic trace (i.e. current based). A first level of resistance can be reached with a binary balancing of the data content when accessing sensitive data in memory elements. A complementary mean is a shielding against electromagnetic analysis (i.e. coating in package material).

The use of complex techniques, such as logic scrambling, bus interleaving or dummy logic, for a resistance to a structural analysis of the device layout is of no interest if the means used to lead the analysis are destructive and the acquaintance of the disclosed information not usable to tamper with another device of the same class.

12.4.5.2 Tamper Detection

The tamper detection is based on the evidence of a physical attack resulting in physical sense. This evidence can be based on active means like sensors – temperature, voltage, ionization – or passive means like embedded conductive grid in device package for the detection of open or shorts created by an opening attempt. The appropriate selection of the means must be suited to the targeted security level based on FIPS 140-1 specification.

12.4.5.3 Tamper Response

The tamper response must erase the content of the volatile elements, namely sequential elements and RAM memories. The erasing of the encrypted secret stored in the FLASH can be considered with the appropriate sector erase command. In this case it should be noted that the efficiency of the tamper response is limited by the latency of the flash erase sequence. The tamper response must be adapted to the possibility of the system to reuse or not the confidential data after this response. This tamper response can be managed on two levels:

- Zeroization with the reset of the confidential data contained in the device.
- Activation of an erasure routine of the protected RAM content and asynchronous reset of the sequential elements of the processors and the secure peripherals.

The erasure of the RAM can be software based (controlled by the processor) or hardware based with the use of the BIST controller if it exists. The PLL will be forced in free run mode to maintain the system clock whatever the possible forgery of the external clock source.

The possibility to keep a status flag of the tampering attempt, voluntary or non-voluntary, can be of main importance for the secure system to take a decision after the zeroization sequence.

It does not matter whether the tampering attack leads to the definitive destruction of the device preventing its further functional operation or whether the tampered data will be renewed anyway when reactivating the secure protocol. Again, this is true only if the acquaintance of the disclosed information does not allow tampering of another device of the same class.

12.5 Secure Platform in Digital Base Band Controller/MODEM

The Digital Base Band controller (DBB)/MODEM is the real central security device of mobile phones. This device has to support secure e-transaction and secure e-billing applications.

It is, application wise, a closed environment, which means that it is not allowed to download on it new native applications. The phone manufacturer itself certifies the applications present on the DBB (often called legacy applications). Nevertheless, it is possible to download on the DBB data such as media files (video, audio – for example ringing melodies). The DBB will then handle some multimedia tasks, so it will have limited e-content applications support.

The DBB central processor is generally already loaded with communication protocol and IO management, so to avoid bandwidth limitation due to cryptographic calculations, hardware accelerators could be preferred for:

- Asymmetrical key generation
- Symmetrical key generation
- Data and key encryption
- Hashing

According to the level of security required by the application, the generated keys that must be stored off-chip can be sent to the SIM/smart card, or to the external flash. In any case, the key stored will be covered (or wrapped) with a device permanent root key, which will never be exported from the DBB.

12.6 Secure Platform in Application Platform

The application platform will be added in hi-end phones as a complement to the DBB controller, to handle complex hi-speed multimedia tasks, such as streaming audio or video.

The application platform, compared to the DBB, is an open environment, which means that it is possible to download data and applications on it. Some of the downloaded applications could involve financial transactions, so this device will have to support secure e-transaction plus e-content applications.

Since the application platform is an open environment, special attention needs to be paid to the application download, installation, certification, registry and control. The secure platform software component will handle these features.

The application platform offers generally high processing power, and it offers flexibility in terms of software implementation of cryptographic algorithms. As long as the processing time is acceptable at the application level, the software solution is interesting.

According to the level of security required by the application, the generated keys that must be stored in permanent storage areas can be sent to the external flash. In any case, the key stored will be covered (or wrapped) with a device permanent root key, which will never be exported from the application platform device.

For any data exchange between the application platform and the DBB, a secure communication must be established by using Internet Layer Security Protocol (IPSec) or Transport Layer Security (TLS, SSL, WTLS), or by using proprietary protocols. It should be possible, once this secure link is established, to transfer for example certificate and keys from the application platform to DBB.

12.7 Conclusion

Mobile commerce can only happen in a trusted environment; you put your money in a bank because you trust the bank. A trusted environment can only exist when the appropriate legal framework and proven technologies are in place and advertised as such. Content download, transactions, and billing have to be reliable and fair to users, merchants, banks, as well as content authors and owners. No system is unbreakable though. We just believe that the modern PKI techniques we described, implemented with smart key management and a combination of software and hardware, are up to the "trusted security" challenge: hacking attempts would be far too costly, or even unreasonable, both in terms of time and money.

13

Biometric Systems Applied To Mobile Communications

Dale R. Setlak and Lorin Netsch

13.1 Introduction

Many modern electronic services and systems require a reliable knowledge of the identity of the current user, as an integral part of their security protection [1]. Examples include secure access to automated banking services, access to media services, access to confidential or classified information in the workplace, and security of information within handheld devices. The results of a breach of security can be costly both to the customer and the providers of services or systems. For wireless devices to take on significant roles in these security-conscious applications, the devices must provide the mechanisms needed for reliable user identification. Among the applications of reliable user identification, wireless or handheld devices present a unique challenge due to limited size, power and memory constraints. Furthermore, the nature of the challenge grows when we consider the scope of mobile device penetration into the worldwide marketplace, where the user identification system must function reliably for a huge number of people with a wide range of user demographics and in widely diverse operational environments.

At its core, security in information and communication systems encompasses those processes that: (1) determine what commands the current user may issue to the system and (2) guarantee the integrity of both the commands and the subsequent system responses as they propagate through the system.

Reliable user identity recognition is the necessary first step in determining what commands can be issued to the system by the current user. It involves collecting enough personal data about the current user to confidently link him to a specific set of system permissions and privileges. In current systems, that linkage is often made in the form of a unique user ID (e.g. name, username, account number, social security number, etc.).

Figure 13.1   Data related to the user ID

Figure 13.1 illustrates this relationship from an information structure viewpoint. Identity recognition deals with the left half of the relationship illustrated in Figure 13.1. As human beings, we typically identify a person using three classes of personal identification information:

- Something he/she has – a badge, ID card, letter of introduction, etc.
- Something he/she knows – a password, code word, mother's maiden name, etc.
- Physical characteristics – height, weight, eye color, voice, face, fingerprint, etc. These are sometimes called biometrics.

The identity recognition process collects these types of data about the current user and compares them to data about users that have been previously collected and stored. The process generally takes one of two forms: (1) *Verification*, where the user enters a specific user ID and the system simply corroborates or denies the claimed identity by comparing the live data to that stored for the specific claimed identity, and (2) *identification*, where the system collects live data and searches its entire stored collection of data for all users to find the identity of the current user. *Verification* is a simpler process computationally than identification and is preferred when the operational requirements permit. *Identification* is used where user convenience is paramount, or when the user cannot be trusted to enter the correct user ID. Both of these forms are discussed in this chapter.

A variety of different means exist to provide the information needed to establish and confirm the person's identity. Each method has its own strengths and weaknesses. A much used method involves assigning each user a unique account number, which provides the claimed identity, and either assigning or allowing a user to specify a Personal Identification Number (PIN) which provides the confirmation of the identity. The drawback of using a PIN is that once knowledge of the PIN is compromised, it becomes an immediate and ongoing breach of security. Further, each separate account a user accesses requires memorizing a separate PIN, resulting in a proliferation of PINs. Those who use many different automated services will find it difficult to remember several different PINs. Systems based on biometric data can avoid the use of passwords and PINs entirely. Automated systems that can avoid passwords and PINS will be more secure and easier to use.

In this chapter we describe in detail two popular biometric user verification/identification technologies, speaker verification and fingerprint verification/identification. These techniques were chosen because of their feasibility and their suitability for the mobile handset environment. Other technologies, such as signature analysis and retinal scan may be inconvenient for a small handheld device and are hence not discussed. In general, the type of biometric measure used will depend on the level of security needed and ability to sample the characteristics.

13.2 The Speaker Verification Task

To solve the above-mentioned problems, ideally one desires a system that verifies a person's identity based upon unique characteristics that each individual possesses. The use of a person's voice for verification provides an attractive biometric measure. Talking is perceived as a natural means of communicating information. The only equipment needed in proximity to the person is a microphone to provide the voice information. This equipment is inexpensive and, for many wireless and personal devices, the microphone and A/D front end are already in place.

The weaknesses of using speech to verify a person's identity include the ability of an impostor to make repeated attempts to gain access, or recording of the user's voice. However, recording of a user's voice can be remedied by prompting the user to say different phrases each time a system performs verification. Even then, we observe that some people "sound alike" and so it is possible at times for voice verification to fail. It is extremely important to design the voice verification system to address the unique challenges and maximize performance.

There are two major challenges in the use of voice verification, both dealing with the random nature of the audio signal. The first challenge is the variation of the speech signal. A speaker's voice obviously varies based upon the words spoken. The way a speaker says a word also varies. The rate of speech for each word is subject to systematic change (for example, the speaker may be in a hurry). The acoustics of speech for each word also vary naturally, due to context, health of the speaker, or emotional state. Additionally, acoustics are systematically changed by the characteristics of the transducer and channel used during collection of the speech. This is especially true in the case of equipment in which the transducer may change between verification attempts. The second challenge is contamination of the speech signal by additive background noise. Since the environment in which the voice verification system will be used is not known beforehand, algorithms must be able to cope with the corruption of the speech signal by unknown noise sources.

In addition to the technical challenges provided by the signal, there are practical design issues that must be addressed. Embedded applications must be concerned with the amount of speaker verification measurement data that must be collected and stored for each user. As the amount of user-specific data stored increases, the verification performance increases. Further, if the embedded application is to be accessed by multiple users, the amount of speaker-specific data that must be stored to represent all speakers will obviously increase. Since the amount of data stored must be kept to a minimum, there will of necessity be a trade-off between performance and resource requirements. This will impact on the selection of the type of verification methodology used. Therefore, the efficiency of the speaker verification measures is important. An identity verification system needs a compact representation of the user-specific voice information for efficient storage and rapid retrieval.

13.2.1 Speaker Verification Processing Overview

The processing of speaker verification involves two steps. These are illustrated in Figure 13.2.

The first step, enrollment, is shown in the upper part of the figure. It consists of gathering speech from a known speaker and using the speech to extract characteristics that are unique to the speaker. These characteristics are stored by the system along with the speaker's identity for later use during verification. The technical challenge presented during enrollment is to find features of speech that are applicable to the voice verification task, minimize the amount

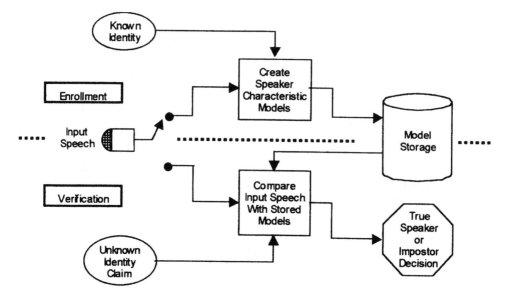

Figure 13.2   Speaker verification block diagram

of storage necessary for each speaker, and provide robust performance in the intended environment.

The second step is the actual verification process. In this step, shown at the bottom of the figure, the system first requests that the speaker claim an identity. This may be performed by many different means, including entering the ID by keypad, or by voice. The system then confirms that it has stored speech characteristics corresponding to the claimed identity. If stored information is available, the system prompts the speaker to say a verification utterance, and then uses the speech to decide if the identity of the speaker is the same as the claimed identity. The challenge presented by verification is to define a metric to be used in a pattern matching process that provides an accurate measure of the likelihood that the verification utterance came from the claimed identity.

13.2.1.1 Types of Voice Verification Processing

There are basically two types of voice verification processing, text-independent and text-dependent voice verification. Text-independent voice verification attempts to verify the claimed identity of speakers from a sample of their voice in which they are free to say whatever they desire. Text-dependent verification, on the other hand, requires that each speaker say a known utterance, often from a restricted vocabulary. Text-dependent verification may also include verification in which the system prompts the speaker to say a specific word or phrase.

Text-dependent verification provides valuable *a priori* information of the expected acoustic signal that may be exploited to optimize performance. The most important benefit (also an added requirement) of a text-dependent verification system is that one may model the utterance statistically both in terms of acoustic sounds and temporal course. The system can specify the acoustic characteristics of the utterance used for verification, thereby ensuring proper acoustic coverage of the signal to ensure the desired level of performance. In addition, text-dependent verification may be used to provide some increased security by constraining the speech to an utterance known only by the true speaker.

Text-independent verification is easier for the speaker to use, since there is no need to memorize an utterance or repeat a prompted phrase. It also can be easier to and more efficient to implement, since it is not required to keep track of the exact temporal course of the input utterance. However, it will normally be necessary to collect longer durations of speech in order to obtain the level of verification performance desired.

13.2.1.2 Measurement of Speaker Verification Performance

One of the most difficult tasks in the design of a speaker verification system is estimating performance that translates well to the intended application. Typically this is done in the laboratory using data from a speech database representing speech from a large number of speakers. The verification system enrolls models of speech specific to each speaker. Then the verification system implements a scoring procedure resulting in a set of "true speaker" likelihood measures (likelihoods in which the models used for verification and test speech come from the same speaker) and "impostor" likelihoods (likelihoods in which the models used for verification and test speech come from different speakers). Verification performance is measured by how well the method separates the sets of "true speaker" likelihoods and "impostor" likelihoods.

Verification performance may be reported in several ways [3]. A commonly used method constructs a plot of two curves. One curve, called the Type I error curve, indicates the percentage of "true speaker" likelihoods above a threshold. The second curve, called the Type II error curve, is a plot of the percentage of "impostor" likelihoods below the threshold. An example is shown in Figure 13.3A.

Often performance is quoted as a number called the Equal Error Probability Rate (EER) which is the percentage at the point where the Type I and Type II curves intersect, indicated by the dot in Figure 13.3A. A verification system operating at this point will reject the same percentage of "true speakers" as "impostors" that it accepts. Another method of reporting performance is to plot percentages of Type I performance versus Type II performance on a log-log plot [2]. This results in an operating characteristic curve as shown in Figure 13.3B.

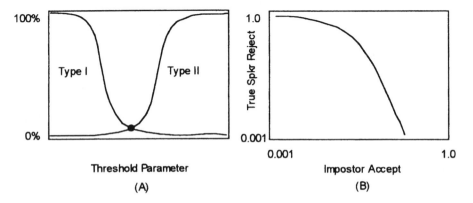

Figure 13.3   (A,B) Methods of measuring speaker verification performance

This type of curve indicates that a verification system may operate over a variety of conditions, depending on the relative costs of rejecting a true-speaker verification attempt or accepting an impostor verification attempt.

Note that these performance measures reflect the average performance over the entire population. They do not indicate speaker-specific performance, and it is possible that verification system failures may be correlated to specific speakers.

13.2.1.3 Acoustic Features for Speaker Verification

Characteristic models of the speech signal may be derived in many ways. A common and computationally tractable method of modeling speech acoustics breaks the speech signal into short segments (termed "frames") and assumes that the speech signal is stationary during each frame [4]. Modeling methods then construct a vector of speech parameters that describe the acoustics of the speech signal contained in the frame. Many methods exist to derive vectors of parameters that describe each frame of speech. However, virtually all speech processing systems utilize some form of spectral energy measurement of the data within the frame of the speech signal as a basis for modeling. Operations applied to the spectrum of the frame result in the parameter vector for the frame. The goal of constructing the parameter vector is to capture the salient acoustic features of speech during the frame that may be useful in pattern matching metric measures, while in some way filtering out the characteristics that are unimportant. The justification for a spectral basis of the speech vector representation is found both in the mechanism of auditory reception and the mechanism of speech production [4,5].

Linear Prediction
One commonly used method of describing the spectrum of the frame of speech is linear predictive analysis [6,7]. It can be shown that the vocal tract resonances can approximately be modeled as an all-pole (autoregressive) process, in which the location of the poles describe the short-term stationary position of the vocal tract apparatus. This method is used as a starting point for many speech feature generation algorithms. The linear prediction model is given by

$$G \cdot H(z) = \frac{G}{1 - \sum_{k=1}^{P} a_k z^{-k}}$$

Here G is a gain term, $H(z)$ is the vocal tract transfer function. The linear predictor parameters, a_k, are determined by first breaking the speech signal into frames, and then calculating the autocorrelation of each frame for 10–15 lags, then applying an algorithm such as the Durbin recursion. Such calculations are efficiently performed in Digital Signal Processor (DSP) hardware. The resulting linear predictor parameters are usually used as a basis for more complex feature representations, which may use the autoregressive filter to calculate spectral energies in non-linearly spaced bandpass filter segments. Overall frame energy parameters and frame difference values may also be included, resulting in parameter sets of 20–30 elements. Since these components are correlated, there is usually some form of linear transformation of the components which is aimed at whitening the resulting feature vector, and reducing the number of parameters. Such transformations result in final feature vectors having 10–20 elements.

Cepstral Features

Another common representation is based on cepstral features [3,8]. As illustrated in Figure 13.4, this signal processing method is based on a direct measurement of the spectrum of the signal using a Fast Fourier Transform (FFT). This is followed by calculation of energy magnitudes in about 20–30 non-linearly spaced bandpass filter segments, non-linear processing by a logarithmic function, and subsequent linear transformation by the discrete cosine transform to reduce correlation of parameters. Again, energy and difference values of the components are usually added to form the final feature vector, which is typically in the range of 10–30 elements in size.

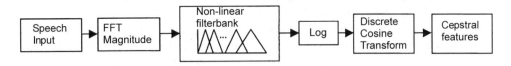

Figure 13.4   Cepstral processing

13.2.1.4 Statistical Models for Measurement of Speech Likelihood

Speaker verification is based on determining the likelihood that the observed speech feature vectors match the parameters for a given speaker. This requires a statistical model of the speech from the speaker. For text-independent speaker verification a typical model is a Gaussian mixture model [9], in which the likelihood of observing a feature vector x from speaker s is given by

$$p(x|s) = \sum_{m=1}^{M} a_{s,m} \cdot N(x, \ \mu_{s,m}, \ v_{s,m})$$

In this equation $N(.)$ is a multivariate Gaussian distribution, $\mu_{s,m}$ is the mean vector for speaker s and Gaussian mixture element m, $v_{s,m}$ is the covariance matrix or variance vector, and $a_{s,m}$ is the mixture weight for the speaker and mixture element. These parameters are estimated during enrollment, and may also be refined during verification if the confidence that the speech came from the true speaker is high enough. During verification the likelihood of all frames of speech are averaged over the duration of the utterance and the result is used to make the decision.

Text-dependent verification is more complicated since the input speech features must be matched to statistical models of the words spoken. A well-known method to do this uses Hidden Markov Models (HMMs) as a statistical model of words or sub-word units [10]. A representative HMM is shown in Figure 13.5. Here the model of speech consists of several states, illustrated by circles. Between the states are transitions, shown by lines, which have associated probabilities. Each state has an associated Gaussian mixture model, which defines the statistical properties of the state of the word or sub-word HMM model for the speaker. The transitions indicate the allowed progression through the model, and the transition probabilities indicate how likely it is that each transition will take place. Each of the parameters of

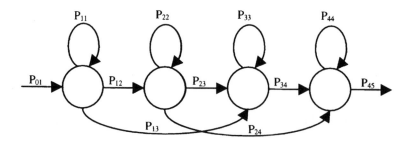

Figure 13.5   A representative HMM

the Gaussian mixture and the transition probability parameters for each state may be estimated during enrollment of the speaker.

Verification is performed by determining the best likelihood of the input speech data frames spoken by the speaker constrained by allowable paths through the HMMs that define the words spoken. Calculation of the likelihood uses a Viterbi search algorithm [10]. This type of processing is similar to speech recognition processing, except that the utterance is known *a priori*. Since the utterance is known, the resource requirements needed to implement verification are not nearly as large as those needed for speech recognition.

The likelihood measure *p(x|s)* may vary significantly with changes in audio hardware or environment. To minimize the impact of these effects on verification performance, some form of likelihood normalization is usually performed [11]. This involves calculating an additional likelihood of the signal x given some confusable set c, where the set c may be chosen as speakers with likelihoods close to speaker s, or as some global set of speakers. The likelihood measure used for making the true speaker or impostor decision is given by

$$L(s|x) = \log(p(x|s)) - \log(p(x|c))$$

13.2.2 DSP-Based Embedded Speaker Verification

Embedded implementations of speaker verification place resource constraints on the verification algorithms.

One of the restrictions is the amount of storage memory available for speaker characteristic parameters. In most verification applications there are usually limited enrollment speech data, which would be inadequate to train a large number of parameters reliably. Therefore, it is possible and necessary to reduce the number of parameters. Storage resources are often conserved by reducing the number of Gaussian mixture components to one or two. With sparse data, the variance parameters of the Gaussian mixture components are sensitive to estimation errors, and so often variance parameters are shared. A trade-off between performance and storage can also be made by reducing the size of the feature vectors representing speech. For text-dependent verification, the number of states in each HMM may be reduced. Per state transition probabilities of the HMMs are often fixed. To reduce storage of transition probabilities further, a systematic structure is often assumed. For the HMM shown in Figure 13.5, if transition probabilities are defined for returning to a state, going to the next sequential state, and skipping a state, then the number of parameters needed to represent transition

Table 13.1   Verification resources example

Verification task	ROM	RAM	MIPS	EER
Long distance telephone, ten continuous digits	8K program	1K search	8	2.1%

probabilities for the model is reduced to three. If a text-dependent verification system contains a set of speaker-independent word HMMs that serve as the basis for allowable verification phrases, then the parameters of the speaker-independent HMMs may be used. For example, it may only be necessary to estimate the parameter $\mu_{s,m}$, while obtaining all other parameters from the speaker-independent HMMs. Other parameters, such as the variance estimates of the Gaussian mixture components may be shared within a model, or even between models. Using these simplifications, typical verification models for text-dependent embedded applications require about 100 parameters per spoken word. However, more complex signal processing algorithms have been developed that retain performance with as low as 20 parameters per spoken word [12].

Program memory storage necessary for speaker verification will depend on the particular speaker verification algorithm used. For a typical text-dependent application, the speaker verification code will be a small addition to the speech recognition code. Processing requirements for the front-end feature processing will be similar to the speech recognition code. Text-dependent recognition requires calculation of the maximum likelihood path through the sequence of HMMs making up the spoken phrase. However, unlike speech recognition applications, speaker verification uses the *a priori* knowledge of the spoken phrase. This implies that processing resources will be less than those reported for speech recognition. As an example, as shown in Table 13.1, except for front-end feature processing, the resources for text-dependent speaker verification using ten digit phrases will be about one-tenth of that reported for digit recognition as reported in Chapter 10.

13.3 Live Fingerprint Recognition Systems

13.3.1 Overview

The ability to implement fingerprint ID systems in mobile devices hinges on the confluence of two technology developments: the recent commercial availability of very small, low power, high quality fingerprint sensors and the introduction of a new generation of fast, powerful DSPs into mobile devices.

In this section we review the engineering elements of designing fingerprint systems into the next generation mobile devices. We briefly characterize the unique aspects of mobile fingerprint systems, develop the concept of operations for mobile fingerprint systems, and then examine the critical performance metrics used to control the system design and ensure its adequacy. The fingerprint system is then decomposed into its basic elements. Each of these is described along with some possible design approaches and implementation alternatives. Lastly, we describe a prototype system architecture based on the Texas Instruments' OMAP architecture, and discuss the design and implementation of a demonstration system constructed using this architecture.

13.3.2 Mobile Application Characterization

13.3.2.1 End-User Benefits

Live fingerprint recognition on mobile devices makes basic security and device personalization convenient for the user. Entering usernames, passwords, or PIN numbers into portable devices is inconvenient enough that most people today don't use the security and personalization functions in their portable devices. With live fingerprint recognition, a single touch of the sensor device is all that is required to determine the user's identity, configure the device for personal use, or authorize access to private resources.

13.3.2.2 Expected Usage Patterns

A portable device typically has a small group of between one and five users. When an authorized user picks up the device and presents his/her finger to the sensor, the device should recognize the user and immediately switch its operation to conform to his/her profile.

13.3.2.3 Unique Aspects of the Application

Mobile devices require fingerprint sensors that are significantly smaller than any previously used. This requirement propagates into two aspects of the fingerprint system design. The first challenge is to build an adequate quality sensor small and light enough for mobile devices. The second challenge comes from the fact that smaller sensors generate images of smaller sections of skin. This means less data is available for comparison than with the larger sensors typically used for fingerprint recognition. To successfully match smaller fingerprint images the sensor must generate higher quality and more consistent images, and the matcher algorithm must be designed to take advantage of the higher quality data.

Alternatively, some systems require the user to slide his finger slowly across the sensor, to increase the area of finger surface imaged. This motion is called swiping. While this approach generates imagery of a larger area of skin, it seriously distorts the skin and has significant operational and performance liabilities.

The prototype application discussed later in this chapter uses an AuthenTec AES-4000 sensor with a sensing area just under 1 cm^2. Systems using even smaller sensors are under development at several fingerprint system suppliers.

13.3.3 Concept of Operations

The operational concepts underpinning most fingerprint authentication systems revolve around three classes of user events: enrollments, verifications, and identifications. Each of these event classes is described below from a high-level process view. The procedures underlying these processes are discussed later in this chapter.

13.3.3.1 Enrollment

Enrollment is the process of authorizing a new person to use the mobile device. In a typical scenario, the owner of the device authorizes a person to use the device by: authenticating himself/herself to the device as the owner, creating a new user profile with the desired privileges on the device, and then training the device to recognize the new user's fingerprints. Typically the system is trained to recognize two or three fingers for each person in case injury makes one finger unavailable.

The process of training the fingerprint system to recognize a new finger can be broken down logically into the following steps:

- Collection of system training data samples
- Feature quality analysis
- Template generation
- Template storage

Collection of Training Data Samples

The system collects several views of a finger, prompting the new user to lift and replace their finger on the fingerprint sensor several times. Each finger placement is considered as one view. Each view may consist of a sequence of image frames that taken together define a view.

13.3.3.2 Feature Quality Analysis

The collected samples (called views) are analyzed to extract the features that will be used for matching. The system then assesses the quantity and quality of matchable feature data present in the views, and estimates the probable robustness of that feature data. The results of this analysis determine whether the system can use this set of views for enrollment, or if more, or better, data are needed. If the data are insufficient, the system may request more views of the same finger or request the new user to present a different finger.

Template Generation

If the data is sufficient for enrollment, the system assembles the best of the available data and formats it into a template that will be used as the reference for subsequent matching of this finger.

Template Storage

The resulting template is then stored under an appropriate encryption scheme for recall during subsequent verification and identification operations. Templates can be stored on any media that can hold digital data. On mobile devices templates are typically stored in flash memory.

13.3.3.3. Verification (Claimed Identity Verification)

Verification is the process of authenticating a claimed user identity. A verification event occurs when: (1) a user indicates his/her identity to the system (usually by typing in a username) and (2) the system verifies the claimed identity by comparing the user's live fingerprint to the template stored for that username. This type of comparison is often called a one-to-one comparison because only one stored template is compared to the live fingerprint.

Verification processes generally require significantly less computational horsepower to perform than identification processes, and may be more reliable. However, verification is generally less convenient for the user as the username must be entered manually. Given that user convenience is a primary requirement for fingerprint systems on mobile devices, verification processes are probably inappropriate and identification processes (discussed in the next section) are preferred. In situations where only one person uses a device (which may be a significant percentage of devices) the identification process essentially devolves to a simple verification, so the performance penalty is minimal.

Data Collection
The system typically collects one view of the finger, which may consist of a sequence of image frames. For extremely small sensors, it may be necessary to collect multiple views of the finger to accumulate enough data to perform the fingerprint match.

Feature Analysis
The collected images are analyzed using various forms of pattern recognition algorithms to extract the features to be used for matching.

Matching to a Template
The data from the live finger is compared to the stored template for the claimed identity and a probability that the claimed identity is true is estimated from the match results. The system returns a binary result. The claimed identity is either true or false.

13.3.3.4 Identification (Unassisted Identification)

Identification is the process of finding the current user's identity from a list of possible identities, based solely on the user's live fingerprint. Identification processes do not require the user to enter a username or any other co-joined authentication. Instead, a single touch of the fingerprint sensor is sufficient. Identification processes typically require significantly more computational power than verification. Additionally, in identification processes both the accuracy and the latency of the process are not constant, as they are functions of the size of the reference dataset being searched.

From the process perspective, identification is similar to verification with two notable exceptions: (1) no username is entered, and (2) the system must perform an indexed search of all of the possible enrolled templates to find the matching template if it exists in the dataset. The result of the process is either the selected ID or an indication that the presented finger is not in the dataset.

The identification process, with its one-step usage paradigm, is significantly better suited to convenient personalization than the verification process.

13.3.4 Critical Performance Metrics

13.3.4.1 Biometric Performance
Biometric performance measures evaluate how well the system does the job of recognizing and differentiating people. At the system level, there are two generally accepted classes of problems to be avoided in these recognition systems. The first class of problems occurs when the system cannot acquire a reasonable quality image of the finger. These usability problems are mostly associated with the sensor itself, and are called "failure to acquire" errors. In some systems available today, acquisition failure errors dominate the behavior of the system. The second class of problems occurs when the system has adequate imagery but makes an error in

performing the pattern recognition. This second class of problems is more generally associated with the pattern matching software, and can be categorized in classical terms as false accept errors (Type 2) and false reject errors (Type 1).

Usability – Ability to Acquire

Mobile communication devices are rapidly becoming a ubiquitous part of our everyday environment. As such, the fingerprint systems will have to work for everyone's fingers. Failure to operate with fingers that fall outside of a norm – such as elderly, sweaty or dry fingers – will not be acceptable. The systems will have to work in a wide range of environments; not just the office and the car, but also the tennis court, the garage, and the ski lodge. Many fingerprint sensors are extremely sensitive to the condition of the finger skin that they must image. Some sensors available today successfully image young healthy fingers, but are unable to image elderly fingers or fingers with dry skin. Some are unable to function in environments more demanding than a clean office. And yet, some sensors can adequately handle all of these conditions.

The Ability-to-Acquire metric measures a system's ability to capture usable fingerprint images over the range of population demographics, usage patterns, and environmental conditions appropriate for that intended application. It can be represented as the expected percentage of successful finger imaging events over the ranges appropriate for a particular application.

For general-purpose mobile communications and information devices we believe that the fingerprint system's ability to acquire fingerprints should be in the range of 99.99%, over a population demographic that represents the entire world population and includes both clean and slightly contaminated fingers, and over a wide range of both indoor and outdoor environments.

Identification/Verification Accuracy

Identification and verification accuracies are usually represented as the percentage of identification/verification events in which the system delivers an inappropriate response; either incorrectly rejecting a finger that should match (false reject), or incorrectly accepting a finger that should not match (false accept). Identification accuracy and verification accuracy (while using the same type of error metrics) are best treated for this discussion as two different kinds of specifications that are associated with two different implementations of fingerprint authentication systems, as discussed earlier in this chapter.

Verification Accuracy

There are two classes of measurement traditionally used to quantify identification/verification accuracy. These are the False Accept Rates (FARs) and the False Reject Rates (FRRs). We believe that mobile communications device applications when used in verification mode require FARs of 0.1% or less (which is sometimes considered similar to the probability of someone intelligently guessing a user-selected four-digit PIN).

For this type of live fingerprint recognizer, false reject errors come in two varieties. Sometimes a valid user will place his finger on the sensor and be initially rejected, but will immediately try again, repositioning the finger on the sensor, and be accepted. This is called a nuisance reject. It is a special case that only occurs in live fingerprint systems, where the user can retry immediately. Our informal experience suggests that nuisance reject rates exceeding the 5–7% range can degrade the user experience and should be avoided. The second and far more serious type of false reject is the denial-of-service event. In this type of event, the system fails to accept a valid user's fingers even after several retry attempts. Clearly, for user satisfaction reasons, the denial-of-service reject should be avoided. We believe that denial-of-service FRRs of 0.1% or less will be required for ubiquitous deployment.

Identification Accuracy

False Accept Errors

When a system performs an identification event, it must in essence compare the live fingerprint to all of the templates that are considered possible matches for that live finger. For comparison, when an non-enrolled finger is presented to the system, a verification function compares the finger to only one template. Hence it has only one opportunity to make a false accept error. In contrast, when an identification system faces that same non-enrolled finger, it must compare the finger to all of the templates in the dataset. Hence it has as many opportunities to make a false accept error as there are templates in the dataset. It is customary to treat the verification FAR as the base metric for a biometric system. The identification FAR can then be estimated for practical systems with small databases as the product of the verification FAR and the size of the template database.

For a mobile information device, let's assume that five users enroll two fingers each for a total of ten templates. If the verification FAR of the system is in the range of 0.1–0.01% then the identification FAR will be in the range of 1–0.1%.

Confused ID Errors

Systems that perform unassisted identification may mistakenly match enrolled person A's finger to enrolled person B's template. This is a special case of a false accept error that is sometimes called a confused ID error. In systems like mobile communication devices where convenient personalization is a key aspect of the application, confused ID errors may be more problematic than other forms of false accept errors. This is because confused ID errors can occur anytime the device is used, while false accept of a non-enrolled person can only occur when an unauthorized person tries to use the device. Confused ID errors also differ significantly from false accept of non-enrolled fingers in that confused ID errors are effectively denial-of-service events to a valid user.

Confused ID error rates as discussed here are specified as the probability that the system will ever generate a confused ID error based upon a specific dataset size. This is in contrast to the FAR and FRR that are specified on a percentage-of-events basis.

Figure 13.6 shows an order of magnitude estimate of the relationship between template set size and the probability of a confused ID error for unassisted identify systems built using matching subsystems with three different verification FARs: 0.1, 0.01, and 0.001%. A matcher having approximately a 1 in 10,000 verification FAR will have approximately a 1 in 1000 probability of confusing someone's fingerprint at some point if four people enroll one finger each.

False Reject Errors

When a system is constructed to perform real-time identification against a dataset of more than just a couple templates, it is often not practical to completely compare the live finger to each template in the dataset. It takes too long and/or requires too much processing horsepower. Practical systems use indexes and filters to select a small list of probable match candidates from the larger dataset. The full matching process is performed only on these

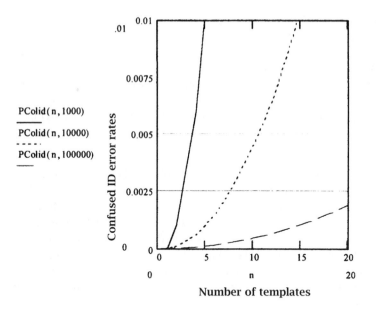

PColid(n , 1000)

PColid(n , 10000)

PColid(n , 100000)

Figure 13.6 Confused ID error rates vs. template dataset size

selected candidates. The indexing and filtering processes are another source of error since they may mistakenly prevent the correct template from being evaluated – leading to a false reject error. Hence the FRRs of identification functions are usually significantly greater than those of verification systems even if they are using the same final match algorithm.

There is no rule of thumb that can be used to predict the identification FRR from the verification false reject rate because the performance depends on the quality of the indexing and filtering implementation, and the degree of latency reduction that the indexing and filtering attempt to achieve.

13.3.4.2 Measuring Biometric Performance and Understanding Test Results

Given the importance of biometric performance to the success of fingerprint recognition applications, it would be useful to have standardized test methods and specification data that could be used by engineers to predict the biometric performance they could expect of a particular product in the field. While progress is being made, it is unfortunately not yet practical to characterize biometric systems in this general way. This section discusses the problems involved in biometric testing, and some things to watch for when interpreting and using biometric test results.

Some systems are designed to work well in specific types of environments and applications (like offices or banks). Others are designed to be adjusted so they can function in a wide range of conditions. The best sensors and systems automatically adapt to the situation presented and work well in all situations. Many systems fall somewhere in between. To ensure that a planned system will be appropriate to its user demographics and its environment, it is best to collect test data across the full range of conditions that the expected application will experience.

Sensitivity to the Population Demographics

The demographics of the user population that will use a biometric device can have a huge effect on the biometric performance of the system. User characteristics such as age, gender, occupation, ethnicity, and general health all can have a major impact on the biometric performance of a fingerprint system. Too many underlying factors contribute to these dependencies to allow a detailed discussion here. However, we can note that the error rates of some fingerprint systems can vary by factors of 10 depending on the demographics of the test population.

Sensitivity to the External Environment

Many fingerprint systems are sensitive to the weather and climate – especially the temperature and humidity trends. Systems that work well in Florida during January may fail miserably in Maine during January, and may also fail in Florida during July. Most of this sensitivity is due to climate-induced changes in the users' skin. Testing must be performed across the full range of weather and climate conditions that the device's users will encounter during use. Since it is the reaction of the human skin to long-term exposure to climate that drives the variations seen, it is difficult to use simulated climates for testing of these effects. To reliably assess the effects of weather and climate requires testing at multiple locations and during several seasons.

Sensitivity to Finger Skin Condition and Contamination

Mobile applications of fingerprint systems are very likely to encounter fingers that are not completely clean. Even if the user wipes his finger off before using the fingerprint device, the finger skin often remains contaminated with residues of many different substances. Typical finger contaminants include: cooking oil, sugar, dust, soil, magic marker, chalk, powder, moisture, sweat, etc. It is not reasonable to require a user to wash his hands before using the mobile device; therefore the fingerprint system must be able to read through these types of finger contamination.

Many fingerprint readers suffer severe performance degradation when presented with slightly contaminated fingers. Optical fingerprint readers are particularly degraded by finger contamination and because of this may be inappropriate for mobile applications.

Sensitivity to the Application Environment and User Experience

The performance of fingerprint systems is also sensitive to several aspects of the application environment. The mounting and presentation of the sensor determines how accurately and repeatable the user will be able to place his finger on the sensor. A poorly presented sensing surface will degrade the systems biometric performance significantly.

Systems that allow users to practice placing their finger on the sensor (and give feedback to help improve the placement) before the finger is actually enrolled have been shown to perform better than systems that do not offer this kind of training.

Response Time

Given the expected usage patterns described earlier, fingerprint system must typically complete the authentication task within 1–3 s from the time the user presents their finger.

13.3.4.3 Security and Spoof Prevention

The fingerprint system is one aspect of the overall security system built into the mobile device and its infrastructure. The fingerprint system is responsible for authenticating the user and delivering that data securely to the rest of the system. In mobile devices, the most significant threat likely to be posed against the fingerprint system is spoofing the sensor with a fake finger.

Spoofing is the use of a mechanical structure (that can be placed on the fingerprint sensor) to fool the sensor into thinking that a real finger having the desired fingerprint pattern has been placed on the sensor.

The principal objective of fingerprint authentication systems on mobile devices is to allow the user to conveniently achieve low to medium security levels. The fingerprint systems in these devices would be like the locks on a typical home. Most residential door locks and home security systems make it more difficult for the average person to break into a home, but in fact can be defeated by skilled professional thieves. The advantage of these residential systems is that they are inexpensive and not too difficult to use, while making break-ins significantly more difficult. Fingerprint systems should follow the same paradigm. It should be inexpensive and simple to use, while making inappropriate use of the protected device significantly more difficult. One simple way to quantify this basic concept is to require the cost of defeating the fingerprint system to exceed the value realized by the unauthorized person who defeats it. Looking at the issue from a different perspective; if the mobile device of the future is used to perform financial transactions, its security should be at least equivalent to that of the four-digit PIN used in today's debit card systems.

RF E-field sensors such as AuthenTec TruePrint sensors are not fooled by typical latex or silicon rubber fakes. This type of sensor can detect the complex electrical impedance of the finger skin and reject fakes that do not present the correct electrical characteristics. While these devices can be spoofed, it is extremely difficult.

13.3.5 Basic Elements of the Fingerprint System

13.3.5.1 Overview and Architecture

At an overview level, fully realized biometric user identification functionality on a mobile device can be viewed as containing the following subsystems: (1) fingerprint sensor hardware, (2) a computational platform host, (3) biometric software, (4) system application software, and (5) user application software. These components are illustrated in the form of a reference architecture in Figure 13.7.

The fingerprint sensor detects the presence of a finger and converts the patterns on the finger into an electrical form that can be used by subsequent information processing stages. The biometric authentication services software manages and optimizes the sensor hardware, optimizes the fingerprint image, performs feature extraction and matching, and makes decisions to accept or reject an identity based on the results of the matching. It is this layer that typically involves the heaviest computational workloads.

The system application software is the link between the biometric identification system and the host device resources and operating systems. It provides host specific user interface and user control functions to the biometric ID system. It also allows biometric identification to be used to gain access to basic system resources. It performs functions like user login, applications and data protection, and browser password replacement.

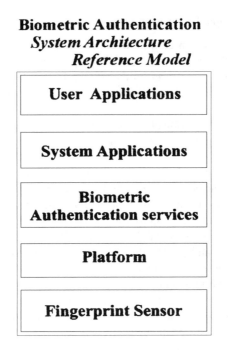

Figure 13.7   Biometric reference architecture

The user application software performs the main functions that the user desires such as voice communications, e-mail, word processing, e-commerce, etc. It uses the identification services to verify the authenticity of the user directing the actions it performs.

13.3.5.2 Fingerprint Sensor Hardware
Candidate Sensor Technologies
Several candidate technologies are currently available for measuring fingerprints within the size and power consumption constraints imposed on mobile wireless equipment. They include optical sensors, thermal sensors, capacitive sensors, and RF E-field sensors.

Optical fingerprint sensors small and thin enough for use in mobile devices can be fabricated using molded plastic fresnel optics and CMOS optical imaging arrays. Their major advantages are physical robustness and industry familiarity. Their principal disadvantages are large size, low image quality, and poor "ability to acquire" finger images in less than optimal conditions.

Thermal array sensors that detect changes in heat flux can be used for fingerprint imaging. In these sensors the finger must be in constant motion to generate a signal, hence they are used by swiping the finger across the surface. Their major advantages are very small size and low cost. Their principle disadvantages are distorted, segmented images caused by the finger swiping motion and poor "ability to acquire" finger images under less than optimal conditions.

Arrays of electronic capacitance sensors can be fabricated into silicon integrated circuits that read fingerprints by measuring differences in local fringing capacitance between pixels. The major advantages of these devices are small size and simple design. Their major disadvantages are physical delicacy, poor image quality, and poor "ability to acquire" finger images in less than optimal conditions.

Silicon fingerprint sensors can be fabricated that couple tiny RF signals into the conductive layer beneath the surface of the finger skin, and then read the shape of this conductive layer using an array of integrated RF sensors. These RF E-field sensors' major advantages are very high image quality, and a high "ability to acquire" finger images in a wide range of less than optimal conditions. Of the sensing technologies practical for mobile device, these are the only ones that read beneath the skin surface. Subsurface imaging makes these devices less sensitive to surface damage and surface contamination than surface reading devices. The major disadvantages are a degree of physical delicacy, and unique control logic that is unfamiliar to the industry.

RF E-field sensors using TruePrint™ technology from AuthenTec, Inc. were used in the prototype system discussed later in this chapter.

Sensor Implementation

Figure 13.8 illustrates the block diagram for a generic fingerprint sensor. All of the blocks may not be present in every sensor, but in most cases they are. The block diagram provides an outline for discussion and comparison of various sensors. In recent silicon sensors, most of the function blocks shown in the diagram are integrated directly onto the sensor substrate.

All sensors start out with some form of signal source that places a signal onto the finger. In optical sensors, the signal source is typically a set of LEDs and a regulated power supply. In thermal sensors the signal source is a heater. In RF E-field sensors, the signal source is a small RF signal generator.

Energy from the signal source is coupled to the finger and the finger's response to that energy is measured by an array of sensor pixels. In optical scanners the pixels are photosensors, in thermal systems the pixels are heat flux sensors. In RF E-field devices the pixels are metal plates that act as tiny field sensing antennas.

The signal from each pixel may be amplified and processed as necessary under pixel, and then multiplexed out of the array by scanning control logic. The result of this scan process is typically an analog signal stream of time multiplexed pixel measurements.

More elaborate signal processing can be applied to the signals once they leave the array prior to conversion from analog signals to digital signals in an A-to-D converter. The output of this converter is a digital data stream representing the sequence of pixel signal values scanned by the multiplexer.

Digital processing may be applied to the data stream to perform tasks not easily handled in the analog circuitry, and to prepare the data for transmission via the digital interface circuitry to a processor.

13.3.5.3 Fingerprint Authentication Software

Introduction

The fingerprint authentication software executes on the host device's processor(s). It operates as a service that responds to a request for fingerprint authentication from a system-level program or an application program.

After receiving a program request the software:

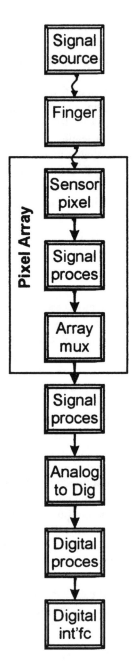

Figure 13.8   Sensor block diagram

- activates the sensor and detects when a finger is placed on the sensor
- collects images from the sensor and manages the user interaction with the sensor
- optimizes the images using various approaches discussed below
- estimates the probability that the live image comes from the same finger as previously stored image data
- determines if the data presented offers sufficient match confidence to declare a verified user identity

The Processing/Accuracy/Latency Trade-Off

Figure 13.9 illustrates a rough rule of thumb for order-of-magnitude estimates of the accuracies and latencies that can be expected from processors of various capabilities.

These plots represent averages of the range of systems that have seen real commercial

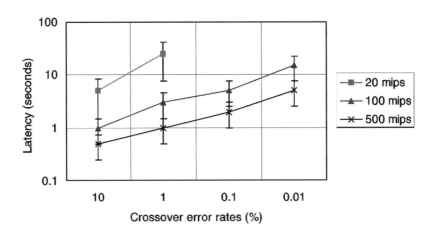

Figure 13.9   CPU accuracy/latency trade-off

deployment from 1997 to 2000. (Note: while specific implementations can vary, these graphs on average appear to remain true for 2001.) The accuracy metric used here is the Equal Error Crossover Rate (EER).

Sensor Control and Image Optimization

Requirements

Different kinds of sensors require different approaches to sensor control and image optimization. Optical fingerprint sensors generally only offer a single operational parameter that can be controlled – the light intensity. These sensors are very simple to control, but cannot adapt to unusual fingers or environmental conditions. Adaptive sensors like the AuthenTec True-Print based sensors are widely adaptable, but require more complex sensor control logic to utilize that flexibility. Strip sensors that are used with a swiping finger motion have very special sensor control and image reconstruction needs that require much more complicated logic than stationary finger sensors.

Approaches

Sensor Adjustment vs. Post Capture Processing

The performance trade-offs illustrated in the above rule of thumb assume a system architecture built around a simple or non-adaptive sensor, which generates large volumes of data that must be optimized through image processing methods before the pattern matching can begin. Industry experience over the last 5 years indicates that several alternative architectures can be built around more flexible adaptable sensors that can generate better quality image data than what has been previously achievable using traditional non-adaptive sensor architectures. The AuthenTec TruePrint™ technology based sensors used in the prototype system fall into this second category of adaptive sensors. These sensors can be adjusted over a wide range of operating points across multiple operational dimensions and operating modes. Systems using these sensors are designed as active closed loop control systems. In these systems, when a finger is placed on the sensor, the image stream is analyzed and the sensor's operating points adjusted on the fly. For each frame, the image quality is analyzed, and the sensor's operating point adjusted toward optimum quality. The result is a better quality image than can typically be acquired from a simple sensor and extensive image processing. From a processor utilization perspective the adaptive sensor systems move some of the processing load from the back end image-processing phase into the real-time sensor control loop. The net result is typically a slightly lower overall computational demand.

These adaptive systems can have somewhat lower CPU utilization than traditional systems for most fingers, but most importantly they can generate high quality images of many fingers that cannot be imaged adequately by traditional systems. This characteristic makes adaptive sensor systems more appropriate for applications that must maintain a very low failure rate across a wide population range and across a wide range of environmental conditions.

Implementation Alternatives

This section will focus on automatic adaptive sensor systems, since they are more appropriate for mobile devices that must work for everyone in a wide variety of environments. The section will discuss the types of logic that can be used for sensor image optimization, as well as their advantages and disadvantages. Also discussed is where the sensor control logic best fits into the overall architecture of a mobile device.

As an example of a highly adaptable sensor, the AuthenTec AES-4000 used in the prototype system (discussed later in this chapter) was used in that system in a configuration that allowed effectively five degrees of freedom in the operating point. The adjustments can be represented as receiver gain, receiver offset, finger excitation voltage, finger excitation frequency, and demodulation phase shift. The first two parameters, gain and offset, allow classic gray-scale normalization of the images directly at the sensor. Each of these first two degrees of freedom has both coarse and fine adjustment capability on the sensor. The latter three parameters are used to adapt to the wide range of electrical behavior seen in human finger skin.

There are several approaches to optimizing the image from a sensor with a wide range of operating points. The choice of approaches is determined by the range of conditions that the system must operate in, the kind of matcher to be used (and the nature of its sensitivities), and the capabilities of the control processor used.

Three of the most common methods of optimization used with flexible fingerprint sensors are:

- Try a fixed sequence of pre-selected operating points
- Use static binary search procedures to find the best operating points
- Use dynamic feedback control regulators to find and track the best operating points

Each of these will be discussed below.

A second aspect arises in characterizing fingerprint sensor control systems. In sensors with multiple configurable operating point control parameters, several approaches can be used to deal with the interactions between control parameters that typically occur in these devices. At one extreme, the interactions between control variables can be ignored, and the controls can be optimized independently by treating them sequentially in time. At the other extreme, the interactions can be modeled and incorporated into a multivariable control algorithm. In this case the system recognizes the control interdependencies and performs simultaneous optimization on the interacting variables. The trade-off here is significantly improved accuracy and repeatability from simultaneous optimization at the price of significantly more complex control logic.

Fixed Sequence of Pre-selected Operating Points

In this optimization strategy a small number of sensor operating points (typically 2–5 points) are pre-selected to span the range of user population and environmental conditions in which the sensor must operate. The control system tries each of these operating points in series until the image quality criterion is reached. In verify and identify operations with local matching, the image quality criterion can be a successful match – the system tries each operating point until a match is achieved. If after trying all operating points no match has been achieved the match attempt is declared false (finger does not match template). When the matching process results cannot be used as a quality criterion (such as during the enroll process, or if the matcher executes remotely from the sensor, or if the matcher is too processing intensive) image quality estimators must be used to identify usable operating points. Estimates of image contrast, brightness, gray scale dispersion, and contrast uniformity have been used as simple image quality estimators.

The advantage of the sequence of pre-selected operating points approach is its simplicity. The disadvantages are low image quality when the small number of operating points selected cannot adequately span the required range of fingers and conditions, and inconsistent images because the system cannot fine-tune itself to specific images.

Static Binary Search Procedures

Static binary search procedures can tune the sensor more accurately to the finger than fixed sequence procedures can. The result is more consistent images over a wider operating range. In this optimization strategy, the image quality is described by a set of measured and/or computed process variables. The control system finds sensor operating parameter values that generate appropriate process variable values using a static binary search of the sensor operating space. The operating point reached at the end of the binary search is used to generate the image used for the biometric processing.

Since binary search procedures assume that the data being searched is stationary, this approach assumes that the finger's imaging characteristics do not change during the period of time the finger is on the sensor and the search is being executed. Unfortunately this assumption is not true in many cases. When a finger is placed on the fingerprint sensor the skin settles slowly onto the sensor surface by slowly changing its shape to match flat sensor surface. The behavior of the skin differs markedly from that of an elastic body, taking on some properties of a viscous fluid. This is sometimes called the finger settling effect. The time it takes the finger skin to settle down onto the sensor surface can vary widely depending on the user's age, degree of callousness, skin moisture content, skin temperature, and finger pressure. This time constant typically ranges from 0.3 to 4 s, and occasionally approaches 10 s for fingers with certain medical conditions. A second non-stationary effect is the accumulation of sweat under the sweat glands during the time the finger is on the sensor. During several seconds of contact, sweat accumulation can significantly change the characteristics of the finger. The non-stationary aspects of finger skin behavior can make binary search techniques less effective and cause wide variations in the quality of the images generated by binary search.

The advantages of binary search procedures include: better quality images – much more finely tuned images than can be achieved with pre-selected operating points, operational simplicity, low computational overhead, and widespread programmer familiarity with binary search procedures. The chief disadvantages are inconsistent image quality resulting from non-stationary finger behavior, and difficulty in adapting binary search to sensors with multiple degrees of freedom.

Dynamic Feedback Control Regulators

The dynamic feedback control regulators approach is similar in some respects to the binary search approach in its use of process variables. However, it applies different, more flexible, methods of selecting the next operating point. The image quality is again described by a set of measured and/or computed process variables. The control system finds sensor operating points that optimize the process variable values using classical analog error regulator algorithms. The system determines the process variable error as the difference between the desired process variable value and its current actual value. Then, using knowledge of the sensor's response to changes in its operating point, the system estimates the value of the operating point parameter that would generate minimum process variable error, and establishes that as the new operating point value. The process is repeated for each image frame, with the regulators constantly readjusting the sensor operating points to adapt to the continuously changing finger characteristics. In these systems several image frames are typically submitted to the matcher until the finger has settled to the point where it matches the template. If no match is achieved after a reasonable period of time (typically 2–4 s) a "does-not-match" result is declared.

Simultaneous Multivariable Control

The most flexible sensors available today adapt to wide ranges of conditions by offering several degrees of freedom within their operating space. Using regulator style controls, several interdependent process variables can be optimized simultaneously using well-understood multivariable control methods. These methods can produce more repeatable image optimization than sequential optimization of individual process variables, and will generally converge faster.

The principle advantages of using regulator style controls are: best quality optimized images, most repeatable images, and operation in wider ranges of conditions by taking advantage of sensors with multiple degrees of freedom. The major disadvantages are the somewhat higher computational loads imposed by regulator systems, more complex control logic, and less programmer familiarity with multivariate analog control design.

Integrating the Sensor Controls into the Architecture

Architecturally there are three places where the intelligence that optimizes the sensor operating point can reside: (1) image analysis and optimization logic can be built into the sensor,

(2) an independent processor can be dedicated to sensor image optimization, or (3) the host system's processor can perform the optimization logic.

The logic to implement a control strategy based on several pre-selected operating points (as well as the image quality estimation that goes along with it) could be implemented directly in the sensor silicon at very little extra cost. Unfortunately these simple control strategies cannot supply adequate image quality across the wide range of user demographics and environmental conditions needed by consumer mobile devices. We believe that fingerprint systems in mobile devices will need full-scale dynamic feedback regulator systems to succeed as ubiquitous commercial devices. This type of system needs the resources of a reasonable sized processor. As a result, it appears to be more cost effective to use the existing host processor for image optimization rather than build a processor into the sensor silicon or incorporate a dedicated processor into the system design for sensor control.

Pattern Matching Algorithms

Input Data

The input data to the pattern matching algorithms is generally in the form of bitmap images. The bitmaps may represent skin areas from less than 1 cm^2 (for the smaller RF E-field devices) to 4 cm^2 (for the smaller types of optical devices).

Approaches

Fingerprint images contain a wealth of information, some of which is stable over time and some of which is less stable. A wide variety of approaches can be used to algorithmically extract the stable information from the images and compare this information across different finger presentations. Different algorithms have different capabilities and accuracies and require different amounts of computational horsepower to achieve those accuracies. An introduction to the most common classes of these algorithms is included in this section.

Algorithms that determine how closely two fingerprint images match can be grouped according to the specific type of information that they use for the comparison. The following discussion will use this grouping to discuss the algorithms. Also included will be an assessment of the characteristics of these approaches when implemented in computerized matching systems for the new generation of fingerprint sensors.

A Shift in Focus

The recent appearance of high-quality sensors capable of producing highly repeatable fingerprint images (such as the AuthenTec TruePrint™ sensors) are opening up a new era in fingerprint matching algorithm approaches.

Algorithms for the previous generation of live fingerprint sensors had to devote large amounts of computing horsepower to filtering and conditioning the raw image, and to attempting to reconstruct the real fingerprint pattern from the often weak and incomplete image data provided by the sensors.

The use of moving-finger placement strategies for image acquisition (rolled finger for law enforcement cards, or swiped finger for strip sensors) introduced large elastic distortions in the images that further complicated the matching process. The matching methods focused on estimating the probability that this weak data in the enrolled templates and in the current image could be mapped back to the same finger. The use of a stationary finger placement on the sensor rather than a rolled placement or swiped placement has eliminated the need to compensate for the large elastic distortions that these acquisition methods introduced. Hence the most recent algorithms focus on accurately matching reliable repeatable images rather than on estimating probable matches from unreliable data.

A Rule of Thumb

One rule of thumb for assessing the value of algorithmic approaches to computerized fingerprint matching is to estimate the amount of differentiation achieved by a method divided by the amount of computation required by the method:

Value = Differentiation/Computation

The following discussion uses an informal, qualitative form of this metric to compare some of the various algorithmic approaches.

Classical Classification

This group of approaches is based on traditional manual methods. It includes approaches such as the Henry system that classifies fingerprints into several types based on the visual patterns created by the ridges in a finger. The types include whorls, arches, loops, tented arches, etc. The human visual perception system performs the complex computations involved in this type of analysis very efficiently.

Classical methods like the Henry system are not often used in computerized matching today because, lacking the efficiencies of the human visual perception system, the differentiation-to-computation ratio of typical implementations is rather low.

Ridge Minutia

Many computerized fingerprint-matching systems identify features in the image called minutia points. Classical minutia points are those locations where either a single ridge line splits (bifurcates) into two ridges, or a ridge line terminates. Ridge minutia have been used in manual fingerprint matching for many years and are well understood. The specific location of the minutia points in fingerprint images has very little correlation with genetic heritage. Even in identical twins, the minutia patterns are different. Figure 13.10 illustrates ridge minutia in a fingerprint.

Figure 13.10   Ridge minutia in a fingerprint image

In computerized systems, these minutia points can be identified within a fingerprint image using several different image analysis techniques. They are then located spatially with respect to each other using physical measurements – e.g. (x, y) coordinates – or using topographic references, e.g. number of ridge transitions between the minutia. The result of this minutia extraction process is a list of minutia points and their spatial relationships for the image under analysis. A comparison is made between two images and a degree of similarity computed by comparing the two minutia lists.

Ridge minutia algorithms have been common in computerized fingerprint matching since the early 1980s. They provide a high degree of differentiation when given full finger images, but require a fairly large amount of computation to extract the minutia points from the image. In minutia systems, the processor workload is not evenly split between the extraction and matching processes. The matching process is much faster than the extraction process. This makes minutia systems useful as an early step in one-to-many matching systems.

Since ridge minutia are small, localized features, small amounts of noise or distortion in the images may hide minutia or simulate minutia, causing false conclusions. Minutia systems are difficult to use by themselves in systems that work with smaller image sizes, because some fingers have such low minutia densities that only a few minutia are seen in the smaller images.

Ridge Flow Direction Vectors

The directional properties of the fingerprint ridges constitute a feature set that can be used for matching. A two-dimensional matrix of vectors representing the ridge direction at various points in the image is called a ridge flow map. Ridge flow maps (Figure 13.11) can be more useful matching features in mobile device systems than they have been in the law enforcement systems of the past.

Ridge flow maps have only limited utility when evaluating images acquired from rolled finger presentations (such as produced for law enforcement using ink and cards) due to the

Figure 13.11   Ridge flow vector matrix superimposed on the fingerprint image

high level of shape distortion produced by rolling the finger across the card. Hence ridge flow maps are not matched directly in most law enforcement systems. However, many modern electronic fingerprint scanners use a slap style of acquisition where the finger is simply placed stationary on the sensor. This method of acquisition minimizes shape distortion and allows the ridge flow map to be used as a reasonable feature set for image matching. (Note that the linear swipe style of the fingerprint sensor does not generate low distortion images and cannot take advantage of this type of ridge flow matching algorithm.)

Ridge flow matching algorithms can generate a moderate degree of differentiation with a moderate amount of computation. For lower accuracy applications, the differentiation-to-computation ratio can be more favorable in ridge flow matching algorithms than in many others. Since the ridge flow direction is a fairly large feature these algorithms are not as susceptible to small amounts of noise as minutia algorithms. However, severely broken images or images with large amounts of twisting or pulling distortion often cannot be accurately matched.

Specific Ridge Pattern

The specific ridge pattern itself can be used for matching directly. Systems that look at the ridge pattern extract the pattern information by removing the fine structure such as ridge width variations, pores, and other small-scale features. The resulting ridge lines can be directly compared using image cross correlation techniques – hence these matchers are often called image correlation matchers. Figure 13.12 is an illustration of a ridge pattern map. Note that the fine detail has been removed to enhance accurate representation of the pattern of the ridges. Ridge pattern correlators can be extremely accurate if they are combined with well-constrained distortion mapping algorithms. The distortion analyzer is needed because even slap prints often exhibit distortion in excess of one ridge width over a 1/4–1/

Figure 13.12   Normalized ridge pattern map – as used in image correlation matching

2 inch distance. The distortion analyzer must be very careful to restrict the distortion to that which is physically probable to avoid warping one fingerprint to falsely match another.

Ridge pattern correlation is another technique that benefits from the low distortion achieved by modern slap sensors when compared to rolled finger sensors and swipe sensors.

Ridge pattern correlators can generate a very high degree of differentiation between fingerprints but require a very large amount of computational energy to do so. These processing intensive algorithms are well suited to performing one-to-one matches in situations that can afford powerful DSPs or Pentium III class processors.

Fine Structure

The previously described approaches to matching fingerprints are all based on large-scale attributes of the ridge structure. Typically these systems attempt to remove small-scale fluc-

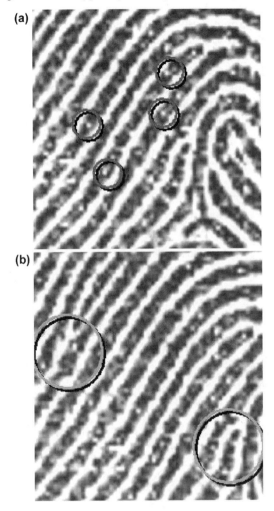

Figure 13.13   (a) Image detail showing strong pore locations. Example pores are circled in violet. The sensor is an AuthenTec AFS-2 250 ppi resolution. (b) Image detail showing areas of strong detail structure near minutia points

tuation from the images before matching the features. When smaller sensors are used, more information must be extracted from smaller areas of skin to provide enough data for differentiation of fingerprints. As a result, methods of using smaller features called fine structure are of considerable interest for use on mobile devices. Two types of fine structure information have been used successfully to aid differentiation in fingerprint matching: pores, and detailed ridge shape. These will be discussed briefly below. These methods require higher image quality and more detailed repeatability from the sensor in order to be useful.

Pores

Pores are the termination of the sweat ducts on the surface of the skin. They can sometimes be seen in optical images and are generally even more pronounced on sensors using electronic imaging techniques. Pores are known to be stable over time and are occasionally used in forensic fingerprint identification. Figure 13.13a shows the pores on an image taken with an AuthenTec RF E-field fingerprint sensor. Notice that strong pores are clearly visible even at 250 ppi resolution using an RF E-field sensor.

Ridge Detail Shape and Thickness Variations

Ridge detail, shape, and thickness variations are other types of fine structure that are fairly stable and can be used for differentiation in fingerprint matching. The detail structure of the area around the core, deltas, and minutia points are sometimes used because these regions are easy to locate and often contain a lot of entropy in the detail. Alternatively areas not containing large-scale structure can be compared for detail structure when a large-scale structure is not available. Figure 13.13b shows some examples of ridge detail structure.

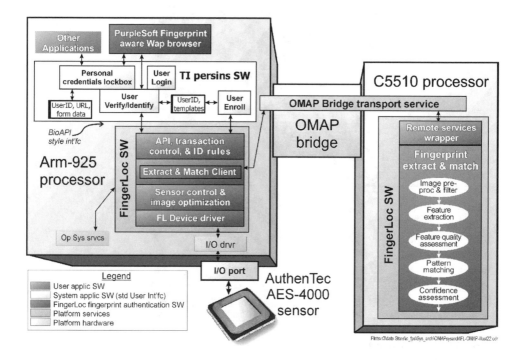

Figure 13.14   Fingerprint authentication prototype architecture for OMAP mobile systems

Multiple Algorithm Matching
Fingerprint matching systems often combine several algorithmic methods, such as those discussed above, to achieve high match accuracy. The best results are obtained when the selected methods use orthogonal features classes. Combining such statistically independent matching algorithms may increase confidence levels beyond that which is practical for a single algorithmic approach.

13.3.6 Prototype Implementation

13.3.6.1 Description
A prototype implementation of fingerprint recognition in a mobile device was constructed and demonstrated at the GSM conference in February 2001. The prototype demonstrated live real-time fingerprint authentication using a very small silicon fingerprint sensor (AuthenTec AES-4000) and a combination microcontroller + DSP development board (Texas Instruments' OMAP development board). The architecture of the prototype system is illustrated in Figure 13.14.

13.3.6.2 Functions

The demo system allowed enrolled users to use their fingerprint to access website accounts that were previously secured by standard personal passwords. For each user, the system stored the user's passwords and the associated website accounts in encrypted internal storage. The data becomes available and is sent to the website when the fingerprint is successfully verified.

Table 13.2   Component operating speeds in the OMAP prototype system and in the production OMAP 3.1 integrated part

System component	OMAP prototype	OMAP 3.1
AuthenTec AES-4000 sensor		
Frame rate (frames/s)	7	7
Comm speed (Kbytes/s)	100	100
ARM-925 processor		
Clock speed (MHz)	64	175
I/O bus speed (MHz)	12	12
C5510 DSP processor		
Clock speed (MHz)	40	200
System authentication latencies: (nominal)		
Image acquisition latency (s)	1	
Fingerprint verification matching (s)	0.5	
Total latency (s)	<2	

13.3.6.3 Architecture

Platform Performance

While the performance of the OMAP system was limited by its board level implementation of the interprocessor bridge, it was sufficient to demonstrate the basic functionality of system.

Component operating speeds in the OMAP prototype system and in the production OMAP 3.1 integrated part are shown in Table 13.2.

13.3.7 Prototype System Processing

13.3.7.1 Sensor Control and Image Optimization

Sensor Scanning and Data Acquisition

Most fingerprint readers have some form of an array of sensitive spots built into a silicon substrate that receives energy from the section of the finger placed over them. These spots called pixels serve as energy transducers; converting light, heat, or electric fields into electrical signals. Scanning a fingerprint requires that each of these transducers be read and the resulting signals measured and converted to digital form for use by a computer. The AuthenTec AES-4000 sensor used in the prototype has over 8000 of these pixels.

Sensor scanning is primarily controlled by logic internal to the AuthenTec AES-4000 sensor. The sensor detects the presence of a finger and begins image streaming. Array scanning logic and timing are internal to the sensor as well as all the timing and sequencing for the RF demodulation, analog data acquisition, and digital-to-analog conversion.

When using sensors that do not have internal scan control logic, the host CPU would control and sequence the detailed scanning and data acquisition process.

Gray Scale Normalization

Gray scale normalization is the process of adjusting the image so that the features of interest can be easily discerned, by positioning the information about those features uniformly across the gray scale range of the image. It is like adjusting the aperture and shutter speed of an optical camera, or the contrast and brightness of a CRT.

In the AuthenTec AES-4000 sensor the gray scale normalization task is split between the on-sensor logic and the host CPU.

Histogram Processing

While scanning the image, the sensor is also accumulating a gray scale histogram of that image. To reduce the time required for gray scale optimization the sensor initially sends only the histogram to the host without the image. The host analyzes the histogram as discussed below in AGC control, and sends the sensor new operating points to improve the gray scale. When the host determines that the histogram is close enough to optimal for use, it tells the sensor to start sending complete images along with the histograms.

AGC control

The AGC controls are implemented as software running in the ARM 925 processor. The current demo runs simplified control logic to reduce the load on the prototype board processor. In order to achieve the acquisition capabilities needed for general purpose mobile devices, implementations designed for full speed integrated OMAP processors will probably execute a more capable control system similar to the three-level hierarchical control system used in some of AuthenTec's PC based systems.

Spatial Filtering

In many systems, the fingerprint image is filtered to emphasize the features of interest and remove elements that will confuse feature recognition. The filters used vary depending on the class of features used for matching and on the type of artifacts and noise introduced by the sensor.

Edge Sharpening and Baseline Variation Removal

Edge sharpening and baseline variation removal are generally performed using some form of two-dimensional spatial bandpass filter. The AuthenTec sensor used in the prototype performs this function on-sensor using parallel processing hardware integrated into the sensor's pixel electronics. Removing undesired signal information at the sensor reduces the required channel bandwidth downstream of the sensor and reduces the processing work-load of the entire system.

Sensors that do not contain integrated spatial filtering often require digital spatial filtering as an early step in the image processing thread.

Artifact Removal

Many types of sensors generate artifacts in their output images that must be removed to avoid confusing the subsequent feature extraction and pattern recognition processes.

Fixed Pattern Noise

Fixed pattern noise is spatial signal variation that is stationary with respect to time. Fixed pattern noise comes primarily in two forms in fingerprint sensors: repetitive scanning noise and random pixel noise. Repetitive scanning noise is generated by differences in the data acquisition paths taken by data from the sensor's structural grouping of the pixels. It is typically visible as row artifacts, column artifacts, or region artifacts. Random pixel noise comes from random variation in the characteristics of the individual pixels. Removal of random pixel noise usually entails construction and maintenance of a detailed pixel perfor-mance map for each individual sensor.

For the prototype system using the AuthenTec AES-4000 sensor, no fixed pattern noise removal was necessary. Some of AuthenTec's higher performance PC based fingerprint systems use a scan noise reduction filter to clarify the images from extremely dry fingers. Similar filters may be used for the high performance systems envisioned for mobile devices.

Data Skew from Finger Movement

Sensors that require finger movement (such as swiping) during scanning must compensate for varying shift in finger position during the scan process. If the sensor is a narrow strip, multiple one-dimensional image strips must be collected. If their data is to be used collectively, spatial relationships between the strips must be established and the strips integrated into a single image. This partial image integration process is sometime called stitching.

The AuthenTec AES-4000 sensor uses a slap presentation rather than a swipe. No finger movement occurs during the scan hence it does not require data skew compensation or image stitching.

Time Varying Noise

Sensors may exhibit time varying noise for a wide variety of reasons. In single image snapshot systems little can be done to correct this. The same is generally true for swipe systems. Multiple image frames from streaming systems like the AuthenTec AES-4000 can be time-integrated to reduce the effects of time varying noise.

The prototype system did not require integration at the image frame level to reduce noise. This type of processing may become more important with lower voltage sensors that operate closer to their noise floor.

Image Feature Optimization

Sensors with very flexible operating characteristics can be adjusted to optimize the clarity of the specific feature type the matcher uses. This behavior can allow the sensor to acquire images of very unusual fingers that require operating points far removed from the typical norm. This type of adaptation requires that the feature extractor provide an estimate of the clarity of the features it is using, and that the sensor has an adjustment mechanism that affects the imaging of those features.

The AuthenTec TruePrint sensor technology offers several patented adaptive mechanisms for clarifying fingerprint ridges. Two of the most useful of these are the demodulation phase adjustment and the excitation signal frequency and amplitude adjustments.

Excitation Signal and Demodulation Phase Shift

The dead skin through which the signals pass in TruePrint sensors is an extremely complicated electrical structure. When presented with an AC excitation signal, both real and imaginary currents can flow, and minute pockets of trapped saline may exhibit electrical-field saturation and hysteresis effects. These characteristics make different fingers respond differently to different excitation frequencies and amplitudes. Optimizing the frequency and amplitude of the excitation signal for best feature clarity can significantly improve the systems ability to acquire unusual fingers.

Synchronous demodulation schemes such as employed in the AuthenTec sensors allow those systems to differentiate the real and imaginary component of the measured signal by shifting the phase of the demodulation clock with respect to the excitation signal. This capability can be used to suppress the effects of surface perspiration on the clarity of the image.

The prototype system implemented a binary search through the demodulation phase shift space to clarify the images of sweaty fingers. The prototype enhanced very dry finger acquisition by adjusting the excitation frequency. It used two excitation frequencies, trying one first and if sufficient image quality could not be obtained trying the second.

13.3.7.1 Pattern Matching Algorithm

The pattern matching algorithm on the prototype system was limited to a single stage, single feature-type algorithm because of the limited computational performance of the OMAP concept demonstration board.

The monolithic silicon implementation of the OMAP architecture is expected to be capable of executing accurate multi-stage, multiple feature-type matching algorithms, that have demonstrated (on other computing platforms) the biometric performance levels we believe are required for mobile communications and information devices.

13.4 Conclusions

In this chapter we have presented two of the most promising techniques for user authentication within mobile handsets. Both speech and fingerprint techniques now exist and can be implemented using compute power that is presently available on modern DSPs. This chapter discussed the technical challenges faced by each method, and the engineering trade-off decisions that must be made in order to obtain the desired level of performance. The use of such technology is now dependent on our ability to integrate it into the handset in a user friendly and efficient manner

Reliable user identification capabilities working together with secure communications capabilities can allow the mobile wireless device to become a highly capable and secure personal controller for the electronic systems that serve us. Whole new classes of applications programs on the mobile device platform can now perform secure remote control of everything from our investment portfolios to the locks on our homes and cars.

References

[1] Jain, A., Hong, L. and Pankanti, S., 'Biometric Identification', *Communications of the ACM*, Vol. 43, No. 2, February 2000, pp. 90–98.
[2] Naik, J.M., 'Speaker Verification: A Tutorial', *IEEE Communications Magazine*, Vol. 28, January 1990, pp. 42–48.
[3] Furui, S., *Digital Speech Processing, Synthesis, and Recognition*, Marcel Dekker, New York, 2001.
[4] Rabiner, L. and Schafer, R.W., *Digital Processing of Speech Signals*, Prentice-Hall, Englewood Cliffs, NJ, 1978.
[5] Allen, J.B., 'Cochlear Modeling', *IEEE ASSP Magazine*, Vol. 2, January 1985, pp. 3–29.
[6] Markel, J.D. and Gray Jr., A.H., *Linear Prediction of Speech*, Springer-Verlag, New York, 1976.
[7] Mammone, R.J., Zhang, X. and Ramachandran, R. 'Robust Speaker Recognition', *IEEE Signal Processing Magazine*, Vol. 13, September 1996, pp. 58–71.
[8] Furui, S., 'Cepstral Analysis Technique for Automatic Speaker Verification', *IEEE Transactions on Acoustics, Speech, and Signal Processing*, Vol. 29, April 1981, pp. 254–272.
[9] Reynolds, D., 'Speaker Identification and Verification using Gaussian Mixture Speaker Models", 1994 ESCA Workshop on Automatic Speaker Recognition, Identification, and Verification, pp. 27–30.
[10] Rabiner, L.R. and Juang, B.H., *Fundamentals of Automatic Speech Recognition*, Prentice Hall, Englewood Cliffs, NJ, 1993.
[11] Liu, C.S., Wang, H.C. and Lee, C.H., 'Speaker Verification using Normalized Log-Likelihood Score', *IEEE Transactions on Speech and Audio Processing*, Vol. 4, January 1996, , pp. 56–60.
[12] Netsch, L.P. and Rao, K.R., 'Robust Speaker Verification using Model Decorrelation and Neural Network Post-Processing', *Proceedings of ISPACS '98*, Melbourne, Australia, November 1998.

14

The Role of Programmable DSPs in Digital Radio

Trudy Stetzler and Gavin Ferris

14.1 Introduction

Existing AM and FM broadcasting have remained relatively unchanged since the 1960s when FM stereo transmission was introduced. Meanwhile, audio recording techniques have undergone tremendous change from traditional analog to high quality digital recording with the introduction of compact disc, and most recently MP3 compressed audio recording to permit music be transmitted via the Internet. Traditional analog broadcasts were originally designed for stationary receivers and suffer from degradation of the received signal when used in a mobile environment with weak signal strength and multipath. The listener typically experiences these deficiencies as pops and dropouts caused by selective fading, distortion of a weak signal caused by a strong interferer or multipath reflections, or bursts of noise caused by electrical interference.

The impact of digital technology on broadcast radio will be as significant as it was for cellular phones. Digital broadcasting offers the opportunity for the broadcaster to deliver a much higher quality signal to home, portable, and automotive receivers. The key features for a Digital Audio Broadcast (DAB) service are to provide near CD quality sound, high immunity to multipath and Doppler effects, spectrum efficiency, and low cost receivers. A digital transmission system also enables a new range of data services to complement the audio programming since it is essentially a wireless data pipe. The new services can include simple text or graphics associated with the program or independent services such as news, sports, traffic, and broadcast websites. These new services will enable broadcasters to compete effectively against products delivered via the Internet and cellular services.

Digital radio can be broadcast over several mediums including RF via transmission towers, satellite delivery, or cable systems. The evolution of a single DAB standard is difficult since broadcasting is regulated by governments and driven by commercial interests. A worldwide allocation of frequency spectrum would simplify receiver design and lower costs, but this is unlikely to occur. There are several DAB technologies either existing or in development today. In 1992, the World Administrative Radio Conference (WARC'92) allocated 40 MHz of spectrum on a worldwide basis in the 1.452–1.492 GHz band for satellite and complementary digital audio broadcast services. The European Telecommunications Standards Institute (ETSI) standardized Eureka 147 as ETS 300401 in 1995. It is the standard for Europe, Canada, parts of Asia, and Australia and is being evaluated in many other countries. It includes operating modes for terrestrial, satellite, and cable delivery. Although Eureka-147 is deployed in several other countries, the United States decided to pursue a different approach, due to frequency spectrum concerns and also partially due to pressure from commercial broadcasters. The US currently has two different DAB systems. The first consists of two satellite systems in the 2.3-GHz S-band and targets national coverage. The second is In-Band On-Channel (IBOC) which broadcasts the digital signal in the sidebands of the existing analog AM/FM signal and will provide local coverage once completed. There is also a second IBOC digital radio standard proposed by Digital Radio Mondiale (DRM) for frequencies below 30 MHz (essentially the AM band). In addition, WorldSpace is a satellite system offering services to Africa, Asia and Latin America using the L-band.

14.2 Digital Transmission Methods

The main problem conventional analog radio suffers from is signal corruption due to channel effects. There are three main categories of effect:

1. Noise: overlaid unwanted signals that have nothing to do with the desired transmission (for example, additive white Gaussian noise).
2. Shadowing, where the wanted signal is attenuated, for example by the receiver going into a tunnel.
3. Multipath fading occurs when delayed signals combine at the receiver with the line of sight signal, if it has not been attenuated.

Delayed signals are the result of reflections from fixed terrain features such as hills, trees or buildings, and moving objects like vehicles or aircraft. The signal delays can vary from 2 to 20 µs and will enhance signal strength at some frequencies while attenuating others by as much as 10–50 dB. When the receiver or its environment is moving, the multipath interference changes with time and creates an additional amplitude variation in the received signal (Figure 14.1) [1]. These channel effects create noise, distortion, and loss of signal in conventional

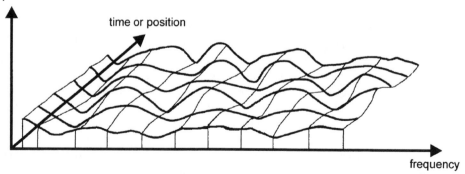

Figure 14.1   Multipath propagation effects on the frequency response of a channel

analog broadcasts. Simply increasing the transmitted power is not a solution because both the direct and reflected signals will increase proportionally, thus preserving the nulls [23].

Since terrestrial digital broadcasting systems operate in the same RF environment as analog systems, they must use a transmission method that reduces the effects of multipath propagation, Doppler spread, and interference. The goal is to develop a system that maintains a sufficient Bit Error Rate (BER), reasonable transmitted power, high data rates, and occupies a small bandwidth.

Digital radio systems (such as the Eureka 147 system described below) employ a number of different techniques to counter the channel effects described above, for example:

1. Forward error correction (FEC): by adding structured redundancy to the signal, the receiver may be able to infer the correct message (or at least, that an error has occurred), despite corruptions imposed by the channel.
2. Wide bandwidth: by utilizing a signal width greater than the channel coherence bandwidth, a degree of frequency diversity is obtained. A modulation system such as Orthogonal Frequency Division Multiplexing (OFDM), described below, is a natural way to utilize a wideband signal.
3. Interleaving of the data across multiple frequencies, to de-cohere frequency specific corruptions.
4. Interleaving of the data across time, to de-cohere temporally-specific corruptions (e.g. lightning or driving through an underpass).
5. Positive utilization of multipath diversity, to reduce the dependence upon any single path, which helps obviate the effects of shadowing.
6. Use of modulation schemes with maximized decision distances between their various valid states, to allow good performance even in the presence of significant noise.

Note that simply changing to a digital system does not by itself solve the multipath problem, although the use of channel coding can significantly mitigate it. Similarly, extending the symbol period of the digital data can reduce the impact of channel interference, Inter-Symbol Interference (ISI), and multipath effects, but entails a reduction in the symbol rate. For a single narrowband signal, a data rate that is sufficiently low to ensure an acceptable BER at the receiver is insufficient for a high quality audio service. One method of obtaining a sufficient data rate is to use Frequency-Division Multiplexing (FDM) where the data is distributed over multiple carriers. Since the data signal occupies a large portion of the bandwidth, there is less chance that the entire signal will be lost to a severe multipath fade on one carrier frequency [1,17,23]. The detailed implementation of the digital transmission will be discussed for the Eureka 147 standard, but similar techniques are used for satellite systems and the proposed terrestrial digital standards for the US.

14.3 Eureka 147 System

14.3.1 System Description

The Eureka 147 DAB standard can be implemented at any frequency from 30 MHz to 3 GHz and may be used on terrestrial, satellite, hybrid (satellite with complementary terrestrial), and cable broadcast networks [24]. The Eureka 147 system uses Coded Orthogonal Frequency Division Multiplexing (COFDM), which is a wideband modulation scheme specifically designed to cope with the problems of multi-path reception. COFDM achieves this by spreading the data across a large number of closely spaced carriers. Since the COFDM carriers are orthogonal, the sidebands of each carrier can overlap and the signals still received without adjacent carrier interference. The receiver functions as a bank of OFDM demodulators, translating each carrier down to DC, and then integrating over a symbol period to recover the raw data. Since the other carriers all translate down to frequencies which, in the time domain, have a whole number of cycles in the symbol period (t_s), the integration process results in zero contribution from all these other carriers. As long as the carrier spacing is a multiple of $1/t_s$, the carriers are linearly independent (i.e. orthogonal). Since any non-linearity causes Inter-Carrier Interference (ICI) and damages orthogonality, all hardware must have linear characteristics [1].

Shown mathematically, the set of normalised signals $\{y\}$ where y_p is the pth element of $\{y\}$, are orthogonal if

$$\int_a^b y_p(t)y_q^*(t)\mathrm{d}t = 1 \quad \text{for } p = q$$

$$= 0 \text{ for } p \neq q$$

where the * indicates the complex conjugate. The use of a regular carrier spacing enables the signal to be generated in the transmitter and recovered in the receiver using the Fast Fourier Transform (FFT).

Although many modulation schemes could be used to encode the data onto each carrier, Phase-Shift Keying (PSK) modulation yields the lowest BER for a given signal strength. In Eureka 147, Differential Quadrature Phase Shift Keying (DQPSK) is used where four phase changes are used to represent two bits per symbol (see Table 14.1) so the symbol rate is half the transmission rate [15].

Table 14.1 DQPSK encoding

Phase change	Encoded Data
0	00
$\pi/2$	01
π	10
$3\pi/2$	11

Multipath interference distorts the received phase of the symbol for each spectral component. As long as the channel is not changing rapidly, successive symbols of any one carrier will be perturbed in a similar manner. Since DQPSK encoding is used, the receiver looks at the difference in phase from one symbol to the next and these errors cancel out, eliminating the need for channel equalization [1].

For mobile receivers, multipath interference changes rapidly and this can cause problems. Since multipath propagation results in multiple reflections at the receiver arriving at different times, it is possible for a symbol from one path (with short delay) to arrive at the same time as the previous symbol from another path (with long delay). This creates a situation known as

ISI and limits the digital system's symbol rate. Eureka 147 overcomes ISI by adding a guard interval of 1/4 of the symbol time to each symbol, which decreases the overall rate [17]. This guard interval is actually a cyclic prefix – in effect, a copy of the last 1/4 of each symbol appended to the front of it, for several reasons. First, to maintain synchronization, the guard interval cannot simply be set to zero. Second, inserting a cyclic prefix that uses data identical to that at the end of the active symbol avoids a discontinuity at the boundary between the active symbol and the guard interval. The duration of the receiver's active sampling window corresponds to the *useful* symbol period, which remains the reciprocal of the carrier spacing and thus maintains orthogonality. The receiver window's position in time can vary by up to the cyclically extended guard interval and still continue to recover data from each symbol individually without any risk of overlap (Figure 14.2) [1,17]. Further, the guard interval can be used to do a channel estimation on a symbol by symbol basis.

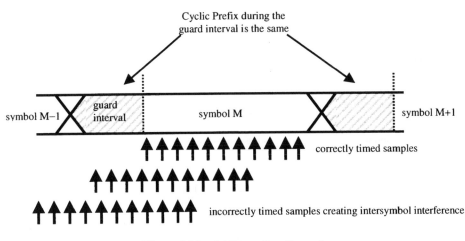

Figure 14.2   Addition of cyclic prefix

For Eureka 147, a special feature called the Single Frequency Network (SFN) is used to increase the spectrum efficiency. A broadcast network can be extended across a country by operating all transmitters on the same radio frequency with the same programming. The transmitters are synchronized in time and frequency using Geographical Positioning System (GPS) clocks. When an identical signal is transmitted from a nearby and a distant transmitter, a receiver would receive two signals – one much delayed compared to the other. However, this case is indistinguishable from a genuine long-delay echo from the nearby transmitter. Provided that the delay did not exceed the guard interval, the receiver would be able to decode the received signal successfully. The guard interval, carrier spacing and operating frequency determine the system tolerance of ISI and therefore the maximum spacing for the transmitters. The carrier separation is a major factor in the immunity of the system to the effects of Doppler spread in mobile receivers [17]. There are four different transmission modes for Eureka 147 modes as shown in Table 14.2 [2].

All the modes have the same spectral occupancy (approximately 1.5 MHz, determined by the number of subcarriers and the spacing between them), and the system operation of each is essentially the same. The choice of transmission modes is dependent on the system imple-

Table 14.2 Definition of the parameters for transmission modes I, II, III and IV

Description	Parameter	Transmission mode I	Transmission mode II	Transmission mode III	Transmission mode IV
Number of OFDM symbols per transmission frame (excluding null symbol)	L	76	76	153	76
Number of transmitted carriers	K	1536	384	192	768
Frame duration	T_F	196608 T, 96 ms	49152 T, 24 ms	49152 T, 24 ms	98304 T, 8 ms
Null symbol duration	T_{NULL}	2656 T, ~1297 μs	664 T, ~324 μs	345 T, ~168 μs	1328 T, ~648 μs
Total symbol duration	T_S	2552T, ~1246 μs	638 T, ~312 μs	319 T, ~156 μs	1276 T, ~623 μs
Useful symbol duration	T_u	2048 T, 1 ms	512 T, 250 μs	256 T, 125 μs	1024 T, 500 μs
Guard interval duration	Δ	504 T, ~246 μs	126 T, ~62 μs	63 T, ~31 μs	252 T, ~123 μs
Nominal maximum transmitter separation for SFN (km)		96	24	12	48
Subcarrier spacing (kHz)	Δf	1	4	8	2
Nominal frequency range		≤375 MHz	≤1.5 GHz	≤3 GHz	≤1.5 GHz

mentation. Transmission mode I is intended to be used for terrestrial SFN and local-area broadcasting in VHF Bands I, II and III. Transmission modes II and IV are intended to be used for terrestrial local broadcasting in VHF Bands I, II, III, IV, V and in L-band. They can also be used for satellite-only and hybrid satellite–terrestrial broadcasting in the L-band. Transmission mode III is intended to be used for terrestrial, satellite and hybrid satellite–terrestrial broadcasting below 3000 MHz. Transmission mode III is the preferred mode for cable distribution since it can be used at any frequency available on cable [2]. (The ability of an OFDM/DQPSK system to operate in the presence of a given frequency shift is directly proportional to the inter-carrier frequency spacing, which in turn is inversely proportional to the number of carriers employed. Hence Mode I is the most sensitive to frequency errors, followed by modes IV, II and III respectively). In a single frequency network environment, the maximum possible separation of transmitters is constrained by the guard interval size; this is approximately 1/4 of the useful symbol length for Eureka 147, hence mode I allows the most widely distributed SFN, followed by IV, II and III respectively, with the last mode being useful only where there is effectively little multipath.

While OFDM with a guard interval minimizes the effects of ISI, multipath interference will still cause the attenuation of some of the OFDM carriers resulting in lost or corrupted data bits. It is important that this does not create any distortions in the audio signal. By using an error-correcting code, which adds structured redundancy at the transmitter, it is possible to correct many of the bits that were incorrectly received. The information carried by one of the degraded carriers is corrected because other information, which is related to it by the error correction code, is transmitted on different carrier frequencies [1].

Eureka 147 uses a channel coding based on a convolutional code with constraint length 7. A kernel using four polynomials is used, with puncturing allowing a variety of less redundant derivative codes to be used. The mother convolutional encoder generates from the vector $(a_i)_{i=0}^{I-1}$ a code word $\{(x_{0,i}, x_{1,i}, x_{2,i}, x_{3,i})\}_{i=o}^{I+5}$, where the codeword is defined by:

$$x_{o,i} = a_i \oplus a_{i-2} \oplus a_{i-3} \oplus a_{i-5} \oplus a_{i-6};$$

$$x_{1,i} = a_i \oplus a_{i-1} \oplus a_{i-2} \oplus a_{i-3} \oplus a_{i-6}$$

$$x_{2,i} = a_i \oplus a_{i-1} \oplus a_{i-4} \oplus a_{i-6}$$

$$x_{3,i} = a_i \oplus a_{i-2} \oplus a_{i-3} \oplus a_{i-5} \oplus a_{i-6}$$

for $i = 0, 1, 2,..., I + 5$.

When i does not belong to the set $\{0, 1, 2, ...I - 1\}$, a_i is equal to zero by definition. The encoding can be achieved using the convolutional encoder shown in Figure 14.3. The octal forms of the generator polynomials are 133, 171, 145 and 133, respectively [2]. This type of coding is conventionally removed at the receiver using a Viterbi decoder, which works best if the errors in the sequence presented to it are "peppered" throughout the input vector, rather than clustered together. To overcome the fact that error-inducing channel effects are likely to show strong frequency and/or time coherence, *interleaving* is applied. Frequency interleaving is used to distribute bit errors associated with a particular range of frequencies within the COFDM spectrum caused by narrow-band interference. Time interleaving is used to distribute errors that affect all carriers simultaneously, such as a rapid reduction of signal strength caused by an overpass. The time interleaving process for the digital radio system covers 16 frames (of 24 ms each) which results in an overall processing delay of 384 ms. Figure 14.4 shows the relationship of the COFDM spectrum in frequency and time domains. The convolutional

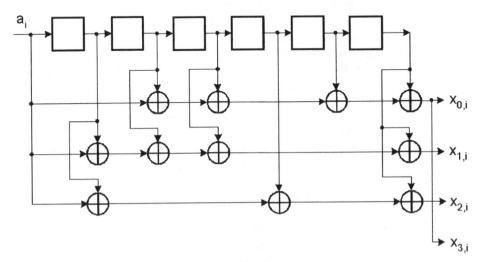

Figure 14.3 Convolutional encoder

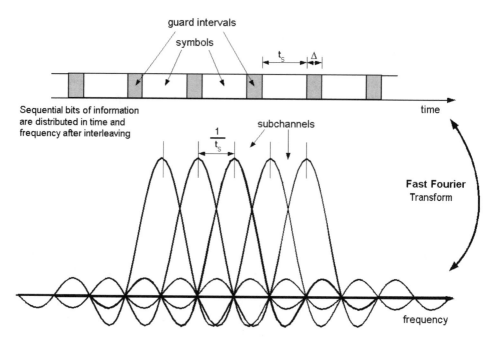

Figure 14.4 Representation of COFDM signal in frequency and time

coding parameters depend on the type of service carried, the net bit rate, and the desired level of error protection. Two error protection procedures are available: Unequal Error Protection (UEP) and Equal Error Protection (EEP). UEP is primarily used for audio since some parts of the audio frame are less sensitive to transmission errors and can tolerate less redundancy than the critical data, such as headers (see Figure 14.5) [2]. EEP is typically reserved for data services since the frame content is unknown, although it can also be used for audio services.

A consequence of the fact that the Eureka 147 transmission system is designed to allow operation as an SFN, and to utilize an efficient, wideband signal, is that it must be able to broadcast several digital services (audio, data, or a combination of both) simultaneously. Therefore, source coding is required to reduce the bandwidth required while maintaining the audio quality. The choice of source coding for Eureka 147 is independent of the choice COFDM for the modulation scheme. Eureka 147 uses MPEG-1/2 Layer II psychoacoustical coding, also known as Masking pattern Universal Sub-band Integrated Coding And Multiplexing (MUSICAM) [14,15].

Perceptual coders are not concerned about the absolute frequency response or dynamic range of hearing, but rather the ear's sensitivity to distortion. MUSICAM relies on the spectral and temporal masking effects of the inner ear. Masking occurs in auditory perception when the presence of one sound raises the threshold required to perceive other nearby sounds. The principle of audio masking is shown in Figure 14.6 [14,15]. The 1 kHz tone raises the audible threshold required for other signal components to curve B (the masking threshold). If a second audio component is present at the same time and close in frequency to the 1 kHz tone, then for the second component to be perceived by the ear, it must be loud enough to exceed the higher masking threshold than it would otherwise need to be if no other sounds

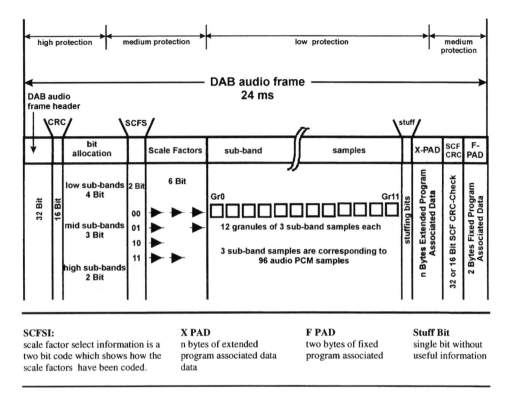

SCFSI:
scale factor select information is a
two bit code which shows how the
scale factors have been coded.

X PAD
n bytes of extended
program associated data
data

F PAD
two bytes of fixed
program associated

Stuff Bit
single bit without
useful information

Figure 14.5   Structure of Eureka 147 audio frame

were present (curve A). The MUSICAM system divides the audio spectrum into 32 equally spaced sub-bands and then requantizes these bands. During the bit allocation process of the requantizing procedure, fewer bits are allocated to components of the spectrum that are effectively masked. This enables a high subjective audio quality to be maintained while conserving valuable bit rate capacity [14,15,18].

The MUSICAM audio coding process can compress digital audio signals to one of a number of possible encoding options in the range 8–384 kbit/s, at a sampling rate of 48 or 24 kHz (if a service can tolerate the limited frequency response). The coding option selected for a given service will depend on the audio-quality required – for example high quality stereo is typically encoded at 128 kbit/s and higher whereas mainly speech based services are encoded at lower rates, typically less than 96 kbit/s. The international standard ISO 11172-3 defines four different coding modes for MPEG 1: stereo, mono, dual channel (two independent mono channels) and joint stereo (where only one channel is encoded for the high frequencies and a pseudo-stereophonic signal is reconstructed using scaling coefficients) [3,14].

The ISO standards only define the format of the encoded data stream and the decoding process. Therefore manufacturers can design their own improved psychoacoustic models and data encoders. In the receiver, psychoacoustic models are not required. The decoder only recovers the scale factors from the bit stream and then reconstructs the original 16-bit Pulse Code Modulation (PCM) samples [15,18,25].

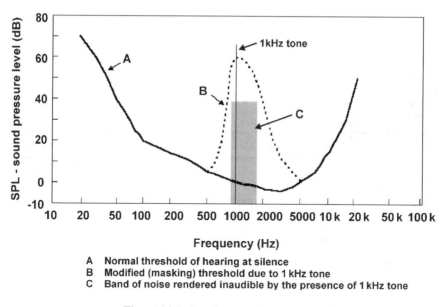

A   Normal threshold of hearing at silence
B   Modified (masking) threshold due to 1 kHz tone
C   Band of noise rendered inaudible by the presence of 1 kHz tone

Figure 14.6 Pyschoacoustical masking

The digital radio data frame format for MPEG audio is shown in Figure 14.5 [2]. In Layer I the audio data corresponds to 384 PCM samples and has an 8-ms frame length. In Layer II the audio corresponds to 1152 PCM samples and has a frame length of 24 ms. The 32-bit header contains information about synchronization, which layer, bit rates, sampling rates, mode, and pre-emphasis. This is followed by a 16-bit Cyclic Redundancy Check (CRC) code. The audio data is followed by ancillary Program Associated Data (F-PAD and X-PAD). The ISO Layer I and II audio data frames contain information about bit allocation to the different sub-bands, scale factors and the sub-band samples themselves [15].

At the receiver, it is possible to transform the MUSICAM audio frames into the more widely commercially adopted, and higher density, MP3 format, without first regenerating the PCM audio. This is possible because the two format share a common filterbank structure, and furthermore, it is possible to drive the MP3 psychoacoustic model using the quantization levels utilized for the MUSICAM frame. More details of this technique are available in Ref. [4].

14.3.2 Transmission Signal Generation

The Eureka 147 transmission signal occupies a bandwidth of 1.536 MHz and simultaneously carries several digital audio and data service components. The gross transmission data rate is 2.304 Mbps, and the net bit rate varies from 0.6 to 1.8 Mbps depending on the convolution code rates utilized across the various services. Each service can potentially have a distinct coding profile if necessary, although the use of half rate coding is widely adopted as a de facto standard in practice. Typically, the useful bit rate capacity is approximately 1.2 Mbps and consists of eight to ten radio services which can include a combination of service components consisting of audio primary and secondary channels, PAD for the audio channel, and packet or streamed data channels. The PAD channel is incorporated at the end of the Eureka 147 audio frame, so it is

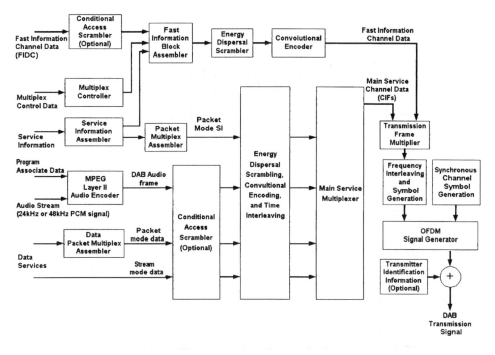

Figure 14.7   Conceptual block diagram

synchronous with the audio. PAD data rates ranges from 667 bps (F-PAD) to 65 kbps (X-PAD), and typical applications are dynamic labels (DLS), graphics, and text information [24].

The Eureka 147 signal-generation path comprises multiple signal-processing blocks for audio and data coding, transmission coding and multiplexing, frequency interleaving, and modulation. Figure 14.7 shows a conceptual block diagram of the Eureka 147 transmission system [2]. Each service signal is coded individually at source level, and then each can have Conditional Access (CA) applied if so desired. The optional CA system operates on both the transmitter and receiver and enables restricted access programming through scrambling/ descrambling, entitlement checking and entitlement management. The data is then scrambled in frequency (energy dispersal block) by a module 2 addition with a Pseudo-Random Binary Sequence (PRBS) defined by the 9th degree polynomial $P(x) = X^9 + X^5 + 1$. The initialization word is applied such that the first bit of the PRBS is obtained when the outputs of all the shift registers stages are set to "1" as shown in Figure 14.8 [2]. This step reduces the possibility of systematic patterns in the data and improves the efficiency of the transmit power amplifier. The data is then convolutionally encoded (with possibly a different rate for each service) and time interleaved as explained previously. Then the services are multiplexed into the Main Service Channel (MSC) according to a pre-determined, but adjustable, multiplex configuration. The multiplexer output is combined with multiplex control and service information, which travel in the Fast Information Channel (FIC), to form the transmission frames in the transmission multiplexer. The fully multiplexed information is then divided into a large number of bit-streams, which are DQPSK modulated onto individual orthogonal carriers generated via an IFFT. This digital signal is then converted to analog and pulse-shaped to limit the bandwidth using raised cosine filters so that the interference from

Initialization word

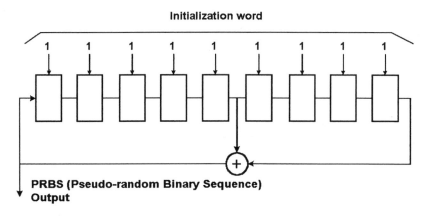

PRBS (Pseudo-random Binary Sequence)
Output

Figure 14.8   PRBS generator

each bit is nulled at the center of other bit intervals, removing in-band ISI (distinct from reflective path ISI, discussed earlier). The signal is then transposed to the appropriate radio frequency band, amplified and transmitted [23,24].

OFDM has a relatively high peak-to-mean ratio, which must be taken into account when considering the upconversion and amplification of the signal to air. The peak-to-mean ratio of a time domain Eureka 147 DQPSK signal is approximately 3:1. Therefore, many modern broadcast transmitters utilize pre-distortion in the digital domain to counterbalance the non-linear performance of the amplifiers used.

Each transmission frame follows a fixed format that allows receivers to synchronize and extract data (see Figure 14.9) [2]. The frame begins with a null symbol (no RF signal transmitted) for coarse receiver synchronization, followed by a phase reference symbol for differential demodulation. Transmitter-Identification-Information (TII) data may also be included. The next symbols are reserved for the FIC and the remaining symbols provide the MSC. The total frame duration is 96, 48 or 24 ms, depending on the transmission mode. Each service

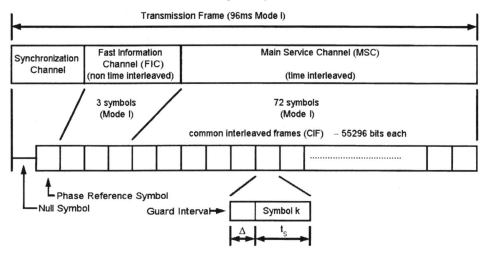

Figure 14.9   Eureka 147 transmission frame

within the MSC is allocated a fixed time slot in a "Common Interleaved Frame" (CIF), which represents 24 ms (55296 bits) of data from each of the subchannels. The transmission frame contains one, two or four CIFs, depending upon the mode. Since there are 72 OFDM symbols for the MSC (in mode I), the data channel (including error correction) is 2.304 Mbps, and each of the four CIFs is mapped to 18 subsequent OFDM signals. This data capacity is organized as Capacity Units (CUs) of 64 bits each, which are purchased by the various services to accommodate their desired subchannel data rate and protection level. Note that for Eureka 147, given a desired subchannel, it is possible for the receiver to determine which symbols it must decode. This feature makes it possible to design highly power-efficient receivers, which can, e.g. shut down parts of the RF during "un-interesting" symbols, or amortize processing in time to reduce the baseband processor's overall clock rate. The FIC contains the Multiplex Configuration Information (MCI) which defines the organization of the subchannels, services, service components, and controls re-configurations. The FIC is not time interleaved to allow rapid access to the MCI for fast tuning of the receiver. The Eureka 147 multiplex can quickly and seamlessly re-allocate the available capacity between services to introduce temporary services or to allow more data capacity at night with lower audio quality – a process known as reconfiguration. The transmitted output spectrum is shown in Figure 14.10 [2].

14.3.3 Receiver Description

14.3.3.1 Architectures and Algorithms
The receiver must perform the inverse of the transmission process just described. Although there are many ways to implement this solution, the fundamental processes and algorithms required are the same. A high-level block diagram of the receive signal path is shown in

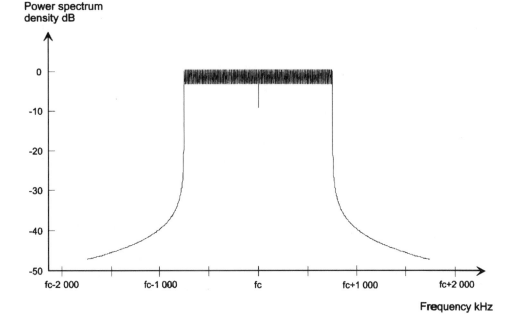

Figure 14.10   Theoretical Eureka 147 transmission signal spectrum (mode I)

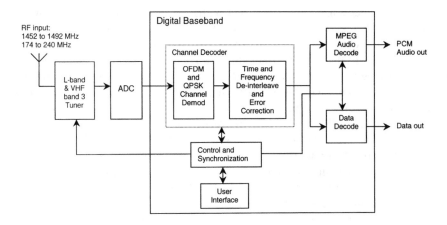

Figure 14.11   Block diagram of Eureka 147 receiver

Figure 14.11. The tuner amplifies the incoming RF analog signal, down-converts it to an Intermediate Frequency (IF), and then filters it prior to the A/D converter. Since this is a non-constant envelope system, the COFDM receiver front-end must be highly linear and requires automatic gain control to accommodate the wide dynamic range required. The ADC digitizes the analog signal so that the channel decoder can transform the signal back to the original Eureka 147 frame. Based on the user's program selection, the appropriate service component is sent to the audio and/or data source decoder. The audio decoder converts the MPEG 1 Layer II audio back to PCM audio for output to a DAC. The data decoder outputs packets or streaming data to an appropriate display or interface device.

The digital baseband of a Eureka 147 receiver requires many digital signal processing techniques. The main algorithms required by the digital baseband shown in Figure 14.11 consist of:

- Conversion to in-phase and quadrature signals at baseband
- OFDM and QPSK demodulation
- Automatic frequency control
- Acquisition and tracking (synchronization)
- Decoding of the FIC
- Time and frequency de-interleaving and Viterbi decoding
- Audio source decoding from MPEG 1 Layer II (MUSICAM) coded audio to linear PCM audio

Demodulation

The key to a COFDM system is its orthogonality. This enables the received data stream to be OFDM demodulated using the Discrete Fourier Transform (DFT). The DFT requires that the sampled signal is repetitive, which is achieved through the addition of the cyclic prefix described earlier. For mode I, the DFT operation samples the full OFDM input to generate 1536 DC components (the desired phase states). Since the DFT integrates over a time period equal to the inverse of the 1-kHz spacing, any harmonics integrate to zero, so only the desired DC voltages (phase states) remain. At the end of the DFT process, the output is the 1536 complex carriers, which are the known phase states.

The DFT is typically implemented in the receiver using the FFT, which demodulates the OFDM symbols to obtain the complex (I, Q) carriers. Common versions of the FFT operate on a group of 2^M samples per integration period and output the same number of frequency coefficients (demodulated data). In order to sample above the Nyquist limit and maintain 2^M samples, the FFT size selected is the next higher 2^M size and the output samples not corresponding to active carriers are set to zero. The capacity of the system remains the same and this also enables the analog filtering requirements to be reduced. The demodulator block must perform different FFT sizes for the four different modes in the Eureka 147 standard, with mode I being the most difficult to implement since the 1536 carriers are spaced at 1 kHz. Modes I, II, III and IV require a FFT size N of 2048, 512, 256, 1024 respectively. In order to maintain real time processing on a symbol basis, the 2048-point FFT must finish within one OFDM symbol period which is 1.246 ms (including the guard interval Δ) for mode I. The FFT operation is defined as [16]

$$X_{k,l} = \frac{1}{\sqrt{N}} \sum_{n=-w}^{N-w-1} x_{n,l} e^{-j2\pi \frac{kn}{N}}$$

$$= \frac{e^{-j2\pi \frac{kw}{N}}}{\sqrt{N}} \sum_{n=0}^{N-1} x_{n,l} e^{-j2\pi \frac{kn}{N}}$$

where: X is the OFDM carrier spectrum; x is the baseband signal; k is the OFDM carrier index $(-K/2...K/2)$ where the $k=0$ carrier is excluded; l is the OFDM symbol index $(0...L)$; n is the sample index current symbol $(0...N-1)$; w is the start of the FFT demodulation window $(0...\Delta - 1)$.

Following the FFT, differential demodulation of the DPQSK carriers is performed via a complex multiplication with the complex conjugate of the previous OFDM

$$Y_{k,j} = X_{k,j} X_{k,j-1}^*$$

This operation requires both the current and previous symbol, so both results must be stored in memory. The resulting differential constellation diagrams are quantized into a soft decision bit stream, in which each ordinate bit has added to it a number of supporting confidence bits. One possible mapping of the demodulated amplitudes to soft decision confidence bits is shown in Figure 14.12. Here the $+3$ represents a "confident" logic 1 and -3 represents a "confident" logic 0. The $+2$, $+1$, -2, and -1 represent progressively less confident logic 1s and 0s, respectively. The 0 level does not represent any logic value. This output is provided to a soft-decision channel Viterbi decoder in order to improve the coding gain.

Synchronization

In order to accurately perform the FFT, the receiver local oscillator must be synchronized to the transmitter local oscillator. As was shown in Figure 14.9, the Eureka 147 system provides two methods for receiver synchronization. The first is the null symbol, in which all the OFDM carriers are turned off (except for occasional transmission of a TII code, which uses a small subset "comb" of carriers, but does not significantly add to the signal energy in the null symbol), and the second is the Phase Reference Symbol (PRS), in which all the carriers have pre-determined phases. The receiver uses a coarse synchronization algorithm to calculate the short term and long term average energy of the baseband signal to find the null symbol and

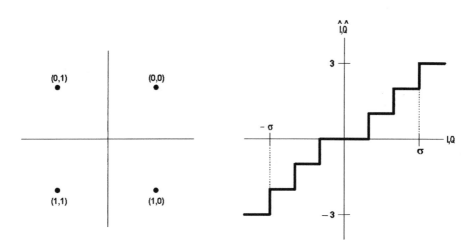

Figure 14.12   Soft Viterbi mapping

generate a timing estimate. This information can also be used to generate an Automatic Gain Control (AGC) value for scaling in the FFT. Since the timing will not be perfect, all of the samples could be displaced by a fixed time offset, which would create inter-symbol interference [17,26].

Once coarse synchronization is obtained, the PRS (containing Constant Amplitude Zero Autocorrelation (CAZAC) sequences) is used for the fine time synchronization and frequency correction. A digital AFC loop can be implemented using the PRS sequences to provide a frequency correction range of approximately $\pm$ 32 carriers [26]. The time synchronization can be performed by correlating the actually received and demodulated PSR with the known transmitted PSR version stored in the receiver

$$h_n = \frac{1}{N} \sum_{k=-K/2}^{K/2} X_{k,1} Z_k^* e^{j2\pi\frac{kn}{N}}$$

$$= \frac{1}{N} \sum_{k=-K/2}^{K/2} Z_k e^{-j2\pi\frac{kw}{N}} Z_k^* e^{j2\pi\frac{kn}{N}}$$

$$= \frac{1}{N} \sum_{k=-K/2}^{K/2} e^{-j2\pi\frac{k(n-w)}{N}}$$

$$= \delta(w)$$

where Z_k is the reference symbol as inserted at the transmitter side; $X_{k,1}$ is the received reference symbol which, without distortion is the same as the transmitted reference symbol Z_k, demodulated by an FFT with offset. In the case of an ideal transmission channel, this time domain correlation results in a single correlation peak δ at the offset position w of the demodulation window. The resulting h_n can be shown to be the Channel Impulse Response (CIR) [16].

For receiver synchronization in a multipath environment, the CIR peak-detection algorithm is one way of adjusting the FFT demodulation window timing to minimize ISI [16]. For a Gaussian or Rayleigh channel, the FFT window is adjusted to the largest CIR peak position, which is usually the first peak. For a Ricean channel with no line-of-sight reception, the window is adjusted relative to the first well formed peak. In the event there is no dominant peak or multiple peaks (due to reflections), the goal is to maximize the impulse energy (area under the CIR curve) which falls within the non-sampling period of the symbol, thereby minimizing the ISI which falls inside the critical sampling period. The PRS provides the phase reference for the following symbol, which carries active data. The frequency information from the sync symbol is also used to correct the down conversion local oscillator and the sampling clock rate of the analog to digital converters as well as the FFT. Since there are many possible algorithms for synchronization with each receiver manufacturer having a preferred implementation, a programmable Digital Signal Processor (DSP) is ideally suited for this application. Fading profiles for the Eureka 147 system are described in Ref. [5].

Equalization

When the OFDM symbols arrive at the receiver, they have been convolved with the time-domain channel impulse response. However, the data are transformed back into the frequency-domain by the FFT in the receiver. Due to the periodic nature of the cyclically-extended OFDM symbol, this time-domain convolution will result in the multiplication of the frequency-domain constellation points of the OFDM signal with the frequency response of the channel. The result is that each subcarrier's symbol will be multiplied by a complex amplitude and phase distortion equal to the channel's frequency response at that subcarrier's frequency [13]. In order to undo these complex gain effects, a frequency-domain equalizer based on the PRS can be used. The frequency-domain one-tap equalizer would consists of a single complex multiplication for each subcarrier (four complex multiplies total), but it could also be integrated with the FFT. For the ideal case of no noise, the equalizer's frequency response is simply the inverse of the channel's frequency response. The use of differential modulation within the Eureka 147 system makes the issue of equalization much simpler, since any phase shift that is purely frequency dependent will be applied to both the target symbol and its prefix, thereby canceling out, and leaving only magnitude issues to be normalized by the equalization processor (which, in turn, given that the Eureka system does not utilize the magnitude information of its carriers directly, will be done primarily to ensure good dynamic range is maintained within the DSP algorithms).

De-Interleaving and Decoding

In order to improve performance in multipath environments, the transmitted OFDM signal was encoded using a convolutional code, and then the encoded bit stream was interleaved in time and frequency. This process must now be reversed in the receiver by removing the interleaving and using a Viterbi decoder to correct any errors in the received signal.

The frequency de-interleaving process performs a XOR function with the PRBS to descramble the energy dispersal of the data. Time de-interleaving and Viterbi decoding are based on a CU of 64 bits, and the main service channel (MSC) contains a maximum of 864 CUs per 24 ms CIF. Time de- de-interleaving is only required for the MSC subchannels, since

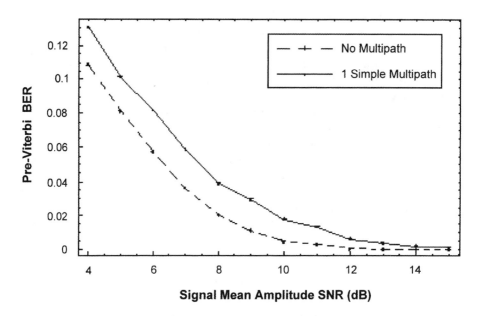

Figure 14.13   Pre-Viterbi bit error rate

the FIC is not time interleaved to allow fast receiver initialization. The data from the MSC subchannel is delayed across 0 to 15 CIFs, creating a reception delay of 15 frames × 24 ms/ frame (0.36 s) in the audio signal [16].

As described previously, Eureka 147 uses a convolutional channel code with constraint length 7. Puncturing enables a variety of code-rates to support various levels of error protection determined by the importance of the transmitted information and the channel characteristics. The Viterbi algorithm is used for decoding, and the decoder usually operates with 2–4 soft decision bits for better performance. Note, however, that Eureka 147 does not provide an external block code (e.g. a Reed–Solomon code), in the way that DVB-T does, which ultimately limits the performance of the system when carrying "pure" data. This issue can be handled either by utilization of a repetitive carousel for data services, the application-level use of a block code, or some combination of the above. Figure 14.13 shows the number of bits in error in a mode I frame into the Viterbi decoder, for a range of Signal-to-Noise Ratios (SNRs). SNR is measured by taking the ratio of the mean amplitude of the wanted signal to the mean amplitude of the unwanted imposed complex Gaussian noise. As can be seen from the graph, the limits of the decoder are around 8 dB SNR in a single path channel and 10 dB SNR in a simple (equal weight) multipath channel. The BER level of 2×10^{-2} is the level where when input to a half-rate Viterbi results in bits in error "breaking through" into the output of the Viterbi. The constellation diagram for a DQPSK signal with 8 dB SNR in complex Gaussian channel with no multipath is shown in Figure 14.14.

Audio Source Decoding
As explained previously, Eureka 147 uses MPEG-1 Layer II (MUSICAM) perceptual audio coding to reduce the data rate required for digital audio broadcast. 1152 samples are used to represent 24 ms of audio signal (at a sampling frequency of 48 kHz). Decoding is accom-

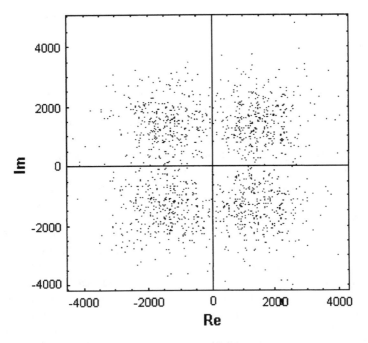

Figure 14.14 Constellation diagram

plished by first separating header information, side information, and sub-band samples and performing a CRC check. The side information containing the bit allocation information, quantization and scaling factor applied to each sub-band by the encoder is then decoded. The samples are requantized by multiplying them with the correct scale factors, which contain all the information necessary to recalculate the masking thresholds. The samples are applied to a synthesis reconstruction filter, which uses this information to recombine the 32 sub-bands into one broadband audio signal (see Figure 14.15 block diagram) [18,19,25]. This sub-band filter is the inverse of the encoding filter. After the sub-band filtering, the output is a linear PCM signal that is converted to an audio signal through a D/A converter. Since the decoder does not contain any psychoacoustic modeling, the encoder psychoacoustic model used in the transmitter can be improved and not affect the receiver's decoder but it will still result in

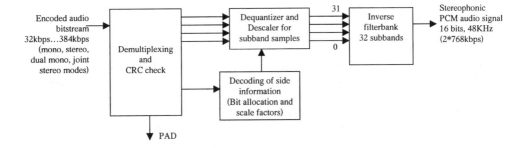

Figure 14.15 Audio MPEG Layer I or II decoder block diagram

improved receiver performance. Layer II coding can use stereo intensity coding, provide for a dynamic range control to adapt to different listening conditions, and uses a fixed-length data word [18,19,25].

Impact of Non-Idealities

Non-idealities in the receiver of an OFDM system will affect performance. These effects include local oscillator offsets, sampling frequency offsets, phase noise, and non-linearities. Some of these effects can be corrected with the timing and frequency correction algorithms described previously. If there is an offset in the synchronization loop, the receiver Local Oscillator (LO) frequency and transmitter LO will be different. The LO offset can be expressed mathematically by multiplying the received time-domain signal by a complex exponential whose frequency is equal to the LO offset amount, which is equivalent to a frequency shift of the received signal spectrum. This frequency shift causes the OFDM subcarriers to lose orthogonality, and the bins of the FFT no longer line up with the peaks of the received signal's sinc pulses. The result is ICI, which occurs when energy from one bin spills over into adjacent bins and this energy distorts the affected subcarriers. Since the zero-crossings of the non-zero subcarrier's spectrum no longer align with the FFT bins, energy from the non-zero subcarrier is spread out among the other subcarriers, with the closest subcarriers receiving the most interference. According to the central limit theorem, the sum of a large number of random processes will result in a signal that has a Gaussian distribution. As a result, the ICI appears as additive Gaussian noise, thus lowering the effective SNR of the system. The effect of a LO frequency offset can be corrected in the carrier tracking loop by multiplying the signal by a correction factor [6,13].

A second source of error is LO phase offset. This can occur in conjunction with or separately from a LO frequency offset error. A LO phase offset occurs when there is a difference between the phase of the LO output and the phase of the received signal. This error can be expressed mathematically by multiplying the time-domain signal by a complex exponential with a constant phase. The result is a constant phase rotation for all of the subcarriers in the frequency-domain, with the constellation points for each subcarrier experiencing the same degree of rotation. A frequency-domain equalizer can be utilized to correct for small phase rotations as long as the constellation points do not rotate beyond the symbol decision regions. Each filter coefficient in the equalizer multiplies its corresponding subcarrier by a complex gain factor (i.e. an amplitude scaling and phase rotation). Larger phase rotations are corrected in the carrier tracking loop [7,13].

Another non-ideal effect that can impact the OFDM system is an offset in the FFT window location. In an ideal system, the receiver processes the FFT data in blocks of N samples at a time, and these samples correspond to the N samples of a single transmitted OFDM symbol. The receiver synchronizes itself with the received signal's OFDM symbol boundaries via a correlation with the PRS. However, inaccuracies in the synchronization process will manifest themselves as an offset in the FFT window location. The result is that the N samples used in the FFT window will not align exactly with the corresponding OFDM symbol. The presence of the guard interval provides some margin to enable a small offset to be present without taking samples from more than one OFDM symbol. However, an offset of even a few samples will cause some degree of distortion. As long as the FFT window location offset does not go beyond an OFDM symbol boundary, the offset appears as a shift in time, which is equivalent to a linearly increasing phase rotation in the frequency-domain constellations. The amount of rotation increases linearly with the subcarrier's FFT bin location. Constellations on low frequency subcarriers are rotated slightly, while constellations on the high frequency subcarriers are rotated proportionately more. FFT window location offsets are corrected by the fine frequency correction loop previously described [6,13].

Another source of error is due to a sampling frequency offset, which occurs when the A/D converter output is incorrectly sampled. Assuming F_S is the sampling frequency $F_S/2$ is the highest allowed frequency in discrete-time. Sampling too fast appears as an increase in the value of $F_S/2$ and results in a contracted spectrum caused by oversampling. Similarly, sampling too slowly appears as a decrease in the value of $F_S/2$ and results in an undersampled expanded spectrum, which can cause aliasing of the spectrum if the error is sufficiently large. Since the expansion or contraction of the spectrum prevents the received subcarriers from aligning with the appropriate FFT bin locations, this results in ICI [6,13]. A sampling frequency offset can be corrected by generating an error term that is used to drive a sampling rate converter or correct the sampling clock. Sampling clock inaccuracies may also be corrected in the digital domain, through the use of interpolation filters. (Note that another unwelcome effect of an inaccurate sampling clock will be bad frame timing, which can produce significant issues within the audio rendering subsystem.)

Phase noise will be added to the signal through the RF to IF frequency-conversion stage in the tuner. The local oscillator used in the conversion is typically in a PLL loop to stabilize it at the desired frequency. The resulting phase noise spectrum will be a function of the PLL for close-in phase noise and the free running oscillator for wideband phase noise. Phase noise is shaped and is primarily concentrated near the center frequency of the signal. This phase noise will be superimposed on the desired signal. An OFDM signal set contains multiple subcarriers, each of which is a smaller percentage of the total frequency bandwidth than in a single carrier system. As a result, phase noise degrades the performance of an OFDM system more than a single carrier system. Phase noise effects in an OFDM system can be separated into two categories, a "common" phase error part and a "thermal-noise-like" part. In the case of "common" phase error, all carriers are rotated by the same angle simultaneously. This results in the rotation of the signal constellation within a given symbol. This error could be corrected by using pilot subcarriers, which are not used in the Eureka 147 system. The "thermal-noise-like" error is different for all carriers. The phase noise sidebands of the local oscillator are superimposed on every carrier. The wanted carrier is correctly demodulated, along with its phase noise sidebands, which cause a random phase rotation. Although in a COFDM system the adjacent carriers fall into the nulls of the demodulator, their phase noise contributions are still demodulated and contribute to the phase error of the desired signal. Since there are many random contributing phase noise components, this error has a Gaussian-like distribution and results in a blurring of the constellation. This effect can be viewed as a form of ICI and a loss of orthogonality [7,13].

14.3.3.2 Implementation

When implementing a practical receiver design, it is important to remember that the Eureka 147 system is designed as a wideband data system that broadcasts near-CD quality audio and data to mobile, portable, and fixed receivers that rely on simple, non-directional antennas. As described previously, the transmission signal uses frequency and time interleaving to improve performance under fading and multipath conditions and allows for optimization of the error protection levels for the channel conditions. Since the approved transmission frequencies are within VHF Band III (174–240 MHz) and L-band (1452–1492 MHz), the receiver must be programmable over a wide range of grid center frequencies. It must also decode the FIC and PAD simultaneously with the audio decode. In addition, the implementation must be low cost and low power since portable radios are necessary for commercial success.

When designing a low cost receiver, the partitioning of functions between the digital baseband and the analog front-end of the receiver is critical. The most common implementation is a super-heterodyne radio-signal path as shown in Figure 14.16 [15]. The incoming L-band and VHF band III RF signals are down-converted to a common IF where the out of band components are rejected by a Surface Acoustic Wave (SAW) filter to ensure the signal stays within the dynamic range of the ADC. The first IF signal is then amplified and converted to a second IF signal, which is digitized by the ADC and fed into the digital baseband section. This simplifies the RF front-end since a programmable DSP performs the signal conversion to baseband In-phase/Quadrature (I/Q), synchronization, COFDM symbol demodulation, de-interleaving, Viterbi forward error-correction, and MPEG-1 Layer II audio decoding.

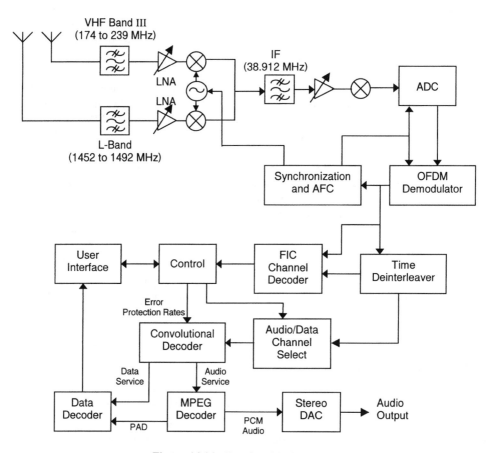

Figure 14.16   Receiver block diagram

Two time concepts exist simultaneously in the digital baseband. The first is the OFDM symbol time and the second is a CIF frame of 24 ms. The symbol time is mode dependent (for example, mode I is 1.246 ms) while the CIF frame time is constant and corresponds to 864 CUs. Each broadcast service is defined in terms of a number of CUs. For example, a stereo audio service with a bit-rate of 160 kbps encoded at UEP protection level 3 (0.5 code rate) needs 128 CUs distributed over three or four symbols [15]. Depending on the transmission mode, the number of CUs per symbol will vary. The time de-interleaving buffer separates the symbol based operations (demodulation, synchronization, frequency de-interleaving) and the time based operations (time de-interleaving, Viterbi, audio/data decode).

Since the quadrature demodulation is performed digitally, it eliminates phase and gain imbalances, thereby removing a significant source of signal distortion. By carefully selecting the sampling rate, the sampling images are equally spaced, and the normal and conjugate images are interleaved allowing the I and Q components to be recovered [27]. This operation is combined with a decimating FIR filter to provide additional adjacent channel filtering, although most of the rejection is provided by the SAW filter. A commonly used real IF is 2.048 MHz, with an 8.192 MSPS ADC clock. The four-times-oversampled data are then down-converted to 0 Hz through the use of a cosine 2.048 MHz and $-j$ sine 2.048 MHz stream, which sampled at 8.192 MHz have very simple 1, 0, -1, 0... formats.

To estimate the DSP processing power required, the Texas Instruments TMS320C5510 Dual MAC programmable DSP is used [28]. It operates at 0.35 mW/MIPS (core + memory) for maximum performance, but can support operation as low as 0.08 mW/MIPS (core + memory) at 0.9 V where MIPS are defined as Millions of Instructions Per Second. Several of the typical algorithms are benchmarked below. For example, a 2048-point, 9-tap complex block FIR requires

$$nx\_samples \times (4 + 2nh\_taps) + 62 = 45118 \text{ cycles}$$
$$\Rightarrow 36.2 \text{ MHz (assuming one } 1.246 - \text{ms frame)}$$

OFDM demodulation places strict requirements on the tuning accuracy of the receiver, since (as discussed previously) even small errors can cause a loss of orthogonality resulting in ICI. The tuning accuracy required for Eureka 147 is $\pm 5\%$ of the carrier spacing for a carrier spacing to interference ratio of approximately 20 dB, which corresponds to ± 50 Hz for transmission mode I [20,29]. Most receivers also include timing synchronization to reduce the effects of multipath interference. A programmable DSP is ideal for fulfilling these strict requirements on frequency and timing accuracy for the synchronization process.

One of the more computationally intensive functions is the OFDM demodulation. In order to operate on a symbol basis, the OFDM demodulation and DQPSK symbol mapping must complete in less than 1.246 ms. Mode I is the most difficult, requiring a 2048-point complex FFT. For the TMS320C5510 this FFT requires

$$5\left(\frac{nx}{2}\right) + 5\left(\frac{nx}{2}\right)(\log_2(nx) - 3) \times 1.15 + 10 \times \left(\frac{nx}{4}\right) = 57344 \text{ cycles}$$

since this must finish within one frame of 1.246 ms $\Rightarrow$ 46.02 MHz

The FFT is performed in place to conserve memory, so the required buffer size is: 2K $\times$ 16 + twiddle factors. Since the DQPSK requires a complex multiplication with the previous symbol, a second 2K $\times$ 16 buffer is required.

Next, the FIC is decoded and the CRC verified with the polynomial $G(X) = X^{16} + X^{15} + X^5 + 1$. The FIC is not time-interleaved so that the receiver can access the configuration and service information data more quickly. Since this information is repeated for each transmission frame, time interleaving is not as critical as it is for the MSC. The receiver uses the FIC to select the wanted subchannel. If the entire ensemble is decoded, the memory required for de-interleaving is 1–2 Mbits (depending on the number of Viterbi soft decisions). However, the frame structure of Eureka 147 enables service component selection prior to de-interleaving. This means that the receiver only needs to demodulate and decode the OFDM symbols carrying the required service and the FIC information, instead of processing the entire transmission frame [17]. This is critical for portable receivers since it enables power savings both in the RF section (through power-down modes) and the baseband (through decreased FFT and Viterbi processing requirements), which extends battery life. There is also a substantial reduction in the time de-interleaver memory requirements. This enables the memory to be integrated with the rest of the channel decoder, further decreasing the power and also decreasing the system cost. The required memory for a half code rate, 384 kbps service with 2 soft decision bits is

(16 CIFs) $\times$ (24 ms/CIF) $\times$ (384 kbps) $\times$ (1/code rate) $\times$ 2 bits = 589824 bits $\approx$ 74 kbytes

By continually decoding the FIC, which contains the Eureka 147 signal organization, the receiver can follow any dynamic multiplex reconfigurations sent by the broadcaster, guaranteeing undisturbed reception of the audio or data stream. However, instantaneous switching to a non-selected service is not possible since it takes 0.36 s to fill the time de-interleaver.

The Viterbi decoder estimates the source message given a soft encoded, time de-interleaved observation sequence. The Viterbi is a trellis decoder with constraint length 7, which means there are 64 states, and rate 1/4, with depuncturing used to obtain other rates. Using a data rate of 384 kbps, 64 states/bit, and a half-rate coding, we can estimate the TMS320C5510 cycles as (assuming three cycles per Butterfly, depuncturing of 12 cycles and traceback of ten cycles)

384 kbps $\times$ (12 + 3 $\times$ 64 + 10) = 82.176 Mcycles

$$\Rightarrow 41\% \text{ loading of a 200 MHz processor}$$

Although the TMS320C5510 has Add-Compare-Select hardware to assist with the Viterbi butterfly calculations and transition registers for the traceback, it does not contain a complete hardware Viterbi accelerator. The energy dispersal calculation is an XOR and does not require significant processing capabilities.

The audio decoding algorithm for the MPEG-1 Layer II is shown in Figure 14.17 [2]. With respect to computational effort, the sub-band synthesis filter is the most demanding portion of this algorithm. The filter implements a 32-point Inverse Modified Discrete Cosine Transform (IMDCT) followed by a windowing process to transform the reconstructed sub-band samples from the frequency domain into the time domain [18,19]. Even with this filter, the MPEG decoder takes ~15–20 MHz on the TMS320C5510. The data decoder merely formats the data for output and does not require significant processing.

Although the channel coding techniques provide a level of error correction, residual errors can still be present in the received encoded audio stream. The receiver can apply several different error concealment techniques to mask these errors. For frame errors, muting of non-reliable or non-decodable frames or repeating of the previously received frame in place of the

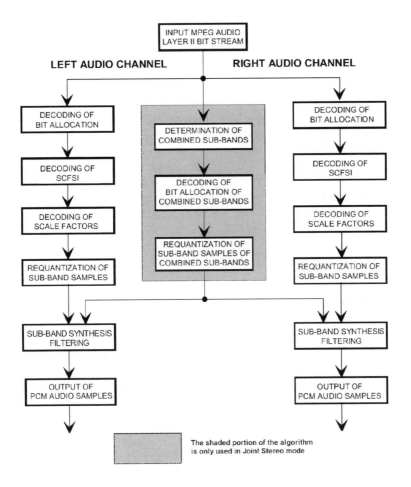

Figure 14.17   General MPEG Audio Layer II stereo decoder flow chart

non-decodable frame are common techniques. In the case where only the scale factor is corrupted, the previously sub-band scale factors can be used since the change from sub-band to sub-band is very small. Alternatively, interpolation from the previous value to the suspect value could also be used [21].

This analysis has focused on the major signal processing tasks to be performed, such as the OFDM demodulation, time de-interleaving, Viterbi decoding, and audio decoding. In addition to these tasks, several smaller signal processing tasks exist and there are also some intermittent operations, such as control of the VCXO and user interface. The DSP can perform all of the tasks in the Eureka 147 receiver, including the overall system control. Therefore, a RTOS (in our case DSP BIOS) is required to handle the CPU interrupts and task scheduling. DSP BIOS schedules both the real-time tasks such as OFDM demodulation, Viterbi decoding, audio decoding, and the non-real time tasks such as the user interface and system control. The input processing OFDM stage of the DSP is symbol based and finishes within one transmission frame. This simplifies the control required and reduces the internal buffering. Interrupts are used to provide a simple user interface for controlling the radio functions such as tuning, volume, equalization, and display. Since the entire acquisition, tracking, and decode loops are in software, the programmable DSP approach provides flexibility for improvements of the synchronization functions and for new features.

The programmable DSP based Eureka 147 receiver also supports an open software API structure. This allows the separate development of each part of the system and also allows users to easily add additional modules or change existing modules. The software provides an extensive command set allowing access to all aspects of the Eureka-147 signal processing pipeline and associated data, as well as peripheral control. This enables implementation of a variety of user interfaces on the DSP. The modular software also allows the user to tailor the implementation to meet his/her exact needs in terms of number of channels and additional features versus power consumption. The receiver's program can be stored in external non-volatile memory or in on-chip ROM, and initializes the DSP. Even if the program is hard-coded into the DSP, the receiver can still be customized for each application via external non-volatile memory. Since the receiver can be easily reprogrammed, this simplifies product development and enables product upgrades.

14.3.3.3 Why Programmable DSP?

If digital radio is to be successful, there must low cost portable receivers available for consumers. The traditional approach has been to use dedicated ASIC gates for the channel decoding with a programmable source decoder and microcontroller. However, new DSP processors are designed to support the computational intensive real-time signal processing algorithms typically found in communications systems. They have adapted instructions, memory architectures and data paths to execute these algorithms efficiently and with low power consumption. Also, the main drivers for the DSP processors in the cellular market are cost and power consumption. Since the algorithms for digital radio are very similar to cellular, these processors can be leveraged for digital radio applications. Many programmable communications DSPs also have accelerators or co-processors for Viterbi decoding, since this is one of the most computationally intensive functions required. The DSP can also be used to implement the man–machine interface, saving the cost of a separate microcontroller.

A flexible and standard software API standard, together with the reconfigurability of the receiver software itself, means that the programmable DSP approach is well suited for new service and feature development. Since new data applications are constantly being developed, the flexible platform is ideal for situations where the output data format may need to be tailored or an additional source decoder added to the platform. In addition, the programmable DSP allows flexibility as the standard evolves, for example the addition of a new transmission mode. A programmable approach also allows competing receiver manufacturer's to include their own intellectual property and to customize the solution for their implementation.

Another benefit of the programmable approach is the ability to implement other features on the same product. For example, the digital radio could be integrated with a solid state audio player (MP3 player), with the additional cost being only the RF front-end and ADC since both solutions could be implemented on the same programmable DSP.

Furthermore, the DSP approach makes it possible for a single baseband part to handle multiple radio standards. For example, some of the RF chipsets utilized for Eureka 147 are capable of tuning into the VHF band-II range, where analogue FM signals are carried. With a change in code load, a DSP solution can rapidly be produced to process this information and generate audio output. Furthermore, it is possible using this paradigm to incorporate multiple digital baseband standards on the same chip, such as Eureka 147 and the IBOC standard briefly discussed below.

Finally, in an emerging market such as digital radio, there is actually a cost and power advantage available to the system developer through the use of a programmable DSP. This is because custom ASIC solutions will be produced to a certain geometry at a relatively high cost, which must in most business models be largely recouped in sales before the next generation (smaller process size, lower cost, higher integration) can be developed. In an emerging market, these volumes can be challenging to produce. In contrast, commercially available DSPs such as the TMS320C5000 series already have a large, established market (in 2.5G cellular phones, modems, etc.) and so enjoy the significant benefits of lower cost, large production runs, and significant testing, documentation, and support.

14.4 IBOC

In-band systems permit broadcasters to simultaneously transmit digital audio signals in the same radio band as existing analog broadcasts. Digital signals are inherently more immune to interference, thus a digital receiver is able to operate in the presence of the analog signals. However, it is more difficult for existing analog receivers to filter out the digital carriers. In leveraging the transmission efficiency of digital radio signals, a low-power digital signal is used to maintain existing coverage areas for digital receivers while allowing analog receivers to reject the interfering signal [23].

The IBOC DSB system for operation below 30 MHz proposed by iBiquity Digital (ITU 6/63-E) is designed to operate in both a "hybrid" and "all-digital" mode [8]. The hybrid mode broadcasts the identical program material in both analog and digital format with the digital carriers located underneath and in the sidebands of the analog signal. In the all-digital mode, the existing analog signal is removed and the number and power of the digital carriers is increased to allow enhanced services.

In addition to the improved AM audio quality, the IBOC DSB system also provides three types of data services. First, there is a low, fixed-rate service that is similar to the existing Radio Broadcast Data System (RBDS). Second, there is an adjustable-rate service that operates at a fixed rate, for a pre-determined period. The broadcaster has the option of adjusting the data rate of this service by either reducing the encoded audio bit rate to allow increased data throughput, or by dynamically allocating specific groups of digital subcarriers among parity, audio, and data services. The third type is opportunistic variable-rate data services whose rates are tied to the complexity of the encoded digital audio. The audio encoder dynamically measures audio complexity and adjusts data throughput accordingly, without compromising the quality of the encoded digital audio [8].

The IBOC DSB system is similar to Eureka 147 in that it uses a codec to compress the audio data rate, Forward Error Correction (FEC) coding and interleaving for robustness, and an OFDM based modem. However, IBOC DSB includes a blending feature that provides a smooth transition from the digital to either the existing analog signal or a back-up digital signal in the event of interference. The IBOC DSB system uses Advanced Audio Coding (AAC) supplemented by Spectral Band Replication (SBR) technology that delivers high quality stereo audio within the bandwidth constraints required. Advanced FEC coding and special interleaving techniques spread burst errors over time and frequency to assist the FEC decoder in its decision-making process [8].

For systems operating below 30 MHz, grounded conductive structures can cause rapid changes in amplitude and phase that are not uniformly distributed across the digital carriers. To correct for this, the IBOC DSB system equalizes the phase and amplitude of the digital carriers to ensure proper recovery of the digital information [8].

As discussed previously, the OFDM transmission system naturally supports FEC coding techniques that maximize performance in a non-uniform interference environment and allows the power of each carrier to be individually adjusted to meet the analog FCC emissions mask. Due to the limited bandwidth available in the AM system, the system uses QAM instead of the DQPSK modulation used in Eureka 147. QAM allows more bits to be transmitted per carrier, which is required to maintain "FM-like" stereo audio quality in the given bandwidth [8].

The IBOC DSB system employs time diversity between two independent transmissions of the same audio source to provide robust reception during outages typical of a mobile environment. This several second time delay in the audio stream is in addition to the time interleaving in the digital signal. The interleaving in the primary digital path provides robustness but at the expense of acquisition time. During tuning, the blend function initially relies on the instantly acquired analog (or digital) back-up signal and then transitions to the full digital signal when it is acquired. Once acquired, blend allows transition to the back-up signal when the digital signal is corrupted. When a digital signal outage occurs, the receiver blends seamlessly to the back-up audio that will not experience the same outage due to the time diversity [8].

The hybrid mode IBOC DAB simultaneously broadcasts identical programming in both an analog and a digital format and targets areas where it is necessary to provide for a rational transition from analog to digital broadcasts. The MF hybrid spectrum is shown in Figure 14.18 [8]. The current US MF band allocation plan assigns stations of 20 kHz of the total bandwidth, with stations interleaved at 10-kHz spacings. The hybrid IBOC DSB signal is comprised of the ±4.5-kHz analog MF signal (present day AM) and digital carriers distributed across a 30-kHz bandwidth. The digital carriers under the analog signal are in quadrature and set at a level that is sufficient to ensure reliable digital service and low enough to avoid objectionable interference to the host broadcast.

The all-digital mode allows for enhanced digital performance after deletion of the existing analog signal. As shown in Figure 14.19, the all-digital mode increases the power of the quadrature carriers that were previously under the analog signal, and adds a low-bit-rate, digital back-up and tuning channel. The additional power in the all-digital waveform increases robustness, and the "stepped" waveform is optimized for performance under strong adjacent channel interference [8].

A functional block diagram of the hybrid MF IBOC DSB transmitter is shown in Figure 14.20 [8]. The audio source from the Studio Transmitter Link ("STL") provides an L + R monaural signal to the analog MF path and a stereo audio signal to the DSB audio. The DSB path digitally encodes the audio, inserts FEC codes, and interleaves the protected digital audio stream. The bit stream is then combined into a modem frame and OFDM modulated to produce the DSB baseband signal. In parallel, the time diversity delay is introduced in the analog MF path and passed through the station's existing analog audio processor and combined the digital carriers in the DSB exciter. This baseband signal is converted to

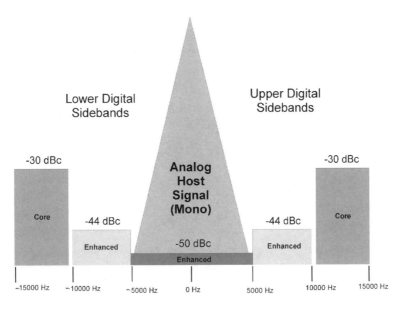

Figure 14.18 Hybrid MF IBOC DSP power spectral density

magnitude ρ and phase ϕ for amplification in the station's existing analog transmitter. A similar approach is used for the all-digital system except the analog transmission path does not exist [8].

The MF IBOC receiver functional block diagram is shown in Figure 14.21 [8]. The signal is received by a conventional RF front-end and converted to IF, similar to existing analog receivers. The difference is that after filtering at IF, the signal is sampled by an ADC and digitally down-converted to baseband in-phase and quadrature signal components in the DDC block. The hybrid signal is then split into analog and DSB components. The analog component is demodulated to produce a digitally sampled audio signal. The DSB signal is synchronized and demodulated into symbols, which are deframed for subsequent de-interleaving and FEC decoding. The audio decoder processes the resulting bit stream to produce the digital stereo DSB output. This DSB audio signal is delayed by the same amount of time as the analog signal was delayed at the transmitter to enable blending. The audio blend function blends the digital signal to the analog signal if the digital signal is corrupted and is also used to quickly acquire the signal during tuning or reacquisition. The receivers typically use tuned circuits to filter out adjacent channels and intermodulation products. Due to the high Q of these tuned circuits, short pulses can be stretched into longer interruptions via ringing. A noise blanker senses the impulse and turns off the RF stages for the short duration of the pulse, effectively limiting the effects of ringing on the analog sound quality. Short pulses have a minimal effect on the digital data stream. An all-digital receiver would be similar except the analog functions are not performed [8].

Clearly, because of the wider FM bandwidth, a FM in-band system is much easier to implement. The details of new FM system created by the merger of USA Digital Radio and Lucent Digital Radio to form iBiquity Digital have not been released. However, it will still use a similar approach to the terrestrial digital audio broadcast systems already discussed. The original USA Digital Radio proposed spectrum is shown in Figure 14.22 [9]. The digital

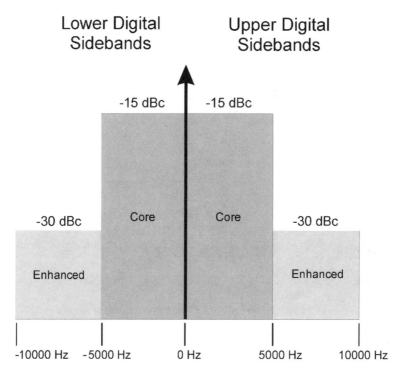

Figure 14.19   All-digital MF IBOC DSP power spectral density

sidebands occupy frequencies higher than approximately ± 129 kHz to ± 199 kHz around the channel center frequency. In an all-digital mode, the center analog signal is deleted and the power of the digital sidebands is increased within the emissions mask. The number of OFDM carriers is increased so that each sideband occupies a 100-kHz bandwidth. The extra 30 kHz

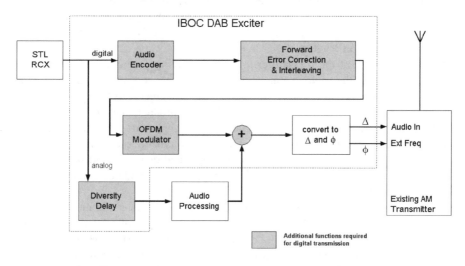

Figure 14.20   Hybrid MF IBOC DSB transmitter block diagram

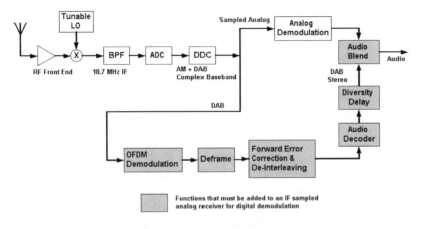

Figure 14.21   Hybrid MF IBOC DSB receiver block diagram

carries back-up audio, additional auxiliary services or additional error correction data. Low-level carriers are also added to the -100 kHz to $+100$ kHz region to support enhanced auxiliary services or multichannel sound [9,10].

The IBOC system has advantages in that it complies with the existing regulatory statutes and no new frequency spectrum is required. Existing analog AM and FM receivers are designed to filter out the digital sidebands, so they will continue to receive the analog broad-

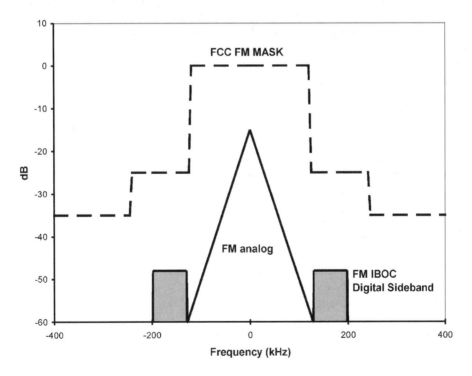

Figure 14.22   Hybrid FM IBOC DSB signal

casts without interference. New digital radio receivers are designed to receive both analog and digital broadcasts, allowing the receivers to switch to analog where there is no digital coverage or when the digital signal is lost. The digital signal provides improved frequency response, lower noise, and less distortion for the same coverage areas. IBOC is also attractive to broadcasters since the signal is in a spectrum they have already licensed and requires minimal upgrades to their existing equipment [23].

14.5 Satellite Systems

For digital radio services that are available anywhere in the continental US, two companies (Sirius Satellite Radio and XM Satellite Radio) have developed systems for digital audio broadcasting from satellites. These systems are known as Satellite Digital Radio Service (S-DARS) and feature direct satellite near-CD quality radio transmissions to any subscriber with a receiver. Although these systems could transmit to home and portable radios, they are intended primarily for mobile users. Miniature satellite dishes that are a few inches in diameter are mounted on the car's roof or trunk to receive the satellite signal, which is then sent to the receiver head unit inside the car. Both services were granted adjacent licenses by the FCC in the Wireless Communications Service (WCS) band. XM Satellite Radio is located at 2332.5–2345 MHz and Sirius Satellite Radio is located at 2320–2332.5 MHz. Both offer 100 channels of programming [23].

To provide a high quality mobile service, the Sirius S-DARS system uses spatial, time, and frequency diversity as well as receiver multipath equalization. The systems uses three satellites that are in a figure "8" ground track over North America and part of South America. Since the satellites are not geostationary, three satellites are required and they are separated by 8 hours. Each satellite can cover the continental US for approximately 16 hours, so at any one time, two of the three satellites are providing transmissions to the receivers. This orbital configuration allows very high elevations angles not possible with geostationary satellites, and this decreases the effects of foliage and signal loss due to buildings. Time diversity in the satellite transmission is essential for preventing signal losses due to obstructions such as overpasses. Therefore, a 4-s delay is introduced between the two active satellites and the earlier stream is stored at the receiver to use if the signal is briefly blocked. The system also uses terrestrial repeaters to avoid outages in the major cities caused by tall buildings blocking the satellite signal or in areas such as long tunnels. The system requires 105 terrestrial repeaters in 46 cities. Frequency diversity is assured by broadcasting the terrestrial, and two satellite signals on separate frequency bands. The block diagram of the satellite receiver is shown in Figure 14.23. The receiver also incorporates a six-tap equalization filter to decrease multipath effects. The system uses PAC audio compression technology to compress the audio bitstream while maintaining CD quality audio at the receiver. Time Division Multiplexing (TDM) is used for the satellite transmissions since multipath effects are less of an issue. COFDM is still used for the terrestrial broadcast repeater network [11].

XM Satellite Radio uses a slightly different approach. They chose to use two geostationary satellites, one over the east coast and one over the west coast of the US, to avoid the changing outage spots associated with satellites in elliptical orbits. However, due to the lower elevation angle, the system requires over 1500 terrestrial repeaters, which is significantly more than the Sirius system. The FCC has mandated that a common standard be developed so that S-DARS receivers can receive programming from both companies [23].

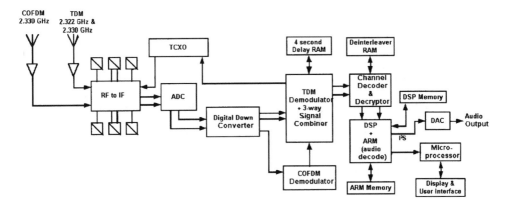

Figure 14.23   Sirius satellite receiver block diagram

WorldSpace is another deployed S-DARS system, but it is targeting Africa, Asia and Latin America. Since the targeted receivers are typically stationary, the system design is less complex enabling lower cost and lower power receivers. The main difference is there is only one TDM channel in the system. The system still uses QPSK modulation, half-rate Viterbi and 255/233 Reed–Solomon FEC. The framing structure is similar to Eureka 147 since a TDM frame contains: the Master-Frame Preamble (MFP), which contains synchronization information; the Time-Slot-Control Channel (TSCC), which describes the structure of the Prime-Rate-Channel (PRC) data; and 96 16-kbps PRCs. The PRCs can be combined in groups of up to eight PRCs to offer services from 16 to 128 kbps. The WorldSpace system uses MPEG-2.5 Layer 3 for audio compression which supports audio-sampling rates of 8–48 kHz. In order to preserve audio quality at low sampling frequencies, MPEG-2.5 Layer 3 compression uses Huffman encoding. It also supports continuous bit-rate variation from 8 to 128 kbps for increased flexibility in the audio transmissions [12,22].

14.6 Conclusion

Regardless of the approach chosen, digital radio systems are better than the current analog systems in several ways. First, they are more tolerant of multipath effects and provide a higher quality audio signal. Second, they also enable new digital broadcast services such as traffic, weather, travel information, websites, and video. Third, digital signals are much more power efficient to transmit than analog signals. While a typical FM analog broadcast would require 100,000 W of power, a digital broadcast would require ~1000 W for the same coverage area.

With the availability of faster, low power programmable DSPs, most of the demodulation and decoding required for the radio can be performed in software. This offers distinct advantages when dealing with multiple standards and different countries. While different RF frontends may be required, the digital baseband portion could remain the same, with just a different program load. Another advantage of a programmable platform is that digital radio will be more than just a radio. This means integration with devices such as solid st1ate audio players and cellphones. To the extent that all the software can be run on the same platform, this integration can be achieved very cost effectively. In addition, as VLSI trends continue to smaller, faster, low-power devices, the software solution will be able to migrate easily to the next generation device. This will enable more features and greater power savings.

References

[1] Shelswell, P., The COFDM Modulation System, The Heart of Digital Audio Broadcasting, BBC Research and Development Report, BBC RD 1996/8.

[2] ETSI, ETS 300 401, *Radio Broadcasting Systems; Digital Audio Broadcasting (DAB) to Mobile, Portable and Fixed Receivers*, Second Edition, May 1997, European Telecommunications Standards Institute, France.

[3] MPEG-1 Audio Layer II specification, ISO/IEC 11172-3 and MPEG-2 Audio Layer II, ISO/IEC 13818-3.

[4] Ferris, G. and Woodward, M., *MPEG Audio Transcoding White Paper*, Institution of Electrical Engineers, IEE, Savoy Place, London, http://www.radioscape.com.

[5] CENELEC EN50248, Characteristics of DAB Receivers, November, 1999.

[6] Stott, J.H., The Effects of Frequency Errors in OFDM, BBC Research & Development Department Report No. RD1995/15.

[7] Stott, J.H., 'The Effects of Phase Noise in COFDM', *EBU Technical Review*, Summer 1998.

[8] International Telecommunication Union (ITU) Radiocommunication Study Groups Document 6/63-E, 25 October 2000, Task Group 6/6, Draft New Recommendation ITU-R BS.[DOC. 6/63], System for Digital Sound Broadcasting in the Broadcasting Bands Below 30 MHz.

[9] FCC RM-9395 Petition for Rulemaking to the United States Federal Communications Commission for In-Band On-Channel Digital Audio Broadcasting, USA Digital Radio Filing, 7 October 1998.

[10] NRSC DAB Subcommittee Evaluation of USA Digital Radio's Submission to the NRSC DAB Subcommittee of Selected Laboratory and Field Test Results for its FM and AM Band IBOC System, Report from the Evaluation Working Group, Dr. H. Donald Messer, Chairman (as adopted by the Subcommittee on 8 April 2000).

[11] 'S-DARS Implementation', Sirius Satellite Radio, National Association of Broadcasters, *NAB Proceedings*, April 2000, pp. 153.157.

[12] Seminar on the WorldSpace Satellite Direct Digital Audio Broadcast System, Washington, DC, 19 July 1998.

[13] Litwin, L. and Pugel, M., 'The principles of OFDM', *RF Design,* January, 2001.

[14] Bower, A.J., 'Digital Radio The Eureka 147 DAB System', *Electronic Engineering,* April 1998, pp. 55–56.

[15] Mason, S., 'NTL Guide to Digital Radio for Anoraks', http://www.ntl.com/gb/en/guides/anoraks/default.asp.

[16] Huisken, J.A., van de Laar, F.A.M., Bekooij, M.J.G., Gielis, G.C.M., Gruijters, P.W.F. and Welten, F.P.J., 'A Power-Efficient Single-Chip OFDM Demodulator and Channel Decoder for Multimedia Broadcasting', *IEEE Journal of Solid-State Circuits*, Vol. 33, No. 11, November 1998, pp. 1793.1798.

[17] Baily, S., 'A Technical Overview of Digital Radio', BBC Research and Development, September 1999, http://www.bbc.co.uk/rd/projects/dab/index.html.

[18] Noll, P., 'MPEG Digital Audio Coding', *IEEE signal Processing Magazine*, September 1997, pp. 59–81.

[19] Hoeg, W. and Lauterbach, T., *Digital Audio Broadcasting Principles and Applications*, John Wiley & Sons, Ltd., New York, 2001, p. 81, p. 229.

[20] Muschallik, C. 'Influence of RF oscillators on an OFDM Signal', *IEEE transactions on Consumer Electronics*, Vol. 41, No. 3, August 1995, pp. 592–603.

[21] Hoeg, W. and Lauterbach, T., *Digital Audio Broadcasting Principles and Applications*, John Wiley & Sons, Ltd., New York, 2001, pp. 97–99.

[22] Sachdev, D.K. 'The WorldSpace System: Architecture, Plans and Technologies', National Association of Broadcasters, *Proceedings of 51st NAB Conference*, Las Vegas, 1997, p. 131.

[23] Pohlmann, K., *Principles of Digital Audio*, McGraw-Hill, New York, 2000, pp. 541–571.

[24] Tuttlebee, W. and Hawkins, D. 'Consumer digital radio: from concept to reality', *Electronics and Communication Engineering Journal*, December 1998, pp. 263–276.

[25] Pohlmann, K., *Principles of Digital Audio*, McGraw-Hill, New York, 2000, pp. 327–343.

[26] Hoeg, W. and Lauterbach, T., *Digital Audio Broadcasting Principles and Applications*, John Wiley & Sons, Ltd., New York, 2001, p. 228.

[27] Cavers, J. and Stapleton, S., 'A DSP-based alternative to direct conversion receivers for digital mobile communications', *Global Telecommunications Conference and Exhibition*, 1990, Vol. 3, pp. 2024–2029.

[28] www.ti.com.

[29] Taura, K., Tsujishita, M., Takeda, M., Kato, H., Ishida, M. and Ishida, Y., 'A digital audio broadcasting (DAB) receiver', *IEEE Transactions on Consumer Electronics*, Vol. 42, No. 3, August 1996, pp. 322–327.

15

Benchmarking DSP Architectures for Low Power Applications

David Hwang, Cimarron Mittelsteadt and Ingrid Verbauwhede

15.1 Introduction

In recent years, the technological trend toward high-performance mobile communications devices has caused a burgeoning interest in the field of low-power design. Indeed, with the proliferation of portable devices such as digital cellular phones, pagers and personal digital assistants, designing for low-power with high throughput is becoming increasingly necessary.

It is often claimed that a full-custom ASIC will be "lower power" than a programmable approach. This is certainly the case when compared to a general purpose processor, but less apparent when compared to a programmable Digital Signal Processor (DSP) processor [13]. An experiment has been designed to verify this claim for a realistic signal processing application in a low-power environment [2]. The goal is to quantify this claim. A meaningful example, one quite larger than a simple FIR filter or autocorrelation, will for the most part execute signal processing functions but will also include some control code and book-keeping operations encountered in many signal processing applications. A Linear Prediction Coefficient (LPC) speech coder was chosen for this task. It is described in subsequent paragraphs. Secondly, the design methodologies for each design approach, ASIC or programmable DSP, will be explained and compared in design effort.

This chapter will investigate five *signal processing specific platforms*: three programmable DSP processors – the TI C55x, the TI C54x, and the TI C6x; and two signal processing design environments – Ocapi, and AIRT Designer. Each design is optimized to reduce cycle count and power consumption. All five designs will be compared based on energy, area, clock frequency/MIPS and design time.In the first section, a brief introduction to the LPC speech coder algorithm will be given. The main computational bottlenecks are identified. In the second section, the details of the design methodology are given. In the third section, the different signal processing platforms are introduced. Finally, in Section 15.5, the results are presented and compared. The last section gives the conclusions.

15.2 LPC Speech Codec Algorithm

Linguistically, sounds can be divided into two mutually exclusive categories: vowels and consonants. Vowels are produced by periodic vibrations of vocal chords. The period of vibrations is known as the pitch. Hence, excitation of vowels can be approximated simply by an impulse train with a period equal to the pitch. For consonants, the excitation is produced by air turbulence, which is approximated by a White Gaussian Noise (WGN) model [7]. If every frame is classified as voiced (periodic) or unvoiced (noisy), we only need to transmit a single bit indicating voiced/unvoiced and the value of pitch period (in the case of voicing). On the receiving side, excitation can then be modeled by either an impulse train or WGN. This excitation source then passes through gain a stage and a time varying IIR filter that uses the α values as coefficients.

The main building blocks of the LPC encoder are summarized in Figure 15.1.

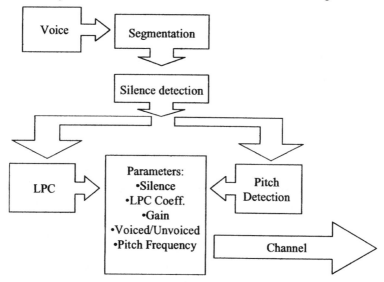

Figure 15.1   LPC speech encoder example

The algorithm consists of the following main steps:

15.2.1 Segmentation

The voice samples are segmented in frames of 240 samples with an overlap of 80 samples between frames. The samples pass through a Hamming window (a raised cosine transfer function).

15.2.2 Silence Detection

Silence is detected by calculating the 0th lag autocorrelation and comparing that value to a threshold value. If silence is detected a bit is set in the parameter list, the gain G is set to 0 and the LPC and pitch detection phases are skipped.

15.2.3 Pitch Detection Algorithm

In order to classify each frame as voiced/unvoiced the autocorrelation function is examined. Indeed, if the frame is voiced, it must be periodic, thus forcing its autocorrelation to be periodic with an identical period. The algorithm used is due to Sondhi [8] and is described below:

1. The frame is low-pass filtered at 1 kHz.
2. A Clipping Level (CL) is set to 30% of the maximum value in the frame.
3. The frame, $x(n)$, is then clipped according to the following equation

$$C[x(n)] = \begin{array}{l} +1 \text{ if } x(n) > C_L \\ -1 \text{ if } x(n) < -C_L \\ 0 \text{ otherwise} \end{array}$$

4. Finally, the autocorrelation function $R(n)$ is computed on the clipped frame $C[x(n)]$ according to

$$R(k) = \sum_{n=0}^{N-k-1} C[x(n)] \times C[x(n+k)]$$

where N is the length of the frame. Since minimum and maximum pitch frequencies for men and women are 80 and 350 Hz, we only need to compute $R(k)$ for k between 22 and 100 inclusive (for an 8-kHz sampling rate).

5. If the largest peak of $R(k)$, max[$R(k)$], satisfies

$$\max[R(k)] = 0.3 \times R(0)$$

the frame is classified as voiced and the index k is transmitted as the pitch period, else the frame is classified as unvoiced.

15.2.4 LPC Analysis – Vocal Tract Modeling

To model the vocal tract, an all-pole function $H(z)$ is assumed

$$H(z) = \frac{G}{\left(1 - \sum_{k=1}^{p} a_k \times z^{(-k)}\right)}$$

where p is the model order, chosen to be 10. The predictor coefficients a_k can be found from solving the linear system

$$\begin{bmatrix} R(0) & R(1) & \dots & R(p-1) \\ R(1) & R(0) & \dots & R(p-2) \\ \dots & \dots & \dots & \dots \\ R(p-1) & R(p-2) & \dots & R(0) \end{bmatrix} \times \begin{bmatrix} a_1 \\ a_2 \\ \dots \\ a_p \end{bmatrix} = \begin{bmatrix} R(1) \\ R(2) \\ \dots \\ R(p) \end{bmatrix}$$

where $R(k)$ is the kth lag autocorrelation function of a frame. The Toeplitz structure of the leftmost matrix can be exploited and the linear system can be solved iteratively with the

Levinson–Durbin recursion [7]. So the main calculations in this step are the computations of the autocorrelation values and the solutions of the Toeplitz system. The set of ten Linear Predication Coefficients (LPC), i.e. the α_i values, as well as the prediction error $E^{(p)}$, which corresponds to the square of the gain G^2, are computed and transmitted for each frame.

15.2.5 Bookkeeping

The last step consists of combining all the calculated values in a vector to be sent over the channel: an indication of silence/no silence, a vector with the α_i values, the gain, an indication of voice/unvoiced frame and the pitch.

15.2.5.1 Compute Intensive Functions

From an implementation viewpoint, the computation intensive modules are the following:

1. Pitch detection. A total of 78 correlations have to be computed – $R(22)$ through $R(100)$. The computations are simplified using the "clipped" coefficients, as explained in step 3 of the algorithm above, to reduce the computational complexity of this module.
2. Levinson–Durbin algorithm. This algorithm requires a lot of computations because it involves 11 correlations as well as an iterative division algorithm.

Execution times on the DSP processor show that around 60% of the cycles are spent on the pitch detection while 25% are spent on the Levinson–Durbin algorithm. The remaining 15% are used for the Hamming window, low-pass filter and miscellaneous memory transfers.

15.3 Design Methodology

This section describes the design methodology for implementing the LPC speech coder on the various platforms. The different steps are illustrated in Figure 15.2. The codec is first designed in floating-point format in MATLAB. The challenge is to efficiently map the software algorithm onto the fixed-point hardware. This involves the conversion of floating-point computations into fixed-point word lengths (along with the ensuing design decisions) as well as the allocation and software mapping/scheduling on the available hardware.

15.3.1 Floating-Point to Fixed-Point Conversion

Since the algorithm executes in real-time on fixed-point hardware, one has to make decisions concerning the internal word lengths of each of the system hardware modules. An inadequate word length can lead to reduced Signal-to-Noise Ratio (SNR), deterioration of sound quality, and clipping. However, a surfeit of word length can create extraneous hardware, leading to wasted area and power.

For some of the platforms (i.e. the TI DSPs), the internal word lengths are fixed to a particular number (i.e. 16 bits). However, on the other platforms, the word lengths can be decided by the designer. There are several criteria which affect the fixed-point word length decision, including recognizable synthesized speech, pitch frequency matching, avoidance of signal overflow/saturation at each point in the algorithm, and avoidance of saturation of the synthesized speech output.

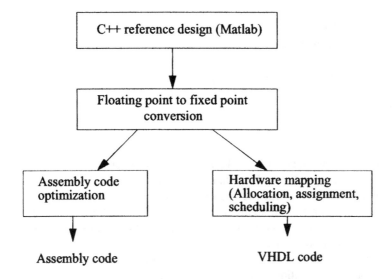

Figure 15.2   Design methodology to map an LPC speech coder on signal processing specific platforms

Of all these factors, the most restrictive criterion is the avoidance of synthesized speech saturation. This particular problem, related to instability (and hence the poles of the system), is inherent to the Levinson–Durbin algorithm. In the fixed-point implementation, the quality of voice is dependent on the number of input bits in a highly non-linear fashion. If the number of bits is insufficient, the algorithm is unstable and clipping occurs. On the other hand, if the number of bits is sufficient, the Levinson–Durbin algorithm is stable and the reconstructed signal is virtually the same as the floating point signal. Hence, by adjusting the word length parameters and checking output saturation, the minimum bit requirements for each module can be found. This iterative refinement was done on the Ocapi and AIRT Designer platforms with the built-in fixed-point C++ libraries. This resulted in varying word lengths according to the modules. Even within one module, the position of the decimal point (Q-format) is adjusted at each point in the algorithm. The hardware modules for the Ocapi implementation vary from 8-bit clipped correlator units to a 24-bit multiplier and a 30-bit accumulator. As an example, the C++ code for the low pass FIR filter at the input of the pitch detection module is given below. Note the fixed-point annotations, $Fix\langle x, y\rangle$ means the data has a length of x bits, with y bits behind the decimal point. This annotated C++ code can be simulated with the fixed-point class libraries and is input into the behavioral level synthesis part of the AIRT design environment.

```
void fir (Fix⟨8,7⟩ b[24], Int⟨9⟩ seg[240], Fix⟨8,3⟩ out[240])
{
  int i, j;
  Int⟨9⟩ state[23];
  for (i = 0; i < 23; i + + ) state[i] = 0;
  for (i = 0; i < 240; i + + )
    {
```

```
    out[i] = Fix⟨8,3⟩(b[0] * seg[i]);
    for (j = 1; j < 24; j + +)
       out[i] = out[i] + Fix⟨8,3⟩(b[j]*state[j-1]);
    }
  for (j = 23; j > 0; j--)
     state[j] = state[j-1];
  state[0] = seg[i];
  return;
}
```

The fixed-point processors (TI 54x, TI 55x, TI 6x) have internal word lengths set to 16 bits for most arithmetic operations. To obtain a fixed-point C++ code suitable for such a processor, it is necessary to rewrite the entire algorithm using 16-bit C arithmetic (i.e. using ANSI C short format). In addition, the code must be heavily modified to exploit the TI Q15 library function. The Q15 format maps each 16-bit word into a fractional two's complement number in the range $[-1,1)$.

After generating suitable fixed-point code using Q15 functions, data scaling needs to be performed to prevent saturation. For example, the autocorrelation function has its maximum value at $R(0)$, which itself has an absolute worst case value of 240 (if every $C[x(n)] = -1$ for all 240 samples). This would require the scaling of each $C[x(n)]$ by 1/240 to ensure that $R(0)$ remains in the range $[-1,1)$. However, this is a pessimistic approach to scaling, as a speech pattern would never be DC. A more optimistic approach for scaling makes sense. If the speech is unvoiced the input will resemble noise and thus the autocorrelation will obtain a maximum value equal to the variance of the $C[x(n)]$ sequence. Therefore, each $C[x(n)]$ will be scaled by a factor of $1/128$ (2^{-7}). This allows for a greater dynamic range, while still keeping $R(0)$ confined to the range $[-1,1)$ for most cases. In the few cases this range is exceeded, $R(0)$ saturates to the boundary points.

15.3.2 Division Algorithm

Division is necessary as part of the Levinson–Durbin algorithm. To illustrate the conversion from floating point to fixed-point and to illustrate the care a designer has to take when making this translation, a specific division algorithm was chosen. Instead of using the more familiar restoring, non-restoring, or SRT division algorithms, a fixed-point Newton–Raphson reciprocation algorithm was chosen [3], because it matches well with a 16-bit fixed-point processor. The goal is to calculate $A = x/y$, in this case x is a 32-bit number and y is a 16-bit number. Using the Newton–Rapson technique, first $1/y$ is calculated and then multiplied by x to produce the final answer. The algorithm works as follows:

1. First, using the MSBs of y, linear interpolation is used to produce a rough estimate of $1/y$, called $z1$.
2. This rough estimate of $1/y$ is used to calculate a new estimate, $z2$ using the following equation: $z2 = z1(2 - y \times z1)$ where y is the 16-bit value of y.
3. Step 2 is repeated until the solution converges to a value z that is accurate in 16-bit resolution. This normally requires three iterations [10].
4. The final answer is calculated using: $A = x \times z$

Using this technique, the cycle count is significantly reduced. A full division is computed within 18 cycle counts.

15.3.3 Hardware Allocation

A|RT Designer [18] assumes a VLIW architecture, where the user is free to choose the data path modules in the architecture. Ocapi [17] gives the user only an environment to specify the architecture and does not impose a particular architecture. This has the advantage that any architecture can be described, but the disadvantage that the designer has to describe all features and details of the architecture.

Thus in both environments, the user allocates the data path modules (ROM, RAM, ALU, MAC, Address Control Units, etc.) necessary to complete the design, as well as designate which modules perform each function in the algorithm code. Hence, by examining processor use statistics and by keenly examining the code structure, one can pinpoint design bottlenecks and alleviate them by reallocation and reassignment.

An example of this can be seen in the iterative design flow of A|RT Designer. The initial design uses the A|RT Designer default minimum hardware allocation. This means that all operations have only one choice of hardware module. Operations on different execution units can still occur in parallel since A|RT uses a VLIW architecture. This implementation requires 8000 cycles to complete one frame. By examining the scheduling load graph, it is found that the autocorrelation function occupies 80% of the processing time and that one autocorrelation iteration takes three cycles where the Address Calculation Unit (ACU) is used every cycle.

Thus, the cycle count can be reduced by 4000 cycles by inserting an additional ACU and a second MAC combined with loop pipelining. By further investigation of the code, one finds that the coefficients for the windowing filter can be reallocated onto a ROM instead of being soft coded and calculated for every frame. This modification reduces the cycle count by another 1000, bringing the total cycle count down to 3000 cycles – a 63% decrease from the original design. These are examples of the design processes required to optimize performance (cycle count) using the various tools.

15.4 Platforms

As mentioned previously, five implementation platforms are under examination, which are described briefly below.

15.4.1 Texas Instruments TI C54x

The TI C54x fixed-point DSP is a signal processor commonly used in cellular phones, digital audio players, and other low-power communications devices [4]. The TI core uses an advanced modified Harvard architecture that maximizes processing power with eight buses (four program data buses and four address buses). The core consists primarily of a 40-bit ALU, a barrel shifter, two accumulators, a 17×17-bit MAC unit and an addressing unit. The program fetch is 16 bits and the instruction length is also 16 bits.

15.4.2 Texas Instruments TI C55x

The TI C55x processor is the most recent DSP in the TMS320C5000 series. It builds on the C54x generation with a one-sixth reduction in power consumption alongside a (maximally) 500% increase in performance [15]. The C55x has additional hardware, including a 17×17-bit MAC, a 16-bit ALU and a total of four 40-bit accumulators. The instruction length is variable between 8 and 48 bits, the program fetch is 32 bits.

15.4.3 Texas Instruments TI C6x

The Texas Instruments' TMS320C6000 series is the line of fixed-point and floating-point processors which emphasize high-performance as the key metric. As such, they are used in base stations and other systems in which bandwidth and processing power are crucial. In our experiment, the C62x processor was chosen, a fixed-point DSP used for multi-channel broadband communications. The core implements a VLIW architecture with eight functional modules [12]. These consist of six parallel 40-bit ALUs and two 16-bit multipliers (with 32-bit outputs). The C62x processor operates at 150–300 MHz and is capable of operating at 1200–2400 MIPS [16].

15.4.4 Ocapi

Ocapi is a C++ based design environment developed by IMEC [14,17]. The Ocapi environment is based upon a library of fixed-point C++ classes that allow the user to fully describe an ASIC at the highest algorithmic and behavioral level. Through different design stages, the C++ code is refined and enhanced with architectural detail. The Ocapi toolset then maps the final code into an RTL level bit-parallel HDL code which next can be synthesized.

15.4.5 A|RT Designer

A|RT Designer is a software environment designed by Frontier Design [5,18]. As with Ocapi, A|RT Designer's purpose is to bridge the gap between the software algorithm design and the hardware implementation. The design is first created in floating-point C and then converted to fixed-point C using a fixed-point library. Simulations with the fixed-point libraries are performed. Upon completion of the fixed-point code, the user directs the software tools to perform resource allocation, resource assignment, and operation scheduling (based upon data interdependencies). A|RT generates synthesizable RTL level code which describes the entire VLIW machine.

15.5 Final Results

In circuit design, a measure of the cost for a particular design can be estimated from the total area. Similarly, on an embedded software platform, cost can be estimated by memory and cycle counts to perform the algorithm. In Table 15.1, the overall area/memory and cycle counts for each platform are summarized.

Table 15.1   Implementation results

	Area–memory	Cycles/frame (K)	Power at min clock[a]	Energy/frame (µJ)	Technology (µm)	Power supply (V)
TMS320C6201 core [12]	16 kB[b]	30	3.3 mW	66	0.15	1.5 core
TMS320VC5410A core [9]	8.7 kB	240	7.2 mW	144	0.15	1.6 core
TI C5510 core [10]	10.2 kB	120	2.64 mW	53	0.15	1.5 core
Ocapi	1.4 mm²	11	107 µW	2.1	0.25	2.5
A|RT	3.2 mm² 2.3 kB ROM 1 kB RAM	3	215 µW	4.3	0.35	3.3

[a] Minimum clock means the minimum clock frequency that needs to be applied to meet the real-time throughput requirement.
[b] This includes only the program code.

15.5.1 Area Estimate

The Ocapi solution is slightly over half the size of the A|RT Designer solution. However, these figures are somewhat deceptive. The reason for this large difference in size is mostly due to the process libraries we had to synthesize for each circuit. For the Ocapi design, a 0.25-µm process was used while for A|RT Designer, a 0.35-µm process was used. Assuming perfect scalability, the A|RT Designer circuit would be only 1.63 mm². This is comparable to the Ocapi design area of 1.4 mm² as one would expect.

An ASIC design can afford to include the minimum amount of memory needed to implement the given applications. For instance for the A|RT design, one small data ROM is provided to store the coefficients of the Hamming window and the coefficients of FIR filters. The size is 264 bytes. The program memory is a ROM of size 2 kB. A data memory of about 1 kB RAM is also included.

For the DSP processor implementations, an estimate is made of what fraction of the available on-chip memory will be used by the program. For the C6x implementation, only the program size is included.

15.5.2 Power Estimate

Power figures for each design are given in Tables 15.1 and 15.2 in units of power at the minimum clock frequency and energy per frame. Power at the minimum clock frequency means that the processor is allowed to run at the lowest clock frequency that still guarantees to meet the real-time constraints. Speech is sampled at 8 kHz, one sample is 8 bit, resulting in 64 kbits/s. One frame is 240 samples but there is an overlap of 80 samples between frames. This results in a 50-Hz frame rate or 0.02 s/frame throughput requirement. The clock frequency of each of the processors is reduced to just meet this throughput requirement.

The power numbers for the ASIC platforms are estimated after synthesis with a standard

Table 15.2 Detailed power estimations

	W	Clock	Duty	mW/MHz	Cycles/frame (K)	Mcycles/s	mW	Energy/frame (μJ)	Technology (μm)	Voltage (V)
TMS320C6204 core [12]	0.44 W	200 MHz	50% high; 50% low	2.175	30	1.5	3.3	66	0.15	1.5 core
TMS320VC5410A core [9]	96 mW	160 MHz	50% MAC; 50% NOP	0.6	240	12	7.2	144	0.15	1.6 core
TI C5510 core [10]				0.44	120K	6	2.64	53	0.15	1.5 core
AIRT [5] – reference design	660 μW	2.5 MHz		0.264	3	0.15			0.25	1.05 I/O + core
AIRT [5] – voltage and tech scaling				3.65	3	0.15	0.55	11	0.35	3.3

cell library but before placement and routing. The power numbers for the DSP processors are estimated by using the power numbers published on the TI application reports and the TI webpage and multiplying these by actual cycle count and correcting them according to the type of instruction [1,9–12].

The ASIC circuits designed with Ocapi and A|RT Designer resulted in the lowest energy per frame with 2.1 and 4.3 μJ, respectively. One should note that the energy for each design refers to the energy of the core and memory and not for the I/O. One should also note that if only a small fraction of the available memory is used, a larger amount of energy will be wasted by accessing memories which are too large for the application. On the other hand, the DSP processors are built in a more advanced technology and can rely on optimized full custom layout compared to synthesized layouts for the ASIC implementations.

The assembly code for the C6x code was generated by C compiler of the TI Code Composer Studio software environment. The original C code with no optimization requires about 80,000 cycles to process a single frame. Successive optimizations of the C code and the use of the *fir_r8* subroutine (assembly optimized FIR code with the loop unrolled eight times) from TI's DSP library reduced the cycle count to about 30,000 cycles per frame. The newest member in the fixed-point C62x series is the C6204. At 200 MHz and 1.5 V, it has a power consumption of 290 mA for the core, at a 50% high, 50% low activity [12]. Thus, if only 30 K cycles are needed to perform one frame, the C6204 processor can either handle 86 channels running in parallel or reduce its clock frequency by this number, resulting in 65 μJ/frame.

The C55x processor has a much lower mW/MIPS or mW/MHz number but does not have the same amount of parallelism as the C62x VLIW architecture. Therefore, it is interesting to note that the final number of 53 μJ/frame is about the same as the C62x. The C54x implementation is almost three times higher. One reason is that more effort has been spent to optimize the code for the C55x. Another reason is that the average instruction on the C55x will execute more operations than the average instruction on the C54x. An example of this is the usage of *fir2* routine in the DSP library which uses the dual MAC instead of a single MAC.

One can argue that making power estimations for ASIC circuits is tricky and error prone before actual processing. Therefore, we have compared our power estimation results with a processed and published processor which is built with the same A|RT environment [5]. This processor consumes 660 μW at 1.05 V running at 2.5 MHz in a 0.25-μm technology. While the application is different, the general architecture is still a VLIW architecture optimized towards the application and also in this case the minimum required clock frequency is very low, which is very beneficial to reduce the power consumption.

This processor has a power consumption of 0.264 mW/MHz. When this number is scaled up according to the scaling rules for short-channel devices [6], a value of 3.65 mW/MHz is obtained. This results in an estimate of 11 μJ/frame instead of the 4.3 μJ estimated from the synthesized results. This is still a factor 5 better than the lowest programmable DSP implementation. A similar comparison cannot be made for the Ocapi environment. The reason is that Ocapi does not have a "built-in" assumption of a basic generic architecture, which makes it hard to compare two designs built with the same environment.

15.6 Conclusions

While large efforts have been made to make programmable DSP processors extremely low power, they still trail in comparison to application specific solutions, by a factor of 5 in this experiment. The above results were obtained in the span of one quarter, indicating the "ease of use" of both the TI programming environment (Code Composer) as well as the design environments, Ocapi and AIRT Designer. The availability of design environments that raise the level of abstraction from Verilog to C++ or from assembly to C, increases the productivity and allows the exploration of design alternatives. It also shortens the design time.

Acknowledgements

The authors would like to thank Alan Gatherer for the careful review of the TI processor power numbers and the EE213A class of Spring 2000. DH acknowledges the support from the Fannie and John Hertz Foundation. This work was partially sponsored by UC-MICRO#00-097.

References

[1] Castille, K., TMS320C6000 Power Consumption Summary, Application Report SPRA486B, November 1999, available from www.ti.com.

[2] Hwang, D., Mittelsteadt, C., Verbauwhede, I., 'Low Power Showdown: Comparison of Five DSP Platforms Implementing an LPC Speech Codec', *Proceedings ICASSP 2001*, Salt Lake City, May 2001.

[3] Koren, I., *Computer Arithmetic Algorithms*, Prentice Hall, Englewood Cliffs, NJ, 1993, pp. 158–160.

[4] Lee, W., Landman, P., Barton, B., Abiko, S., Takahashi, H., Mizuno, H., Muramatsu, S., Tashiro, K., Fusumada, M., Pham, L., Boutaud, F., Ego, E., Gallo, G., Tran, H., Lemonds, C., Shih, A., Nandakumar, M., Eklund, R. and Chen I., 'A 1-V Programmable DSP for Wireless Communications', *IEEE Journal of Solid-State Circuits*, Vol. 32, No. 11, November 1997, pp. 1766–1776.

[5] Mosch, P., van Oerle, G., Menzl, S., Rougnon-Glasson, N., Van Nieuwenhove, K., Wezelenburg, M., 'A 660-mW 50-MOPS 1-V DSP for a Hearing Aid Chip Set', *IEEE Journal of Solid-State Circuits*, Vol. 35, No. 11, November 2000, pp. 1705–1712.

[6] Rabaey, J., *Digital Integrated Circuits: A Design Perspective*, Prentice-Hall, Englewood Cliffs, NJ, 1996.

[7] Rabiner, L., Schafer, R., *Digital Processing of Speech Signals*, Prentice-Hall, Englewood Cliffs, NJ, 1978.

[8] Sondhi, M.M., *New Methods of Pitch Extraction, IEEE Transactions in Audio and Electroacoustics*, Vol. AU-16, No. 2, June 1968, pp. 262–266.

[9] Texas Instruments, TMS320VC5410A, Fixed Point Digital Signal Processor, Document SPRS139A, November 2000, Revised February 2001.

[10] Texas Instruments, TMS320C55x DSP Function Library (DSPLIB), February 2000, pp. 56–57.

[11] Texas Instruments, TMS320C6211 Cache Analysis Application Report SPRA472, September 1998.

[12] Texas Instruments, TMS320C6204, Fixed Point Digital Signal Processor, Document SPRS152A, October 2000, Revised June 2001.

[13] Verbauwhede, I., Nicol, C., 'Low Power DSP's for Wireless Communications,' *Proceedings of the 2000 International Symposium on Low Power Electronics and Design*, pp. 303–310.

[14] Vernalde, S., Schaumont, P., Bolsens, I., 'An Object Oriented Programming Approach for Hardware Design', *IEEE Computer Society Workshop on VLSI*, 1999, Orlando, FL, April 1999.

[15] www.ti.com/sc/docs/products/dsp/c5000/index.htm.

[16] www.ti.com/sc/docs/products/dsp/c6000/index.htm.

[17] www.imec.be/Ocapi/.

[18] www.frontierd.com.

16

Low Power Sensor Networks

Alice Wang, Rex Min, Masayuki Miyazaki, Amit Sinha and Anantha Chandrakasan

16.1 Introduction

In a variety of scenarios, often the only way to fully observe or monitor a situation is through the use of sensors. Sensors have been used in both civil and military applications, in order to extend the field of view of the end-user. However, most current sensing systems consist of a few large macrosensors, which while being highly accurate are expensive. Macrosensor systems are highly sensitive; the entire system can break down even with one faulty sensor. Trends in sensing applications are shifting towards designing networks of wireless micro-sensor nodes for reasons such as lower cost, ease of deployment and fault tolerance. Networked microsensors have also enabled a variety of new applications, such as environment monitoring [1], security, battlefield surveillance [2], and medical monitoring. Figure 16.1 shows an example wireless sensor network.

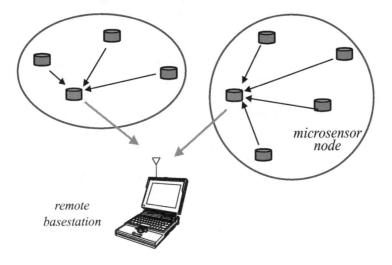

Figure 16.1. Microprocessor networks can be used for remote sensing

Another challenge in wireless microsensor systems is that all nodes are energy-constrained. As the number of sensors increases, it becomes infeasible to recharge all of the batteries of the individual sensors. In order to prolong the lifetimes of the wireless sensors, all aspects of the sensor system should be energy-efficient. This includes the sensor, data conversion, Digital Signal Processor (DSP), network protocols and RF communication. One important low power design consideration is leakage reduction. In sensing applications, often the sensors are idle and waiting for an external event. In such low duty cycle systems, more time is spent in idle mode than active mode, and leakage currents can become large. Therefore, circuit techniques for leakage reduction should be considered.

There are many important differences between the wireless microsensor systems discussed here and their wireless macrosensor system counterparts, that lead to new challenges in low power design [3]. In typical microsensor applications, the number of microsensors will be large which leads to high sensor node densities. As a result, the amount of sensing data will be tremendous, and it will be increasingly difficult to store and process the data. A network protocol layer and signal processing algorithms are needed to extract the important information from the sensor data. Also, the transmission distance between sensors tend to be short (< 10 m) as compared to conventional macrosensors. This leads to lower transmission power being dissipated, and different architectures for computation partitioning will be necessary.

Another important design consideration in microsensors is power awareness, where all sensors are able to adapt energy consumption as energy resources of the system diminish or as performance requirements change. A power-aware node will have a longer lifetime and lend to more efficient sensor systems. Power-aware design is different than low power design, which often assumes a worst case power dissipation [4]. Instead, in power-aware design, the idea is that the system energy consumption should scale with changing conditions and quality requirements. This is important in order to enable the user to trade-off system performance parameters as opposed to hard-wiring them. For example, the user may want to sacrifice system performance or latency in return for maximizing battery lifetime. One property of a well-designed power-aware system, is one that degrades its quality and performance as available energy resources are depleted instead of exhibiting an "all-or-none" behavior. In this chapter, we discuss desirable traits of energy-quality scalable implementation of algorithms.

The application of this chapter involves use of acoustic sensors to make valuable inferences about the environment. Acoustic sensors are highly versatile and can be used in a variety of applications, such as speech recognition, traffic monitoring, and medical diagnosis. An example application is source tracking and localization. Multiple sensors can be used to pinpoint the location of an acoustic source (e.g. vehicle, speaker), by using a line of bearing estimation technique. Another example application is source classification and identification. For example, in a speech application, the end-user may want to gather speech data which can be used for speaker identification and verification. We will explore the networking, algorithmic and architectural challenges of designing wireless sensor networks in the context of these applications.

16.2 Power-Aware Node Architecture

A prototype sensor node based on the StrongARM SA-1100 microprocessor has been developed as part of the MIT micro-Adaptive Multi-domain Power-Aware Sensors (μAMPS) project. In order to demonstrate power-aware DSPs for sensor network applications, networking protocols and algorithms have been implemented and run on the SA-1100.

Figure 16.2 is a block diagram of the sensor node architecture. The node can be separated into four subsystems. The interface to the environment is through the sensing subsystem which for our node consists of two sensors (acoustic and seismic), connected to an Analog-to-Digital (A/D) converter. The acoustic sensor which is primarily used for source tracking and classification is an electret microphone with low-noise biasing and amplification. The seismic sensor is not used for this application, but is also useful for data gathering. For source tracking, a 1-kHz conversion rate of the 12-bit A/D is required. The data is continuously sampled, and stored in on-board RAM to be processed by the data and control processing subsystem.

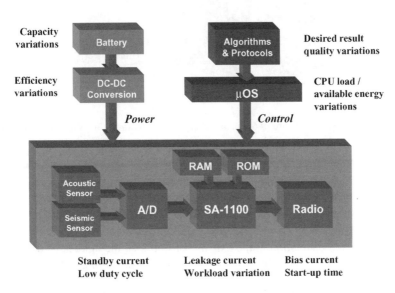

Figure 16.2.   The architectural overview of a sensor node

Depending on the role of the sensor within the network, the data is processed in the data and control processing subsystem. For example, if the sensor is a data-aggregator, then signal processing is performed. However, if the node is a relay, then the data is routed to the communication subsystem to be transmitted. The central component of the data and control processing subsystem is the StrongARM SA-1100 microprocessor. The SA-1100 is selected for its low power consumption, sufficient performance for signal processing algorithms, and static CMOS design. In addition the SA-1100 can be programmed to run at a range of clock speeds from 50 to 206 MHz and at voltage supplies from 0.8 to 1.44 V [5]. On-board ROM and RAM are included for storage of sampled data, signal processing algorithms and the "m-OS". The μ-OS is a lightweight, multithreaded operating system constructed to demonstrate the power-aware algorithms. Figure 16.3 shows a printed circuit board which implements the StrongARM based data and control processing subsystem.

In order to collaborate with neighboring sensors and with the end-user, the data from the StrongARM is passed to the radio or communication subsystem of the node. The primary

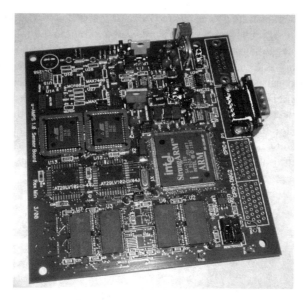

Figure 16.3. Printed circuit board of the data and control processing subsystem of the μAMPS sensor node

component of the radio is a commercial single-chip transceiver optimized for ISM 2.45 GHz wireless systems. The PLL, transmitter chain, and receiver chain are capable of being shut-off under software or hardware control for energy savings. To transmit data, an external Voltage-Controlled Oscillator (VCO) is directly modulated, providing simplicity at the circuit level and reduced power consumption at the expense of limits on the amount of data that can be transmitted continuously. The radio module is capable of transmitting up to 1 Mbps at a range of up to 10 m.

The final subsystem of the node is the battery subsystem. The power for the node is supplied by a single 3.6-V DC source, which can be provided by a single lithium-ion cell or three NiCD or NiMH cells. Regulators generate 5-, 3.3-, and adjustable 0.9–1.5-V supplies from the battery. The 5-V supply powers the analog sensor circuitry and A/D converter. The 3.3-V supply powers all digital components on the sensor node with the exception of the processor core. The core is powered by a digitally adjustable switching regulator that can provide 0.9–1.6-V in 20 discrete increments. The digitally adjustable voltage allows the SA-1100 to control its own core voltage enabling Dynamic Voltage Scaling (DVS) techniques.

16.3 Hardware Design Issues

It is important to accurately estimate the energy requirements of the hardware, so that the sensors are able to estimate the energy requirement of an application, make decisions about their processing ability based on user-input and sustainable battery life, and configure themselves to meet the required goals. For example, based on the energy model for the application and the system lifetime requirements, the sensor node should be able to decide whether a particular application can be run. If not, the node might reduce its voltage using an embedded

DC/DC converter and run the application at reduced throughput or run at the same throughput but with reduced accuracy. Both of these configurations would reduce energy dissipation and increase the node's lifetime. These energy–accuracy–throughput trade-offs necessitate robust energy models for software based on parameters such as operating frequency, voltage and target processor.

16.3.1 Processor Energy Model

The computation or signal processing needed will be performed by the SA-1100 in the software. We have developed a simple energy model for software using frequency and supply voltage as parameters that incorporates explicit characterization of both switching and leakage energy. Most current models only consider switching energy [6], but in microsensor nodes which have low duty cycles, leakage energy dissipation can become large.

$$E_{tot}(V_{dd},f) = NC_L V_{dd}^2 + V_{dd}\left(I_0 e^{\frac{V_{dd}}{nV_T}}\right)\left(\frac{N}{f}\right) \tag{1}$$

where C_L is the average capacitance switched per cycle and N is the number of cycles the program takes to execute. Both these parameters can be obtained from the energy consumption data for a particular supply voltage, V_{dd}, and frequency, f, combination. The model can then be used to predict energy consumption for different supply–throughput configurations in energy-constrained environments, such as wireless microsensor networks.

Experiments on the StrongARM SA-1100 have verified this model. For the SA-1100, the processor-dependent parameters I_0 and n are computed to be 1.196 mA and 21.26 mA, respectively [7]. Then at $V_{dd} = 1.5$ V and $f = 206$ MHz, for several typical sensor DSP routines, C_L is calculated from Equation (1), and E_{tot} is measured from the StrongARM. This C_L is used with our processor energy model to estimate E_{tot} for all possible V_{dd},f combinations. Table 16.1 shows that the maximum error produced by the model was less than 5% for a set of benchmark programs.

A more advanced level of processor energy modeling is to profile the energy for different instructions. It is natural that for different instructions the processor will dissipate different amounts of energy. Figure 16.4 shows the average current drawn from the StrongARM SA-1100 while executing different instructions at $V_{dd} = 1.5$ V. This figure shows that there are variations in current drawn for different classes of instructions (e.g. memory access, ALU), but the differences are not appreciable. Thus, the common overheads associated with all

Table 16.1 Software energy model performance

DSP routines	Meas. energy (mJ)	Model parameters		Error (%)
		N ($\times 10^6$)	C_L	
FFT	53.89	43.67	0.65	1.24
DCT	0.10	0.08	0.66	4.22
IDCT	0.13	0.10	0.66	2.59
FIR	1.23	0.97	0.70	3.28
TDLMS	21.29	17.10	0.71	1.91

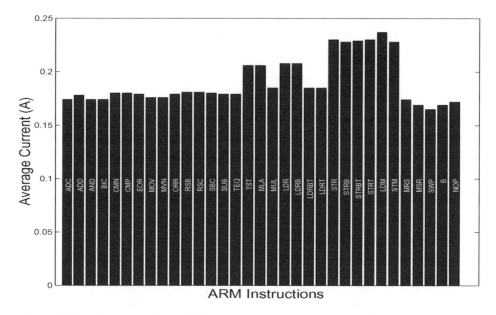

Figure 16.4.   Current profiling of different instructions executed on the StrongARM SA-1100

instructions (e.g. instruction fetch, caches) dominate the energy dissipated per operation. We expect that the variation between instructions will be more prominent in processors that use clock gating. Clock gating is a widely used low power technique where the clock is only enabled for those circuits that are active. Disabling non-active circuits eliminates unnecessary switching, which leads to energy savings.

16.3.2 DVS

Most systems are designed for the worst case scenario. For example, timing is often based on the worst case latency. For energy-scalable systems, where there is a variable computational load, this may not be optimal for energy dissipation. For example, assume in a fixed through-put system, the computation with the worst case latency takes T seconds to compute. Suppose that profiling done on the application shows that most of the time the processor is executing a task which has a computational load half that of the worst case, as shown in Figure 16.5. Since in this case, the number of cycles is halved, so the energy of a system which has fixed voltage supply will have energy savings of 1/2 over the worst case scenario. However, this is not optimal, because after the processor completes the task, it will idle for $T/2$ seconds. A better idea is to reduce the clock frequency by half, so that the processor is active for the entire period, and allows us to reduce the voltage supply by 1/2. According to the processor energy model (Equation (1)) the energy is linearly related to N, the number of cycles of a program and is also related to voltage supply squared. This means that by using a variable voltage supply the amount of energy dissipated is 1/4 that of the fixed voltage supply case. Figure 16.5 also shows a graph comparing E_{fixed} and E_{var} for a variable workload. This graph shows the quadratic relationship between energy and computation when using a variable voltage scheme.

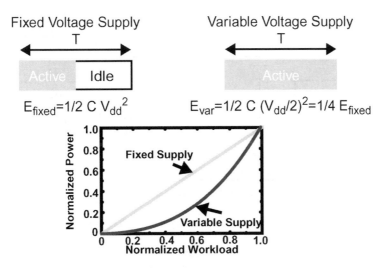

Figure 16.5.   As processor workload varies, using a variable power supply gives quadratic savings

Figure 16.6a depicts the measured energy consumption of an SA-1100 processor running at full utilization. Energy consumed per operation is plotted with respect to the processor frequency and voltage. This figure shows the quadratic dependence of switching energy on supply voltage, and also for a fixed voltage, the leakage per operation increases as the operations occur over a longer clock period. Figure 16.6b shows all 11 frequency–voltage pairs for the StrongARM SA-1100. DVS is a technique which changes the voltage supply and clock frequency of a processor depending on the computational load. It is one technique which enables energy-scalability, as it allows the sensor to change its voltage supply depending on changing requirements. Figure 16.7 illustrates the regulation scheme on our sensor node for DVS support. The µOS running on the SA-1100 selects one of the above 11

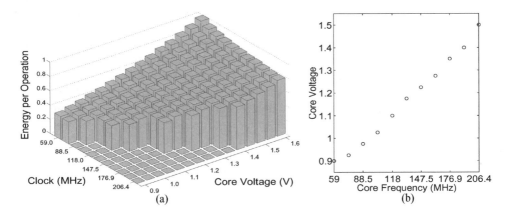

Figure 16.6.   (a) Measured energy consumption characteristics of SA-1100. (b) Operating voltage and frequency pairs of the SA-1100

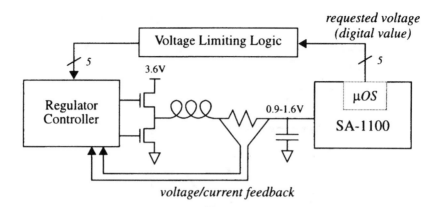

Figure 16.7. Feedback for dynamic voltage scaling

frequency–voltage pairs in response to the current and predicted workload. A 5-bit value corresponding to the desired voltage is sent to the regulator controller, and logic external to the SA-1100 protects the core from a voltage that exceeds its maximum rating. The regulator controller typically drives the new voltage on the buck regulator in under 100 ms. At the same time, the new clock frequency is programmed into the SA-1100, causing the on-board PLL to lock to the new frequency. Relocking the PLL requires 150 ms, and computation stops during this period.

16.3.3 Leakage Considerations

Processor leakage is also an important consideration that can impact the policies used in the network. With increasing trends towards low power design, supply voltages are constantly being lowered as an effective way to reduce power consumption. However, to satisfy the ever demanding performance requirements, the threshold voltage is also scaled proportionately to provide sufficient current drive and reduce the propagation delay. As the threshold voltage is lowered, the subthreshold leakage current becomes increasingly dominant.

We can measure the leakage current from the slope of the energy characteristics, for constant voltage operation. One way to look at the energy consumption is to measure the amount of charge that flows across a given potential. The charge attributed to the switched capacitance should be independent of the execution time, for a given operating voltage, while the leakage charge should increase linearly with the execution time. Figure 16.8a shows the measured charge flow as a function of the execution time for a 1024-point Fast-Fourier Transform (FFT). The amount of charge flow is simply the product of the execution time and current drawn. As expected, the total charge consumption increases almost linearly with execution time and the slope of the curve, at a given voltage, directly gives the leakage current at that voltage. The leakage current at different operating voltages was measured as described earlier, and is plotted in Figure 16.8b. These measurements verified the following model for the overall leakage current for the microprocessor core

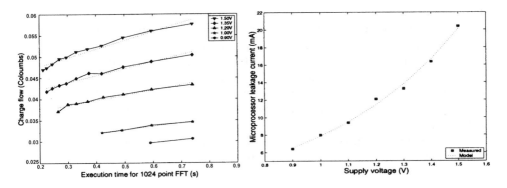

Figure 16.8.   (a) The charge consumption for 1024-point FFT. (b) The leakage current as a function of supply voltage

$$I_{\text{leak}} = I_0 e^{\frac{V_{\text{dd}}}{nV_{\text{T}}}} \qquad\qquad (2)$$

where $I_0 = 1.196$ mA and $n = 21.26$ for the StrongARM SA-1100.

Figure 16.9 shows the results after running simulations of the FFT algorithm on the StrongARM SA-1100 to demonstrate the relationship between switching and leakage energy dissipated by the processor. The leakage energy rises exponentially with supply voltage and decreases linearly with increasing frequency. Therefore to reduce energy dissipation, leakage effects must be addressed in low power design of the sensor node.

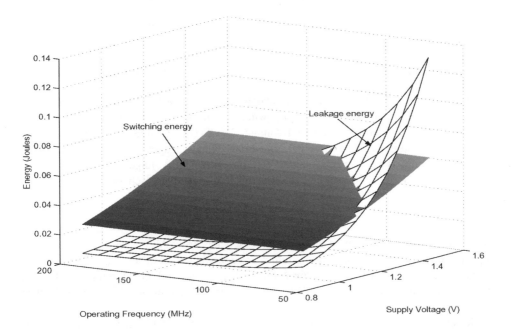

Figure 16.9.   Leakage energy dissipation can be larger than switching energy dissipation

Figure 16.10 shows current consumption trends in microprocessors for low power applications based on the ITRS report [8]. Process technology scaling improves switching speed of CMOS circuits and increases the number of transistors in a chip. To suppress a rapid increase in current consumption, the power supply is also reduced. Therefore, processor operating current increases slightly as shown in Figure 16.10i. The device threshold is reduced to maintain the switching speed at reduced power supply values. As a result, subthreshold-leakage current grows larger with technology scaling as shown in Figure 16.10ii. Thus, the operating current of the processor is strongly affected by the leakage current in advanced technologies. Using leakage control methods can significantly reduce the subthreshold leakage current in idle mode as shown in Figure 16.10iii.

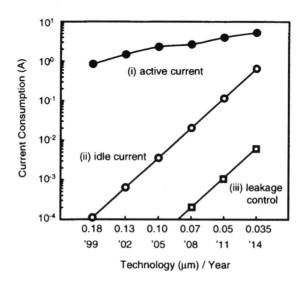

Figure 16.10.   Current consumption trend in microprocessors for portable applications

There are various ways to control the leakage. One approach is to use a high threshold voltage (V_{th}) MOS transistor as a supply switch to cut-off leakage during idle mode. It is represented by the Multiple Threshold-voltage CMOS (MT-CMOS) scheme [9]. Another approach to control leakage involves threshold-voltage adaptation using substrate-bias (V_{bb}) control that is represented by the Variable Threshold-voltage CMOS (VT-CMOS) scheme [10]. A third scheme is the switched substrate-impedance scheme.

16.3.3.1 MT-CMOS

Figure 16.11 shows a diagram of the MT-CMOS scheme. MT-CMOS uses high-V_{th} devices as supply-source switches. Inner logic circuits are constructed from low-V_{th} devices. During active mode, the switches "short" power source (V_{dd}) and ground (GND) with virtual V_{dd} and virtual GND, respectively. The high-V_{th} devices turn off during idle mode to cut-off the leakage current of low-V_{th} devices.

Measurements taken from the 1-V TI DSP [11] show that leakage energy is reduced without any drop in performance. They compared a DSP fabricated entirely in high-V_{th}

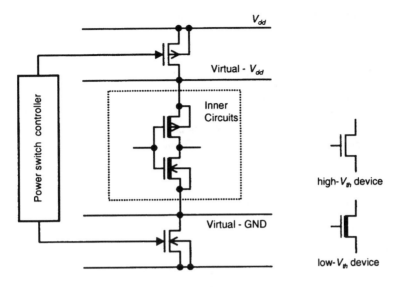

Figure 16.11.   MT-CMOS scheme

with one fabricated using MT-CMOS. At 1 V they were not able to clock the high-V_{th} DSP at the same clock speed as the MT-CMOS DSP, and therefore had to increase the voltage supply. This caused an increase in energy dissipation.

One drawback of using MT-CMOS is that the high-V_{th} supply switch produces a voltage drop between the power lines and the virtual-power lines due to the switch resistance [12]. Consequently, the drop reduces performance of the inner circuits. Therefore, the total width of the high-V_{th} devices must be as large as the inner circuits to reduce the voltage drop. However, large switches require a large chip area.

Another drawback is that power is not supplied to the inner circuits during idle mode and memory circuits, such as registers and caches lose their information in the idle mode. There have been some proposed circuits to hold the memory information. One solution is an intermittent power supply scheme, similar to the refresh process of DRAMs [13]. A balloon circuit is another method that is separated from the low-V_{th} inner devices to keep the inner circuit performance high in active state [14]. A third method is using virtual power rail clamps that set diodes between power lines and virtual power lines. The clamps supply low voltage to the inner circuits during idle state [15].

16.3.3.2 VT-CMOS

The VT-CMOS scheme changes the substrate biases ($V_{bb.p}$ and $V_{bb.n}$, for pMOS-substrate and nMOS-substrate, respectively) depending on a processor's mode. A V_{bb} controller generates V_{dd} and GND during active mode, and pulls $V_{bb.n}$ lower than GND and $V_{bb.p}$ higher than V_{dd} during idle mode. In general, decreasing $V_{bb.n}$ and increasing $V_{bb.p}$ increases V_{th}, which leads to lower subthreshold leakage currents. Similar to the MT-CMOS case, there is a possibility that active operation of the circuits controlled by the VT-CMOS becomes unstable due to the V_{bb} noise. However, a low impedance source to the substrate does not lead to high subthreshold leakage. The width of the V_{bb} switches is increased to prevent V_{bb} noise, but they are still smaller than the V_{dd} switch for the MT-CMOS case.

16.3.3.3 Switched Substrate-Impedance Scheme

The switched substrate-impedance scheme [16] as shown in Figure 16.12 is one solution for reducing V_{bb} noise by using high-V_{th} transistors. This system distributes switch cells as V_{bb} supply switches. The switch cells turn on by signals Φ_p and Φ_n during active mode. For the inner circuits, the switches connect V_{dd} and GND lines with $V_{bb.p}$ and $V_{bb.n}$ lines, respectively. In idle mode, the high-V_{th} switches turn off and the substrate-switch controller supplies appropriate $V_{bb.p}$ and $V_{bb.n}$.

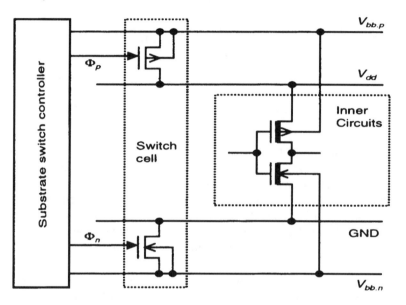

Figure 16.12.   Switched substrate-impedance scheme

There are some drawbacks to using the switched substrate-impedance scheme. One drawback occurs when the impedance of the V_{bb} switch is high, and the inner-circuits substrate lines are floating from power sources during the active mode. Unstable V_{bb}s degrade the performance of inner circuits. Another drawback is that the signals Φ_p and Φ_n have propagation delays, which cause operation errors of the inner circuits when the circuits change from idle to active mode. A feedback cell is adopted to avoid this problem.

This technique has been used in the design of the Hitachi low power microprocessor "Super-H4 (SH4)". The SH4 has achieved 1000 MIPS/W of performance. It has several operating modes. In the active mode, the processor is fully operational and consumes 420 mA current at 1.8 V voltage supply. In sleep mode, the distributed clock signals are disabled, but the clock generators and peripheral circuits are active. The sleep mode current is 100 mA. In standby mode, all modules are suspended including the clock generators and peripherals, and also the V_{bb} controller is enabled. A test chip was built to measure the standby current [16]. Results show that without the switched substrate-impedance scheme enabled, the standby current is 1.3 mA. With the V_{bb} controller on, the standby current of the entire test chip is 46.5 µA, and the overhead due to the V_{bb} controller and substrate leakage is 16.5 µA. Figure 16.13 shows the test chip micrograph of the SH4 with the switched substrate-impedance scheme implemented. The SH4 contains approximately 4M transistors, and the switch cells added for leakage reduction totaled 10K transistors. A uniform layout of one V_{bb} switch every 100 standard cells is employed. In an SRAM, the switches are set along with the word line, and in the data path, they are lined in vertical direction to the data flow. Such a distribution method guarantees the stable and high-speed inner-circuit operation. The substrate-switch controller transistor occupies 0.3% of the chip. The area overhead of the scheme is less than 5%.

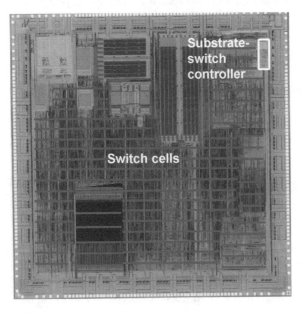

Figure 16.13.   Chip micrograph adapting switched substrate-impedance scheme

16.4 Signal Processing in the Network

As the number of sensors grows larger and larger, it becomes difficult to store and process the data collected from the sensors. Also node densities increase, so that multiple sensors may view the same event. To reduce energy dissipation, the sensors should collaborate with each other, should reduce communicating redundant information and should extract the important information from the sensor data. This is done by having a network protocol layer in order for sensors to communicate locally. Data aggregation should done on highly correlated data to reduce redundancies. By providing hooks to trade-off between computational energy and communication energy, the sensor nodes can be more energy efficient. Commercial radios typically dissipate ~150 nJ/bit and the StrongARM dissipates 1 nJ/bit [17]. In a custom DSP, the energy dissipated can be as low as 1 pJ/bit. Therefore since communication is cheap, it is more energy efficient if we can reduce the amount of data transmitted, by first doing data aggregation. Finally signal processing algorithms are used to make important inferences from the data. This section discusses optimizing protocols for microsensor networks such that signal processing is done locally at the sensor node. We also discuss optimal partitioning of computation among multiple sensors for further energy savings.

16.4.1 Optimizing Protocols

Often, sensor networks are used to monitor remote areas or disaster situations. In both of these scenarios, the end-user cannot be located near the sensors. Thus, direct communication between the sensors and the end-user, as shown in Figure 16.14a, is extremely energy-intensive, since transmission energy scales as r^n (n typically 2–4 [18]). In addition, since direct communication does not enable spatial re-use, this approach may not be feasible for large-scale sensor networks. Thus new methods of communication need to be developed.

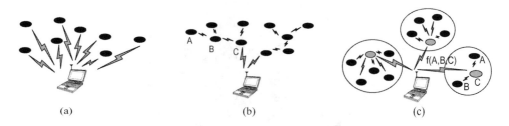

| (a) | (b) | (c) |

Figure 16.14.   (a) Direct communication with basestation. (b) Multi-hop communication with base-station. (c) Clustering algorithm. The grey nodes represent "clusterheads", and the function f(A,B,C) represents the data fusion algorithm

A common method of communication in wireless networks is multi-hop routing, where sensors act as routers for other sensors' data in addition to sensing the environment, as shown in Figure 16.14b [19–21]. Multi-hop routing minimizes the distance an individual sensor must transmit its data, and hence minimizes the dissipated energy for that sensor. One method of choosing routes is to minimize the total amount of transmit power necessary to get data from the node to the base station. In this case, the intermediate nodes are chosen such that the transmit amplifier energy (e.g. $E_{\text{Tx-amp}}(k,d) = \epsilon_{\text{amp}} \times k \times d^2$) is minimized. For example, as shown in Figure 16.14b, node A would transmit to node C through node B if

$$E_{\text{Tx-amp}}(k,d=d_{\text{AB}})+E_{\text{Tx-amp}}(k,d=d_{\text{BC}}) < E_{\text{Tx-amp}}(k,d=d_{\text{AC}}) \tag{3}$$

or

$$d_{\text{AB}}^2+d_{\text{BC}}^2 < d_{\text{AC}}^2 \tag{4}$$

However, multi-hop routing requires that several sensors transmit and receive a particular signal, so this protocol may not achieve global energy efficiency. In addition, the sensors near the end-user will be used as routers for a large number of the other sensors, and their lifetimes will be dramatically reduced using such a multi-hop protocol.

Since data from neighboring sensors will often be highly correlated, it is possible to aggregate the data locally using an algorithm such as beamforming and then send the aggregate signal to the end-user to save energy. Algorithms which can be used for data aggregation include the maximum power beamforming algorithm [22] and the Least Mean Square (LMS) beamforming algorithm [23]. These algorithms can extract the common signal from multiple sensor data.

Figure 16.15 shows the amount of energy required to aggregate data from two, three, and four sensors and to transmit the result to the end-user (E_{local}), as compared to all of the individual sensors transmitting data to the end-user ($E_{basestation}$). As shown in this plot, there is a large advantage to using local data aggregation (LMS beamforming algorithm), rather than direct communication when the distance to the basestation is large.

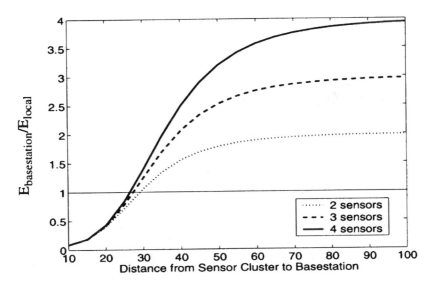

Figure 16.15.   Data aggregation done locally can reduce energy dissipation

Clustering protocols that utilize the energy savings from data aggregation can greatly reduce the energy dissipation in a sensor system. Using Low Energy Adaptive Clustering Hierarchy (LEACH) [24], an energy-efficient clustering protocol, the sensors are organized into local clusters, as shown in Figure 16.14c. Each cluster has a "clusterhead", a sensor that receives data from all other sensors in the cluster, performs data fusion (e.g. beamforming), and transmits the aggregate data to the end-user. This greatly reduces the amount of data that is sent to the end-user and thus achieves energy efficiency. Furthermore, the clusters can be organized hierarchically such that the clusterheads transmit the aggregate data using a multi-hop approach, rather than directly to the end-user so as to further reduce energy dissipation.

16.4.2 Energy-Efficient System Partitioning

One way to improve energy efficiency is to design algorithms which take advantage of the dense localization of nodes in the network. As discussed in Section 16.4.1, closely located sensors have highly correlated data that can be used in signal processing algorithms to reduce communication costs. Another way to reduce energy dissipation is to distribute the computation among the sensors. One application which demonstrates distributed processing is vehicle tracking using acoustic sensors. In this section we present an algorithm that can be performed in a distributed fashion to find the location of the vehicle.

Suppose a vehicle is moving over a region where a network of acoustic sensing nodes has been deployed. In order to determine the location of the vehicle, we first need to find the Line of Bearing (LOB) or direction from which sound is being detected. Using a LOB estimation algorithm, localization of the source can be easily accomplished. Multiple arrays or clusters of sensors determine the source's LOB from their perspective and the intersection of the LOBs determines the source's location. Figure 16.16 shows the scenario for vehicle tracking using LOB estimation.

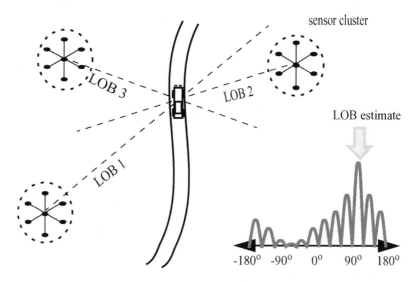

Figure 16.16.   LOB estimation to do vehicle tracking

To do tracking using LOB estimates, a clustering protocol is assumed. First within a cluster individual sensors send their acoustic data to the clusterhead. Multiple clusters determine the source's LOB to be the direction with maximum sound energy from their perspective. At the basestation the intersection point of multiple LOBs will determine the source's location.

To perform LOB estimation, one can use frequency-domain beamforming [25]. Beamforming is the act of summing the outputs of filtered sensor inputs. In a simple delay-and-sum beamformer, the filtering operations are delays or in the frequency domain phase shifts. The first part of frequency-domain beamforming is to transform collected acoustic sensor data from each sensor into the frequency domain using a 1024-point FFT. Then, we beamform the FFT data into 12 uniform directions to produce 12 candidate signals. The direction of the signal with the most energy is the LOB of the source. Figure 16.17 is a block diagram of the LOB algorithm.

The LOB estimation algorithm can also be implemented in two different ways. In the direct technique, each sensor i has a set of acoustic data $s_i(n)$. This data is transmitted to the clusterhead where the FFT and beamforming are performed. This technique is demonstrated in Figure 16.18a. Alternatively we can first perform the FFTs at each sensor and then send the FFT results to the clusterhead. This method is called the distributed technique and is demon-

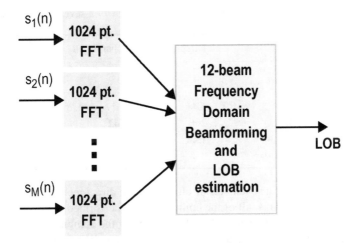

Figure 16.17.   Block diagram of the LOB estimation algorithm

strated in Figure 16.18b. If we assume the radio and processor models discussed previously, then performing the FFTs with the distributed technique has no energy advantage over the direct technique. This is because performing the FFTs at the sensor node does not reduce the amount of data that needs to be transmitted. Thus the communication costs remain the same. However, by adding circuitry to perform Dynamic Voltage Scaling (DVS), the node can take advantage of the parallelized computation load by allowing voltage and frequency to be scaled while still meeting latency constraints.

For example, if computation, C, can be computed using two parallel functional units instead of one, then the throughput is increased by 2. However if the latency is fixed, by instead using a clock frequency of $f/2$, and voltage supply of $V_{dd}/2$, then the energy is reduced by 4 times over the non-parallel case. Equation (1) demonstrates that by reducing V_{dd} yields quadratic savings in energy, but at the expense of additional propagation delay through static logic.

In the DVS enabled sensor node, there is a large advantage to having the computation distributed among the sensor nodes, since the voltage supply can be reduced. Table 16.2

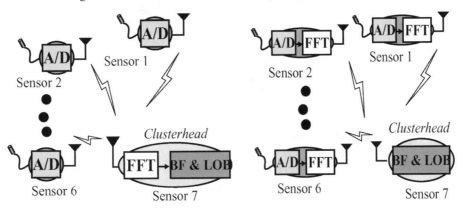

Figure 16.18.   (a) Direct technique: all of the computation is done at the clusterhead. (b) Distributed technique: distribute the FFT computation among all sensors

Table 16.2   Energy results for direct and distributed techniques for a seven-sensor cluster

		Direct	Distributed
Nodes	V_{dd} (V)	–	0.85
	f (MHz)	–	74
Clusterhead	V_{dd} (V)	1.44	1.17
	f (MHz)	206	162
	Latency (ms)	19.2	18.4
	Energy (mJ)	6.2	3.4

shows the energy results for a seven-sensor cluster. In the direct technique, with a computation latency constraint of 20 ms, all of the computation is performed at the clusterhead at the fastest clock speed, $f = 206$ MHz at 1.44 V. The energy of the computation is 6.2 mJ and the latency is 19.2 ms. In the distributed technique, the FFT is parallelized to the sensor nodes. In this scheme, the sensor nodes sense data and perform the 1024-point FFTs on the data before transmitting the FFT data to the clusterhead. At the clusterhead, the beamforming and LOB estimation is done. Since the FFTs are parallelized, the clock speed and voltage supply of both the FFTs and the beamforming can be lowered. For example, if the FFTs at the sensor nodes are run at 0.85 V and 74 MHz clock speed while the beamforming algorithm is run at 1.17 V and 162 MHz clock speed then with a latency of 18.4 ms, only 3.4 mJ is dissipated. This is a 45.2% improvement in energy dissipation. This example shows that energy-efficient system partitioning by parallelism in system design can yield large energy savings.

Figure 16.19 compares the energy dissipated for the direct techniques versus that for the distributed technique as the number of sensors is increased from three to ten sensors. This plot shows that a 20–50% energy reduction can be achieved with the system partitioning scheme.

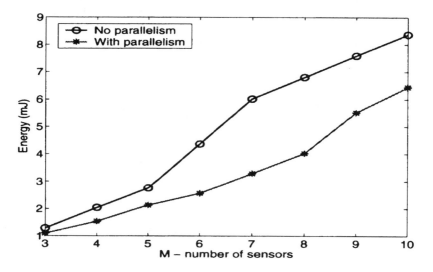

Figure 16.19.   Comparing computation energy dissipated for the direct technique vs. the distributed technique

Therefore it is important to have efficient system partitioning of computation considerations when designing protocols for wireless sensor networks.

16.5 Signal Processing Algorithms

Energy scalability can be achieved by monitoring energy resources, latency and performance requirements to dynamically reconfigure system functionality [25]. Energy–Quality (E–Q) trade-offs have been explored in the context of encryption processors [26]. Energy scalability at the algorithm and protocol levels is highly desirable because a large range of both energy and quality can be achieved by varying algorithm parameters. A large class of algorithms, as they stand, do not render themselves to such E–Q scaling.

Let us assume that there exists a quality distribution $p_Q(x)$, the probability that the end-user desires quality x. Then the average energy consumption per output sample can then be expressed as

$$\bar{E} = \int p_Q(x)E(x)\mathrm{d}x \tag{5}$$

where $\bar{E}(x)$ is the energy dissipated by the system to give quality x. A typical E–Q distribution is shown in Figure 16.20. It is clear that Algorithm II is desirable over Algorithm I because it gives higher quality at lower energies and especially when $p_Q(x)$ is large.

When the quality distribution is unknown, the E–Q behavior of the algorithm can be engineered such that it has two desirable traits. First, the quality on average should be monotonically increasing as energy increases and second, the E–Q curve should be concave downward

$$Q(E_1) \geq Q(E_2) \text{ if } E_1 \geq E_2 \tag{6}$$

$$\frac{\mathrm{d}^2 Q(E)}{\mathrm{d}E^2} \leq 0 \text{ for } 0 \leq E \leq E_{\max} \tag{7}$$

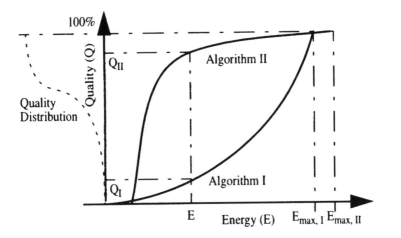

Figure 16.20.   Examples of E–Q curves

where $Q(E)$ is an accurate model of the algorithm's average quality as a function of computational energy and is the inverse of $E(Q)$. These constraints lead to intelligent energy-scalable systems. An $E–Q$ curve that is concave downward is highly desirable since close to maximal quality is achieved at lower energies. Conversely, a system that has a concave upward $E–Q$ curve can only guarantee high quality by expending a large amount of energy.

Algorithmic transformations can be used to improve the $E–Q$ characteristics of a system. For example, we will show that the $E–Q$ curves for both Finite Impulse Response (FIR) filtering and LMS beamforming for data aggregation can be transformed for better energy scalability systems.

16.5.1 Energy–Agile Filtering

FIR filtering is one of the most commonly used DSP operations. FIR filtering involves the inner product of two vectors one of which is fixed and known as the impulse response, $h[n]$, of the filter [27]. An N-tap FIR filter is defined by Equation (8).

$$y[n] = \sum_{k=0}^{N-1} x[n-k]h[k] \tag{8}$$

However, when we analyze the FIR filtering operation from a pure inner product perspective, it simply involves N Multiply and Accumulate (MAC) cycles. For desired $E–Q$ behavior, the MAC cycles that contribute most significantly to the output $y[n]$ should be done first. Each of the partial sums, $x[k]h[n-k]$, depends on the data sample and therefore it is not apparent which ones should be accumulated first. Intuitively, the partial sums that are maximum in magnitude (and can therefore affect the final result significantly) should be accumulated first. Most FIR filter coefficients have a few coefficients that are large in magnitude and progressively reduce in amplitude. Therefore, a simple but effective most-significant-first transform involves sorting the impulse response in decreasing order of magnitude and reordering the MACs such that the partial sum corresponding to the largest coefficient is accumulated first as shown in Figure 16.21a. Undoubtedly, the data sample multiplied to the coefficient might be so small as to mitigate the effect of the partial sum. Nevertheless, on an average case, the coefficient reordering by magnitude yields a better $E–Q$ performance than the original scheme.

Figure 16.21b illustrates the scalability results for a low pass filtering of speech data sampled at 10 kHz using a 128-tap FIR filter whose impulse response (magnitude) is also outlined. The average energy consumption per output sample (measured on the StrongARM SA-1100 operating at 1.44 V power supply and 206 MHz frequency) in the original scheme is 5.12 mJ. Since the initial coefficients are not the ones with most significant magnitudes the $E–Q$ behavior is poor. Sorting the coefficients and using a level of indirection (in software that amounts to having an index array of the same size as the coefficient array), the $E–Q$ behavior can be substantially improved. It can be seen that fluctuations in data can lead to deviations from the ideal behavior suggested by Equation (7), nonetheless overall concavity is still apparent. The energy overhead associated with using a level of indirection on the SA-1100 was only 0.21 mJ which is about 4% of the total energy consumption.

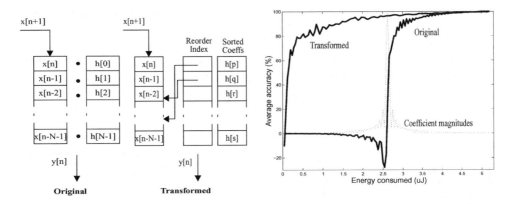

Figure 16.21. (a) FIR filtering with coefficient reordering. (b) E–Q graph for original and transformed FIR filtering

16.5.2 Energy–Agile Data Aggregation

The most-significant-first transform can also be used to improve the E–Q curves for LMS beamforming of sensor data as shown in our testbed in Figure 16.22. In order to determine the E–Q curve of the LMS beamforming algorithm, we perform beamforming on the sensor data, measure the energy dissipated on the StrongARM, and calculate the matched filter (quality) output as we vary the number of sensors in beamforming and as the source moves from location A to B. In Scenario 1, we perform beamforming without any knowledge of the source location in relation to the sensors. Beamforming is done in a preset order $\langle 1,2,3,4,5,6 \rangle$. The parameter we use to scale energy is k, the number of sensors in beamforming. As k is increased from 1 to 6, there is a proportional increase in energy. As the source moves from location A to B, we take snapshots of the E–Q curve, shown in Figure 16.23a. This curve shows that with a preset beamforming order, there can be vastly different E–Q curves depending on the source location. When the source is at location A, the beamforming

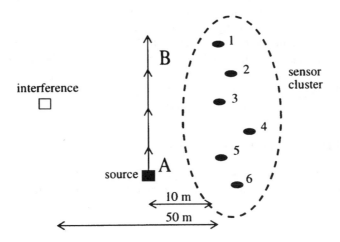

Figure 16.22. A testbed of sensors for vehicle classification

quality is only close to maximum when $k = 5, 6$. Conversely, when the source is at location B, the beamforming quality is close to maximum when $k = 2$. Therefore, since the $E–Q$ curve is highly data dependent, desirable $E–Q$ scalability cannot be guaranteed.

An intelligent alternative is to perform some initial pre-processing of the sensor data to determine the desired beamforming order for a given set of sensor data. Intuitively, we want to beamform the data from sensors that have higher sensor signal energy. We propose the

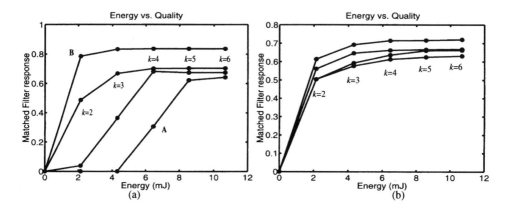

Figure 16.23.   (a) The $E–Q$ snapshot as the source moves from location A to B for the scenario of LMS beamforming with a pre-set order of sensor data. (b) The $E–Q$ snapshot as the source moves from location A to B for the scenario of LMS beamforming with significance ordering of sensor data.

most-significant-first transform, which can be applied to many algorithms to improve $E–Q$ characteristics. To find the desired beamforming order, first the sensor signal energy is estimated from the sensor data. Then the sensor signal energies are sorted using a quicksort method. The quicksort output determines the desired beamforming order.

Figure 16.23b shows the $E–Q$ relationship when an algorithmic transform is used. In this scenario, with the most-significant-first transform, we can ensure that the $E–Q$ graph is concave downward, thus improving the $E–Q$ characteristics for beamforming. However, there is a price to pay in computation energy. If the energy cost required to compute the sensor signal energies and quicksort is large compared to LMS beamforming, the most significant first transform does not improve energy efficiency. For our example, the overhead computational energy was 8.8 mJ, only 0.41% the required energy for two-sensor LMS beamforming when simulated on the SA-1100.

16.6 Signal Processing Architectures

Energy-scalable architectures have been explored to implement power-aware systems. In this section, we introduce architectures for variable-length filtering and variable-precision filtering.

16.6.1 Variable-Length Filtering

The LMS beamforming algorithm block can be implemented using a tapped delay line approach as shown in Figure 16.24. Energy is scaled by powering down the latter parts of the tapped delay line, at the expense of shorter filter lengths which can affect the performance of the LMS algorithm. Increasing the length of the adaptive filter improves the frequency resolution of the signal processing done, thus reducing mean squared error (MSE) and improving performance. However, this comes at the cost of an increase in energy dissipation. In a software implementation on the SA-1100, the number of cycles increases linearly as the filter length is increased. Thus given a specified performance requirement, the latter parts of the tapped delay line can be disabled to reduce the number of processor cycles. This, in turn, reduces the energy dissipated.

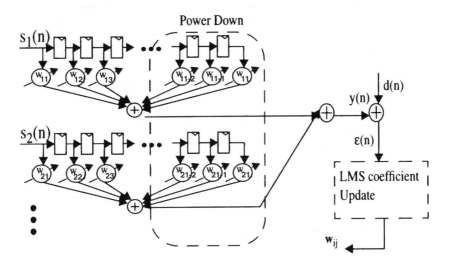

Figure 16.24.   The LMS beamforming algorithm can be implemented using a tapped delay line structure

A simple variable-length filter controller can determine the appropriate filter length by monitoring the (MSE)

$$\text{MSE} = \frac{1}{L} \sum_{n=1}^{L} \varepsilon^2(n) \qquad (9)$$

where the error function, $\varepsilon(n)$, is the mean squared error of the converging processes. A programmable threshold, α, is set and the filter length is set initially to the maximum length, $L_{\max}$. On a frame to frame basis, the filter length is decreased until the MSE is greater than α. Figure 16.25 shows the relationship between matched filter response (quality) and filter length. The optimal filter length is highly data dependent, but in general, a filter that is too short may not provide enough frequency resolution, but a filter that is too long takes longer to converge to the optimal solution.

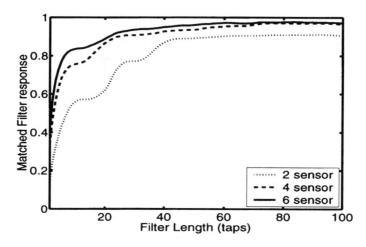

Figure 16.25.   A plot of quality vs. filter length for the LMS algorithm

Measurements from the StrongARM show that a variable-length filtering scheme can be used in conjunction with DVS for further energy savings. This is shown in Figure 16.26. This plot shows that as the program latency increases which corresponds to decreasing clock frequency, there is a large amount of energy savings using a variable voltage supply over using a fixed voltage supply.

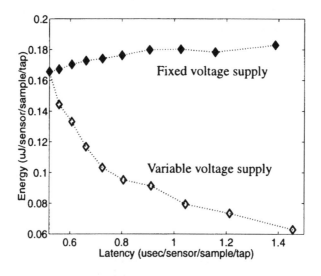

Figure 16.26.   Latency vs. energy for a variable voltage supply on the StrongARM SA-1100

16.6.2 Variable Precision Architecture

When the filter coefficients in an FIR filtering scheme are fixed, the flexibility offered by a dedicated multiplier is not required. Distributed Arithmetic (DA) is a bit-serial, multiplier-less technique that exploits the fact that one of the vectors in the inner product is fixed [28]. All possible intermediate computations (for the fixed vector) are stored in a Look-up Table (LUT) and bit slices of the variable vector are used as addresses for the LUT. A four-tap DA based FIR filter is shown in Figure 16.27.

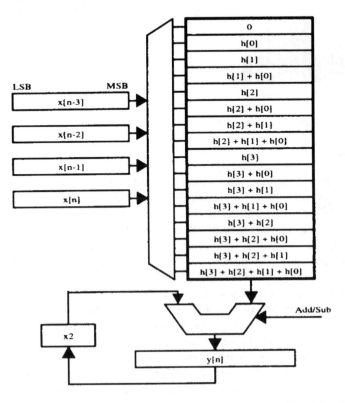

Figure 16.27.   Implementation of a four-tap filter using DA

The MSB first implementation using variable precision architecture will have desirable energy–quality characteristics. By processing the MSB first, then the most significant values are processed first. In the MSB first implementation of Figure 16.27, it has been shown in [29] that each successive intermediate value is closer to the final value in a stochastic sense. Let us assume that the maximum precision requirement is M_{max} and the immediate precision requirement is $M \leq M_{max}$. This scales down the energy per output sample by a factor M/M_{max}. Lesser precision implies that the same computation can be done faster (i.e. in M cycles instead of M_{max}).

DVS can also be used in conjunction with our variable precision architecture for further energy savings. By switching down the operating voltage such that we still meet the worst case throughput requirement (i.e. corresponding to one output sample every M_{max} cycles when operating at V_{max}), quadratic energy savings are achieved. Figure 16.28 illustrates the precision distribution for typical speech data. Notice that the distribution peaks around $M = 4$ which implies that a large fraction of the speech processing can be done using a DA for variable precision processing and reduced energy dissipation.

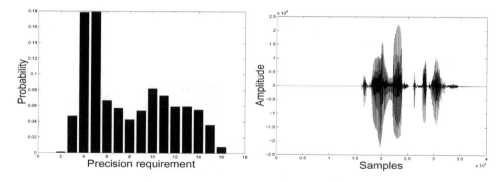

Figure 16.28.   Probability distribution of precision requirement

16.7 Conclusions

In this chapter, many new design challenges have been introduced in the field of wireless microsensor networks. A great deal of signal processing will happen at the sensor node, so it is important that the sensor node have a very energy-efficient DSP. For example, the DSP should be designed with leakage reduction due to the low-duty cycle nature of the applications. Also the DSP should allow for dynamic voltage scaling, so that for variable computational loads, the voltage and clock frequency can be dynamically reduced for further energy reduction.

Other important design considerations are the networking, algorithmic and architectural aspects of the microsensor system. In this chapter, it is shown that a clustering network protocol, which allows for sensor collaboration is highly energy-efficient. Also, optimal system partitioning of computation among the sensors in a cluster is explored. Also introduced in this chapter, are ways to implement signal processing algorithms such that they have desirable energy–quality characteristics. This is verified through energy–agile filtering and energy–agile beamforming. Finally, energy-scalable architectures are presented, which enable power-aware algorithms. Two examples of energy-scalable architectures are variable filter-length architectures and variable precision distributed arithmetic units.

As technology continues to advance, one day it will become possible to have these compact, low cost wireless sensors embedded throughout the environment, in homes/offices and ultimately inside people. With continued advances in power management, these systems should find more numerous and more impressive applications. Until that day, there is a rich set of research problems associated with distributed microsensors that require very different solutions than traditional macrosensors and multimedia devices. Energy dissipation, scalability, and latency must all be considered in designing network protocols for collaboration and information sharing, system partitioning and low power electronics.

References

[1] Karn, J., Katz, R. and Pister, K., 'Next Century Challenges: Mobile Networking for Smart Dust', *Proceedings of ACM MobiCom '99*, August 1999.
[2] Asada, G., Dong, M., Lin, T.S., Newberg, F., Pottie, G. and Kaiser, W.J., 'Wireless Integrated Network Sensors: Low Power Systems on a Chip', *Proceedings of ESSCIRC '98*, 1998.

[3] Estrin, D., Govindan, R., Heidemann, J. and Kumar, S., 'Next Century Challenges: Scalable Coordination in Sensor Networks', *Proceedings of ACM MobiCom '99*, August 1999.

[4] Bhardwaj, M., Min, R. and Chandrakasan, A., 'Power-Aware Systems', *Proceedings of the 34th Asilomar Conference on Signals, Systems, and Computers*, November 2000.

[5] Advanced RISC Machines Ltd., *Advance RISC Machines Architectural Reference Manual*, Prentice Hall, New York, 1996.

[6] Tiwari, V. and Malik, S., 'Power Analysis of Embedded Software: A First Approach to Software Power Minimization', *IEEE Transactions on VLSI Systems*, Vol. 2, December 1994.

[7] Sinha, A. and Chandrakasan, A., 'Energy Aware Software', *VLSI Design 2000*, Calcutta, India, January 2000.

[8] http://public.itrs.net/files/1999_SIA_Roadmap/Home.htm

[9] Mutoh, S., Douseki, T., Matsuya, Y., Aoki, T. and Yamada, J., '1-V high-Speed Digital Circuit Technology with 0.5 μm Multi-Threshold CMOS', *Proceedings of IEEE International ASIC Conference and Exhibition*, 1993, pp. 186–189.

[10] Kuroda, T., Fujita, T., Mita, S., Nagamatu, T., Yoshioka, S., Sano, F., Norishima, M., Murota, M., Kako, M., Kinugawa, M., Kakumu, M. and Sakurai, T. 'A 0.9-V, 150-MHz, 10-mW, 4-mm^2 2D Discrete Cosine Transform Core Processor with Variable-Threshold-Voltage Scheme', *ISSCC Digest of Technical Papers*, 1996, pp. 166–167.

[11] Lee, W., Landman, P.E., Barton, B., Abiko, S., Takahashi, H., Mizuno, H., Muramatsu, S., Tashiro, K., Fusumada, M., Pham, L., Boutaud, F., Ego, E., Gallo, G., Tran, H., Lemonds, C., Shih, A., Nandakumar, M., Eklun, R.H. and Chen, I.C., '1-V Programmable DSP for Wireless Communications', *IEEE Journal of Solid-State Circuits*, Vol. 32, No. 11, November 1997.

[12] Mutoh, S., Douseki, T., Matsuya, Y., Aoki, T., Shigematsu, S. and Yamada, J., '1-V power supply high-speed digital circuit technology with multithreshold-voltage CMOS', *IEEE Journal of Solid-State Circuits*, Vol. 30, No. 8, August 1995, pp. 847–854.

[13] Akamatsu, H., Iwata, T., Yamamoto, H., Hirata, T., Yamauchi, H., Kotani, H. and Matsuzawa, A., 'A Low Power Data Holding Circuit with an Intermittent Power Supply Scheme for Sub-1-V MT-CMOS LSIs', *Symposium on VLSI Circuits Digest of Technical Papers*, 1996, pp. 14–15.

[14] Shigematsu, S., Mutoh, S., Matsuya, Y., Tanabe, Y. and Yamada, J., 'A 1-V High-Speed MTCMOS Circuit Scheme for Power-Down Application Circuits', *IEEE Journal of Solid-State Circuits*, Vol. 32, No. 6, 1997, pp. 861–869.

[15] Kumagai, K., Iwaki. H., Yoshida, H., Suzuki, H., Yamada, T. and Kurosawa, S., 'A Novel Powering-Down Scheme for Low V_t CMOS Circuits', *Symposium on VLSI Circuits Digest of Technical Papers*, 1998, pp. 44–45.

[16] Mizuno, H., Ishibashi, K., Shimura, T., Hattori, T., Narita, S., Shiozawa, K., Ikeda, S. and Uchiyama, K., 'A 18-μA-Standby-Current 1.8-V, 200-MHz Microprocessor with Self Substrate-Biased Data-Retention Mode', *ISSCC Digest of Technical Papers*, 1999, pp. 280–281.

[17] Nord, L. and Haartsen, J., *The Bluetooth Radio Specification and the Bluetooth Baseband Specification*, Bluetooth, 1999–2000.

[18] Rappaport, T., *Wireless Communications: Principles and Practice*, Prentice Hall, NJ, 1996.

[19] Meng, T. and Volkan, R., 'Distributed Network Protocols for Wireless Communication', *Proceedings of the IEEE ISCAS*, May 1998.

[20] Shepard, T., 'A Channel Access Scheme for Large Dense Packet Radio Networks', *Proceedings of the ACM SIGCOMM*, August 1996, pp. 219–230.

[21] Singh, S., Woo, M. and Raghavendra, C., 'Power-Aware Routing in Mobile Ad Hoc Networks', *Proceedings of the Fourth Annual ACM/IEEE International Conference on Mobile Computing and Networking (MobiCom '98)*, October 1998.

[22] Yao, K., Hudson, R.E., Reed, C.W., Daching, C. and Lorenzelli, F., 'Blind Beamforming on a Randomly Distributed Sensor Array System', *IEEE Journal on Selected Topics in Communications*, Vol. 16, No. 8, October 1998.

[23] Haykin, S., Litva, J. and Shepherd, T.J., *Radar Array Processing*, Springer-Verlag, Berlin, 1993.

[24] Heinzelman, W., Chandrakasan, A. and Balakrishnan, H., 'Energy-Efficient Communication Protocol for Wireless Microsensor Networks', *Proceedings of HICSS 2000*, January 2000.

[25] Nawab, S.H., Oppenheim, A.V., Chandrakasan, A.P. and Winograd, J., 'Approximate Signal Processing', *Journal of VLSI Signal Proceedings of Systems*, Vol. 15, No. 1, January 1997.

[26] Goodman, J., Dancy, A. and Chandrakasan, A.P., 'An Energy/Security Scalable Encryption Processor Using an Embedded Variable Voltage DC/DC Converter', *IEEE Journal of Solid-State Circuits*, Vol. 33, No. 11, November 1998.

[27] Oppenheim, A. and Schafer, R., *Discrete Time Signal Processing*, Prentice Hall, NJ, 1989.

[28] White, S., 'Applications of Distributed Arithmetic to Digital Signal Processing: A Tutorial Review', *IEEE ASSP Magazine*, July 1989.

[29] Xanthopoulos, T., 'Low Power Data-Dependant Transform Video and Still Image Coding', Ph.D. Thesis, Massachusetts Institute of Technology, February 1999.

17

The Pleiades Architecture

Arthur Abnous, Hui Zhang, Marlene Wan, Varghese George, Vandana Prabhu and Jan Rabaey

Rapid advances in portable computing and communication devices require implementations that must not only be highly energy efficient, but they must also be flexible enough to support a variety of multimedia services and communication capabilities. The required flexibility dictates the use of programmable processors in implementing the increasingly sophisticated digital signal processing algorithms that are widely used in portable multimedia terminals. However, compared to custom, application-specific solutions, programmable processors often incur significant penalties in energy efficiency and performance. The architectural approach presented in this chapter involves trading off flexibility for increased efficiency. This approach is based on the observation that for a given domain of signal processing algorithms, the underlying computational kernels that account for a large fraction of execution time and energy are very similar. By executing the dominant kernels of a given domain of algorithms on dedicated, optimized processing elements that can execute those kernels with a minimum of energy overhead, significant energy savings can potentially be achieved. Thus, this approach yields processors that are domain-specific.

In this chapter, a reusable architecture template (or platform), named Pleiades [1,2], that can be used to implement domain-specific, programmable processors for digital signal processing algorithms will be presented. The Pleiades architecture relies on a heterogeneous network of processing elements, optimized for a given domain of algorithms, that can be reconfigured at run time to execute the dominant kernels of the given domain. To verify the effectiveness of the Pleiades architecture, prototype processors have been designed, fabricated, and evaluated. Measured results and benchmark studies will be used to demonstrate the effectiveness of the Pleiades architecture.

17.1 Goals and General Approach

The approach that was taken in developing the Pleiades architecture template, given the overall objective of designing energy-efficient programmable architectures for digital signal processing applications, was to design processors that are optimized for a given domain of signal processing algorithms. This approach yields domain-specific processors, as opposed to general purpose processors, which are completely flexible but highly inefficient, or application-specific processors, which are the most efficient but very inflexible. The intent is to develop a processor that can, by virtue of its having been optimized for an algorithm domain, achieve high levels of energy efficiency, approaching that of an application-specific design, while maintaining a degree of flexibility such that it can be programmed to implement the variety of algorithms that belong to the domain of interest.

Algorithms within a given domain of signal processing algorithms, such as CELP based speech coding algorithms [3,4], have in common a set of dominant kernels that are responsible for a large fraction of total execution time and energy. In a domain-specific processor, this fact can be exploited such that these dominant kernels are executed on highly optimized hardware resources that incur a minimum of energy overhead. This is precisely the approach that was taken in developing the Pleiades architecture.

An important architectural advantage that can be exploited in a domain-specific processor is the use of heterogeneous hardware resources. In a general-purpose processor, using a heterogeneous set of hardware resources cannot be justified because some of those resources will always be wasted when running algorithms that do not use them. For example, a fast hardware multiplier can be quite useful for some algorithms, but it is completely unnecessary for many other algorithms. Thus, general-purpose processors tend to use general-purpose hardware resources that can be put to good use for all types of different algorithms. In a domain-specific processor, however, using a heterogeneous set of hardware resources is a valid approach, and must in fact be emphasized. This approach allows the architect a great deal of freedom in matching important architectural parameters, particularly the granularity of the processing elements, to the properties of the algorithms in the domain of interest. Even within a given algorithm, depending on the particular set of computational steps that are required, there typically are different data types and different operations that are best supported by processing elements of varying granularity, and this capability can be provided by a domain-specific design. This is precisely one of the key factors that makes an application-specific design so much more efficient than a general-purpose processor, where all operations are executed on processing elements with pre-determined architectural parameters that cannot possibly be a good fit to the various computational tasks that are encountered in a given algorithm.

Our overall objective of designing energy-efficient programmable processors for signal processing applications, and our approach of designing domain-specific processors can be distilled into the following architectural goals:

- Dominant kernels must be executed on optimized, domain-specific hardware resources that incur minimal control and instruction overhead. The intent is to increase energy efficiency by creating a good match between architectural parameters and algorithmic properties.

- Reconfiguration of hardware resources will be used to achieve flexibility while minimizing the energy overhead of instructions. Field-Programmable Gate Arrays (FPGAs), for example, do not suffer from the overhead of fetching and decoding instructions. However, the ultrafine granularity of the bit-processing elements used in FPGAs incurs a great deal of overhead for word-level arithmetic operations and needs to be addressed.

- To minimize energy consumption, the supply voltage must be reduced aggressively. To compensate for the performance loss associated with reducing the supply voltage, concurrent execution must be supported. The relative abundance of concurrency in Digital Signal Processing (DSP) algorithms provides a good opportunity to accomplish this objective.

- The ability to use different optimal voltages for different circuit blocks is an important technique for reducing energy consumption and must be supported. This requires that the electrical interfaces between circuit modules be independent of the varying supply voltages used for different circuit modules.
- Dynamic scaling of the supply voltage is an important technique to minimize the supply voltage, and hence energy consumption, to the absolute minimum needed at any given time and must be supported.
- The structure of the communication network between the processing modules must be flexible such that it can be reconfigured to create the communication patterns required by the target algorithms. Furthermore, to reduce the overhead of this network, hierarchy and reduced voltage swings will be used. The electrical interface used in the communication network must not be a function of the supply voltages of the modules communicating through the network.
- In order to avoid the large energy overhead of accessing large, centralized hardware resources, e.g. memories, datapaths, and buses, locality of reference must be preserved. The ability to support distributed, concurrent execution of computational steps is the key to achieving this goal, and it is also consistent with our goal of highly concurrent processing for the purpose of reducing the supply voltage.
- A key architectural issue in supporting highly concurrent processing is the control structure that is used to coordinate computational activities among multiple concurrent hardware resources. The control structure has a profound effect on how well an architecture can be scaled to match the computational characteristics of the target algorithm domain. The performance and energy overheads of a centralized control scheme can be avoided by using a distributed control mechanism. Ease of programming and high-quality automatic code generation are also important issues that are influenced by the control structure of a programmable architecture.
- Unnecessary switching activity must be completely avoided. There must be zero switching activity in all unused circuit modules.
- Time-sharing of hardware resources must be avoided, so that temporal correlations are preserved. This objective is consistent with and is in fact satisfied by our approach of relying on spatial and concurrent processing. Point-to-point links in the communication network, as opposed to time-shared bus connections, should be used to transmit individual streams of temporally-correlated data.

17.2 The Pleiades Platform – The Architecture Template

In this section, a general overview of the Pleiades architecture will be presented. Additional details and architectural design issues will be presented and discussed in the following sections. Architectural design of the P1 prototype and the Maia processor will be presented subsequently.

The Pleiades architecture is based on the *platform template* shown in Figure 17.1. This template is reusable and can be used to create an *instance* of a domain-specific processor, which can then be programmed to implement a variety of algorithms within the given domain

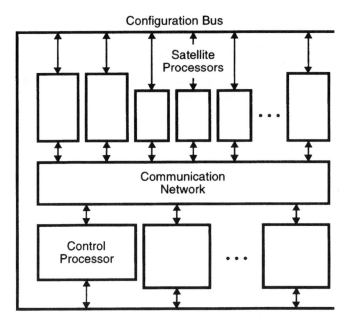

Figure 17.1.   The Pleiades architecture template

of interest. All instances of this architecture template share a fixed set of control and communication primitives. The type and number of processing elements in a given domain-specific instance, however, can vary and depend on the properties of the particular domain of interest.

The architecture template consists of a *control processor*, a general-purpose microprocessor core, surrounded by a heterogeneous array of autonomous, special-purpose *satellite processors*. All processors in the system communicate over a reconfigurable communication network that can be configured to create the required communication patterns. All computation and communication activities are coordinated via a distributed data-driven control mechanism. The dominant, energy-intensive computational kernels of a given DSP algorithm are implemented on the satellite processors as a set of independent, concurrent threads of computation. The rest of the algorithm, which is not compute-intensive, is executed on the control processor. The computational demand on the control processor is minimal, as its main task is to configure the satellite processors and the communication network (via the configuration bus), to execute the non-intensive parts of a given algorithm, and to manage the overall control flow of the algorithm.

In the model of computation used in the Pleiades architecture template, a given application implemented on a domain-specific processor consists of a set of concurrent communicating processes [5] that run on the various hardware resources of the processor and are managed by the control processor. Some of these processes correspond to the dominant kernels of the given application program and run on satellite processors under the supervision of the control processor. Other processes run on the control processor under the supervision of a simple interrupt-driven foreground/background system for relatively simple applications or under the supervision of a real-time kernel for more complex applications [6]. The control processor configures the available satellite processors and the communication network at run-time to construct the dataflow graph corresponding to a given computational kernel directly in the hardware. In the hardware structure thus created, the satellite processors correspond to the nodes of the dataflow graph, and the links through the communication network correspond to the arcs of the dataflow graph. Each arc in the dataflow graph is assigned a dedicated link through the communication network. This ensures that all temporal correlations in a given stream of data are preserved and the amount of switching activity is thus minimized.

Algorithms within a given domain of applications, e.g. CELP based speech coding, share a common set of operations, e.g. LPC analysis, synthesis filtering, and codebook search. When and how these operations are performed depend on the particular details of the algorithm being implemented and are managed by the control processor. The underlying details and the basic parameters of the various computational kernels in a given domain vary from algorithm to algorithm and are accommodated at run-time by the reconfigurability of the satellite processors and the communication network.

The Pleiades architecture enjoys the benefit of reusability because (a) there is a set of predefined control and communication primitives that are fixed across all domain-specific instances of the template, and (b) predefined satellite processors can be placed in a library and reused in the design of different types of processors.

17.3 The Control Processor

A given algorithm can be implemented in its entirety on the control processor, without using any of the satellite processors. The resulting implementation, however, will be very inefficient: it will be too slow, and it will consume too much energy. To achieve good performance and energy efficiency, the dominant kernels of the algorithm must be identified and implemented on the satellite processors, which have been optimized to implement those kernels with a minimum of energy overhead. Other parts of the algorithm, which are not compute-intensive and tend to be control-oriented, can be implemented on the control processor. The computational load on the control processor is thus relatively light, as the bulk of the computational work is done by the satellite processors.

In addition to executing the non-compute-intensive and control-oriented sections of a given algorithm, the control processor is responsible for *spawning* the dominant kernels as independent threads of computation, running on the satellite processors. In this capacity, the control processor must first configure the satellite processors and the communication network such that a suitable hardware structure for executing a given kernel is created. The satellite processors and the communication network are reconfigured at run-time, so that different kernels are executed at different times on the same underlying reconfigurable hardware fabric. The functionality of each hardware resource, be it a satellite processor or a switch in the communication network, is specified by the configuration state of that resource, a collection of bits that instruct the hardware resource what to do. The *configuration state* of each hardware resource is stored locally in a suitable storage element, i.e. a register, a register file, or a memory. Thus, storage for the configuration states of the hardware resources of a processor are distributed throughout the system. These configuration states are in the memory map of the control processor and are accessed by the control processor through the reconfiguration bus, which is an extension of the address/data/control bus of the control processor.

Once the satellite processors and the communication network have been properly configured, the control processor must initiate the execution of the kernel at hand. This is accomplished by generating a request signal to an appropriate satellite processor which will trigger the sequence of events whereby the kernel is executed. After initiating the execution of the kernel, the control processor can either halt (to save power) and wait for the completion of the kernel, or it can start executing another computational task, including spawning another kernel on another set of satellite processors. This mode of operation allows the programmer to increase processing throughput by taking advantage of coarse-grain parallelism. When the execution of the kernel is completed, the control processor receives an interrupt signal from the appropriate satellite processor. The interrupt service routine will determine the next course of action to be taken by the control processor.

17.4 Satellite Processors

The computational core of the Pleiades architecture consists of a heterogeneous array of autonomous, special-purpose satellite processors. These processors are optimized to execute specific tasks efficiently and with minimal energy overhead. Instead of executing all computations on a general-purpose datapath, as is commonly done in conventional programmable processors, the energy-intensive kernels of an algorithm are executed on optimized datapaths, without the overhead of fetching and decoding an instruction for every single computational step.

Kernels are executed on satellite processors in a highly concurrent manner. A cluster of interconnected satellite processors that implements a kernel processes data tokens in a pipelined manner, as each satellite processor forms a pipeline stage. In addition, each satellite processor can be further pipelined internally. Furthermore, multiple pipelines corresponding to multiple independent kernels can be executed in parallel. These capabilities allow efficient processing at very low supply voltages. For bursty applications with dynamically varying throughput requirements, dynamic scaling of the supply voltage is used to meet the throughput requirements of the algorithm at the minimum supply voltage.

As mentioned earlier, satellite processors are designed to perform specific tasks. Let us consider some examples of satellite processors:

- Memories are ubiquitous satellite processors and are used to store the data structures processed by the computational kernels of a given algorithm domain. The type, size, and number of memories used in a domain-specific processor depend on the nature of the algorithms in the domain of interest.
- Address generators are also common satellite processors that are used to generate the address sequences needed to access the data structures stored in memories in the particular manner required by the kernels.
- Reconfigurable datapaths can be configured to implement the various arithmetic operations required by the kernels.
- Programmable Gate Array (PGA) modules can be configured to implement various logic functions, as needed by the computational kernels.
- Multiply-Accumulate (MAC) processors can be used to compute vector dot products very efficiently. MAC processors can be useful in a large class of important signal processing algorithms.
- Add-Compare-Select (ACS) processors can be used to implement the Viterbi algorithm

efficiently. The Viterbi algorithm is widely used in many communication and storage applications.

- Discrete Cosine Transform (DCT) processors can be used to implement many image and video compression/decompression algorithms efficiently.

Observe that while most satellite processors are dedicated to performing specific tasks, some satellite processors might support a higher degree of flexibility to allow the implementation of a wider range of kernels. The proper choice of the satellite processors used in a given domain-specific processor depends on the properties of the domain of interest and must be made by careful analysis of the algorithms belonging to that domain.

The behavior of a satellite processor is dictated by the configuration state of the processor. The configuration state of a satellite processor is stored in a local configuration store and is accessed by the control processor via the reconfiguration bus. For some satellite processors, the configuration state consists of a few basic parameters that determine what the satellite processor will do. For other satellite processors, the configuration state may consist of sequences of basic instructions that are executed by the satellite processor. Instruction sets and program memories for the latter type of satellite processors are typically shallow, as satellite processors are typically designed to perform a few basic operations, as required by the kernels, very efficiently. As such, the satellite processors can be considered weakly programmable. For a memory satellite processor, the contents of the memory make up the configuration state of the processor.

Figure 17.2 shows the block diagram of a MAC satellite processor. Figure 17.3 illustrates how one of the energy-intensive functions of the VSELP speech coder, the weighted synthesis filter, is mapped onto a set of satellite processors.

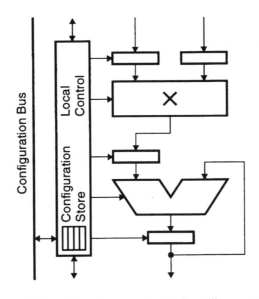

Figure 17.2. Block diagram of a MAC satellite processor

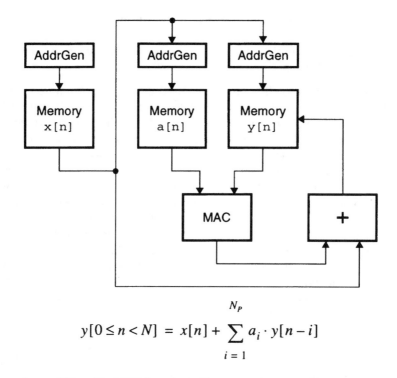

$$y[0 \leq n < N] = x[n] + \sum_{i=1}^{N_P} a_i \cdot y[n-i]$$

Figure 17.3.   The VSELP synthesis filter mapped onto satellite processors

17.5 Communication Network

In the Pleiades architecture, the communication network is configured by the control processor to implement the arcs of the dataflow graph of the kernel being implemented on the satellite processors. As mentioned earlier, each arc in the dataflow graph is assigned a dedicated channel through the communication network. This ensures that all temporal correlations in a given stream of data are preserved, and the amount of switching activity is reduced.

The communication network must be flexible enough to support the interconnection patterns required by the kernels implemented on a given domain-specific processor, while minimizing the energy and area cost of the network. In principle, it is straightforward to provide the flexibility needed to support all possible interconnection patterns for a given set of processors. This can be accomplished by a *crossbar* network. A crossbar network can support simultaneous, non-blocking connection of any of M input ports to any of N output ports. This can be accomplished by N buses, one per output port, and a matrix of $N \times M$ switches. The switches can be configured to allow any given input port to be connected to any of the output buses. However, the global nature of the buses and the large number of switches make the crossbar network prohibitively expensive in terms of both energy and area, particularly as the number of input and output ports increases. Each data transfer incurs a great deal of energy overhead, as it must traverse a long global bus loaded by N switches.

In practice, a full crossbar network can be quite unnecessary and can be avoided. One reason is that not all output ports might be actively used simultaneously. Some output ports might in fact be mutually exclusive of one another. Therefore, the number of buses needed can be less than the number of output ports in the system. Another practical fact that can be exploited to reduce the complexity of a full crossbar (and other types of networks, as well) is that not all input ports need to be connected to all available output ports in the system. For example, address generators typically communicate with memories only, and there is no need to allow for the possibility of connecting the address inputs of memory modules to the output ports of the arithmetic units. This fact can be used to reduce the span of the buses and the number of switches in the network.

The efficiency of data transfers can be improved by taking advantage of the fact that most data transfers are local. This is a direct manifestation of the principle of locality of reference. Instead of using buses that span the entire system, shorter bus segments can be used that allow efficient local communication. Many such architectures have been proposed, particularly for use in multiprocessor systems [7]. These topologies provide efficient point-to-point local channels at the expense of long-distance communications. One simple scheme for transferring data between non-adjacent nodes is to route data tokens through other intervening processors. This increases the latency of data transfers, but keeps the interconnect structure simple. An additional drawback is that the latency of a data transfer becomes a function of processor placement and operation assignment. As a result, scheduling and assignment of operations become more complicated, and developing an efficient compiler becomes more difficult.

The mesh topology has been particularly popular in modern FPGA devices. The mesh structure is simple and very efficient for VLSI implementations. A simplified version of the mesh structure, as used in many modern FPGAs [8], is illustrated in Figure 17.4. To transfer data between non-adjacent processing elements, multiple unit length bus segments can be concatenated by properly configuring the switch-boxes that are placed at the boundaries of the processing elements. Local communication can be accomplished efficiently, and non-

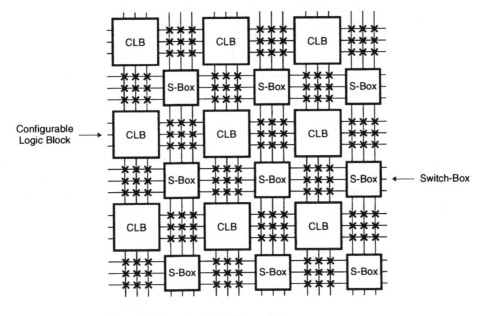

Figure 17.4.   Simple FPGA mesh interconnect structure

local communications can be supported, as well, and the degradation of communication bandwidth with distance, due to the increasing number of switches as more switch-boxes are traversed, is relatively graceful. This scheme has worked quite well in FPGAs, but it is not directly applicable to a Pleiades-style processor because a Pleiades-style processor is composed of a heterogeneous set of satellite processors with different shapes and sizes and the regular two-dimensional array structure seen in FPGAs cannot be created.

The scheme used in the Pleiades architecture is a generalization of the mesh structure, i.e. a generalized mesh [9], which is illustrated in Figure 17.5. For a given placement of satellite processors, wiring channels are created along the sides of the satellite processors. Configurable switch-boxes are placed at the junctions between the wiring channels, and the required communication patterns are created by configuring these switch-boxes. The parameters of this generalized mesh structures are the number of buses employed in a given wiring channel, and the exact functionality of the switch-boxes. These parameters depend on the placement of the satellite processors and the required communication patterns among the satellite processors.

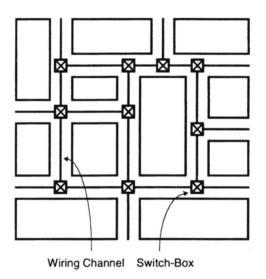

Wiring Channel   Switch-Box

Figure 17.5.   Generalized mesh interconnect structure

An important and powerful technique that can be used in improving the performance and efficiency of the communication network is the use of hierarchy. By introducing hierarchy, locality of reference can be further exploited in order to reduce the cost of long-distance communications. One approach that has been used in some FPGAs, e.g. the Xilinx XC4000 family [8], is to use a hierarchy of lengths in the bus segments used to connect the logic blocks. Instead of using only unit-length segments, longer segments spanning two, four, or more logic blocks are also used. Distant logic blocks can be connected via these longer segments by using far less series switches than would have been needed if only unit-length bus segments were available.

Another approach to introducing hierarchy in the communication network is to use additional levels of interconnect that can be used to create connections among *clusters* of processing elements. An example of this approach is the two-level network structure used in the PADDI-2 multiprocessor [10]. In PADDI-2, a level-1 reduced crossbar network is used to connect nanoprocessors within clusters of four nanoprocessors. A level-2 reduced and segmented crossbar is used to create connections between the clusters. Another example of the application of hierarchy is the binary tree structure used in the Hierarchical Synchronous Reconfigurable Array (HSRA) architecture [11]. In this approach, a binary-tree hierarchy of switch-boxes is used to reduce the cost of communications between distant logic blocks. Local shortcuts are also used to facilitate efficient neighbor-to-neighbor connections, without the need to traverse the tree of switch-boxes.

In the Pleiades architecture, hierarchy is introduced into the communication network by creating clusters of tightly-connected satellite processors that internally use a generalized-mesh structure. Communication among clusters is accomplished by introducing inter-cluster switchboxes that allow inter-cluster communication through the next higher level of the communication network. This is illustrated in Figure 17.6. The key challenge is the proper clustering of the satellite processors and the proper placement of the inter-cluster switch-boxes in order to avoid routing congestions. The proper organization can be found by closely studying the interconnection patterns that occur in the computational kernels of a given domain of algorithms.

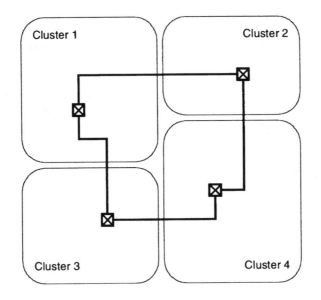

Figure 17.6.   Hierarchical generalized mesh interconnect structure

In addition to the techniques mentioned above, the Pleiades architecture uses reduced swing bus driver and receiver circuits to reduce the energy of data transfers through the network [12,13]. An additional benefit of this approach is that the electrical interface through the communication network is standardized and becomes independent of the supply voltages of the communicating satellite processors. This facilitates the use of dynamic scaling of the supply voltage, as satellite processors at the two ends of a communication channel can run at independent supply voltages.

17.6 Reconfiguration

In the Pleiades architecture, the flexibility needed to support the various kernels of a given domain of algorithms is achieved by the ability to reconfigure the satellite processors and the communication network at run-time, such that a hardware organization suitable for implementing a given kernel is created. This mode of programming is known as *spatial programming*, whereby the act of programming changes the physical interconnection of processing elements, thus creating a new hardware organization, i.e. a particular set of processing elements interconnected in a particular way, to implement a new computation. This is the mode of programming used in FPGAs. Traditional programmable processors rely on *temporal programming*, whereby the behavior of processing elements is altered in time, on a cycle-by-cycle basis, by a stream of instructions, and the underlying hardware organization is fixed.

As mentioned earlier, the behavior of satellite processors and the pattern of interconnections among them is dictated by the configuration state of the satellite processors and the switches in the communication network. Configuring a set of satellite processors or a set of switches in the communication network consists of altering the configuration state of these hardware resources by the control processor via the configuration bus. This is similar to what is done when programming FPGAs. However, in conventional FPGAs such as the Xilinx XC4000 family, reconfiguration is a very slow task that can take milliseconds of time. As a result, run-time reconfiguration is not practical with conventional FPGAs. One basic reason for this shortcoming is that it takes a tremendous amount of configuration information to configure an FPGA. Part of the problem is the bit-level granularity of the processing elements. All details of the logic functions that are needed to implement a particular function must be fully specified. For example, it takes 360 bits of information to configure a Xilinx XC4000E CLB and its associated interconnect switches [14]. The situation is further exacerbated when implementing word-level arithmetic operations, when a great deal of the configuration information is redundant and specifies the same logic functionality for different bits of a datapath. An additional obstacle to run-time reconfiguration is that FPGAs are typically configured in a bit-serial fashion[1]. The PADDI-2 DSP multiprocessor was also configured in a bit-serial manner, and as a result run-time reconfiguration was not practical, but this was not really a limitation for the design, as PADDI-2 was designed for rapid prototyping applications.

In the Pleiades architecture, since hardware resources are configured at run-time, so that different kernels can be executed on the same basic set of satellite processors at different times during the execution of an algorithm, a key design objective is to minimize the amount of time spent on configuring and re-configuring hardware resources. This can be accomplished with a combination of architectural strategies. The first strategy is to reduce the amount of configuration information. The word-level granularity of the satellite processors and the communication network is one contributing factor. No redundant configuration information is wasted on specifying the behavior of individual bit-slices of a multi-bit datapath. This is a direct result of the types of data tokens processed by signal processing algorithms. Another factor is that the behavior of most satellite processors (with the notable exception of PGA-style satellite processors) is specified by simple coarse-grain instructions choosing one of a few different possible operations supported by a satellite processor and a

---

[1] In some recent devices, configuration information can be loaded into the device via a byte-wide bus [14].

few basic parameters, as necessary. For example, a MAC satellite processor can be fully configured by specifying whether to perform multiplication operations or to perform vector dot-product operations. Address generators can be configured by specifying one of a few different address sequences and specifying the associated address generation bounds, steps, and strides, as necessary. As a result, all it takes for the control processor to configure the satellite processors and the communication network is to load a few configuration store registers with the appropriate values.

Another strategy to reduce reconfiguration time in the Pleiades architecture is that configuration information is loaded into the configuration store registers by the control processor through a wide configuration bus, an extension of the address/data/control bus of the control processor. For example, with a 32-bit control processor, such as the ARM9 microprocessor core [15], configuration information can be loaded into the configuration store registers of the satellite processors and the communication network at a rate of 32 bits per cycle.

Another technique to minimize or even eliminate configuration time is to overlap configuration and kernel execution. While satellite processors are busy executing a kernel, they can be configured by the control processor for the next kernel to be executed. When the execution of the current kernel is completed, the satellite processors can start the next kernel immediately by switching to the new configuration state. This can be accomplished by allowing multiple configuration contexts, i.e. multiple sets of configuration store registers. This technique is similar to those used in multi-context and time-multiplexed FPGA devices [16–18]. While one configuration context is active and is used by the satellite processors and the communication network to execute the current kernel, a second passive configuration context is simultaneously loaded by the control processor in preparation for the next kernel. When the execution of the kernel is finished, the new context becomes the active context, and the old context can be loaded with new configuration state in anticipation of the next kernel to be executed. This mode of operation is illustrated in Figure 17.7. An extension of this technique is to allow more than two configuration contexts, at least for some of the satellite processors. These configuration contexts can be pre-loaded when the system is initialized, and there will be no need to reconfigure the associated satellite processor at run-time. This latter technique was used in the address generators of the Maia processor.

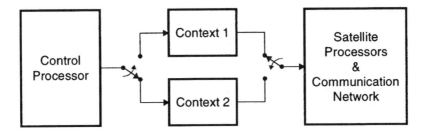

Figure 17.7.   Concurrent reconfiguration and kernel execution

17.7 Distributed Data-Driven Control

Coordination of computation and communication activities among the processing elements of a multiprocessor system is one of the most important architectural design issues, as it has a profound effect on the efficiency of the overall design. This task is performed by a suitable control mechanism. The responsibility of the control mechanism is to provide instructions to the processing elements, i.e. the functional units, the data memories, and the communication network. In doing so, the control mechanism requires control information from the processing elements, indicating their current states to the control mechanism. How instructions are stored, how they are dispatched to the processing elements, and how control information provided by the processing elements is handled are the key issues that must be addressed in designing a control mechanism.

In the Pleiades architecture, computational kernels are executed on the satellite processors in a distributed, concurrent manner. This approach avoids the energy and performance overheads of large, centralized functional units and data memories by replacing global interactions across long distances by more local interactions across shorter distances. This same approach can be applied to the design of the control mechanism. The Pleiades architecture uses a *distributed* control mechanism that employs small local controllers in place of a large global controller.

In a *centralized* control mechanism, a single global controller is responsible for controlling the activities of all processing elements. VLIW and SIMD architectures, for example, use a centralized control mechanism. The conceptual simplicity of this scheme works well when there is a single thread of computation. In a multiprocessor system with multiple processors executing multiple threads of computation, however, a centralized control mechanism loses its conceptual simplicity and becomes quite cumbersome, as the controller has to deal with the combinatorial explosion of control states as the combined states of the individual processing elements are considered together. As a result, developing programs and compilers for architectures that use a centralized control mechanism becomes very complex and difficult. Furthermore, a centralized control mechanism incurs a great deal of energy and performance overhead because instructions to the processing elements and control information from the processing elements are all communicated globally through the central controller. As a result, a centralized control mechanism cannot practically be scaled up to deal with a large number of processing elements because the required bandwidth for distributing instructions and control information and the associated energy overhead and performance penalty can become prohibitive.

In a distributed control mechanism, each processing element has a local controller with a local program memory. As a result, the energy and performance overheads of storing and distributing instructions and communicating control information are greatly reduced as these interactions assume a local nature. With a distributed control mechanism, a computational problem can be partitioned into multiple threads of computation in the most natural way dictated by the problem itself, without the artificial constraints of a centralized control mechanism, and these threads of computation can then be distributed across multiple processing elements or multiple clusters of processing elements. The ability to take such a modular approach eases programming and developing compilers for an architecture with a distributed control mechanism. Another important advantage of a distributed control mechanism is that it can be gracefully scaled to handle multiprocessor systems with a large number of processing elements to tackle increasingly complex computational problems.

The key design issue with a distributed control mechanism is how a local controller coordinates its actions with other local controllers that it needs to interact with during the course of the execution of a given algorithm. One aspects of this problem is that each local controller must somehow determine *when* it can start executing a particular task. The objective here is to synchronize the actions of the controllers, so that computational activities are executed in the correct sequence. This can be accomplished by the exchange of *tokens* of control information among the controllers through the communication network, in the same way that data tokens are exchanged among the processing elements. Arriving control tokens cannot only be used by a controller to determine when to initiate the next computational task, but depending on the control information encapsulated into the control tokens, they can also be used to determine which particular task is to be initiated by the controller.

Minimizing the overhead of control tokens is an important design issue in a distributed control mechanism. An even more fundamental issue is how to map a given algorithm onto processing elements that are controlled in a distributed manner. The approach taken in the Pleiades architecture is to map the dataflow graph of a given signal processing kernel directly onto a cluster of satellite processors interconnected through the communication network. In this approach, a satellite processor directly corresponds to a node or a cluster of nodes in the dataflow graph of a given kernel, and a communication channel through the communication network directly corresponds to an arc in the dataflow graph. Just as in the dataflow graph representation, the execution of an operation in a satellite processor is triggered by the arrival of all required data tokens, i.e. operations are executed in a *data-driven* manner [19]. Thus, data tokens not only provide the operands to be processed by the satellite processor, but they also implicitly provide synchronization information. A *handshaking* mechanism is required to implement a data-driven mode of operation: the arrival of a data token is signaled by a *request* signal from the sending satellite processor, and the acceptance of a data token is signaled by an *acknowledge* signal from the receiving satellite processor (see Figure 17.8).

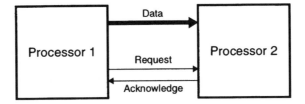

Figure 17.8. Data-driven execution via handshaking

This approach to distributed control is similar to the control mechanism of the PADDI-2 architecture [10] and the DSP architecture proposed by Fellman [20]. As we will soon see, however, the particular control mechanism used in the Pleiades architecture provides additional support for handling common signal processing data structures such as vectors and matrices more efficiently.

The conceptual simplicity and elegance of data-driven distributed control greatly simplify the task of developing programs and compilers for the Pleiades architecture. Extensive prior experience by researchers has demonstrated that dataflow graphs are perhaps the most natural and most effective means to represent signal processing algorithms [21,22]. One of the key strengths of dataflow graphs is that they expose parallelism by expressing only the data dependencies that are inherent to a given algorithm. There is a rich body of knowledge addressing the problem of compiling dataflow graphs onto multiprocessor architectures [23–26].

A data-driven control mechanism has another important benefit: it provides a well-defined and elegant framework for managing switching activity in hardware modules. The handshaking mechanism that is used to implement the data-driven semantics of dataflow graphs can also be used to control switching activity in the satellite processors. When all required data tokens have arrived at a satellite processor, the satellite processor can start executing its task; otherwise, the satellite processor will stay dormant, and no unnecessary switching activity will take place.

17.7.1 Control Mechanism for Handling Data Structures

Distributed execution of an algorithm on multiple processing elements involves partitioning the calculations performed by the algorithm into multiple threads. These threads are then assigned to appropriate processing elements. A convenient first step is to partition the algorithm into *address* calculations and *data* calculations. Address calculations produce memory address sequences that are used to access data structures in the particular manner specified by

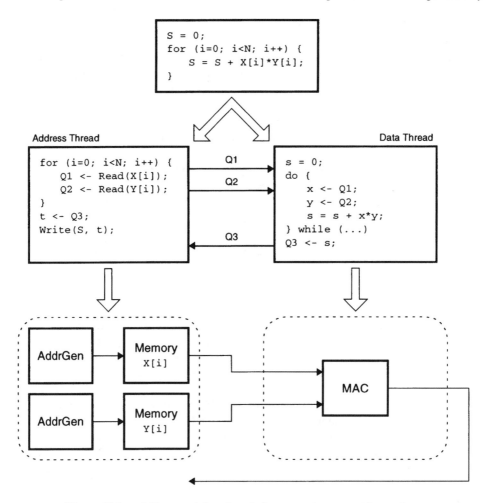

Figure 17.9.   Address and data threads for computing vector dot product

the algorithm. Data calculations process the accessed data structures and produce the desired results. This is illustrated in Figure 17.9 for the vector dot product example. Address calculations involve loop index and memory address pointer calculations. These calculations are mapped onto address generators. The address sequences produced by the address generators are used to access the required data structures (two vectors in this example) from the memory units. The resulting data streams are then communicated to the functional units performing the data calculations (a single MAC unit in this example). The MAC unit must have a way of knowing when the end of a vector is reached. This information will provide the missing condition of the `while()` statement in the data thread in Figure 17.9. One approach is to replicate the loop index calculation of the address thread in the data thread. A better approach that avoids the overhead and inconvenience of replicating the loop index calculation is to let a data stream itself indicate the boundaries of the data structure that it is carrying. This can be done by embedding special control flags that indicate the last element of a sequence into data tokens. The latter approach was taken in the Pleiades architecture.

In the Pleiades architecture, a data stream can be a scalar, a vector, or a matrix. These data types are the most common in signal processing algorithms. The boundaries of vectors and matrices are indicated by special *End-of-Vector* (EOV) flags that are embedded into data tokens. Fig.10 illustrates how this is accomplished. An EOV flag can have one of three values: 0, 1, or 2. The value 1 marks the last data token of a one-dimensional data structure or the last data token of a one-dimensional sub-structure of a two-dimensional data structure. The value 2 marks the last data token of a two-dimensional structure. The value 0 marks all other data tokens. Thus, two additional bits are needed to encode the EOV flag into a data token. Observe that the manner in which the elements of a vector or a matrix are scanned determines how the resulting data stream is delimited with EOV flags. Data structures of

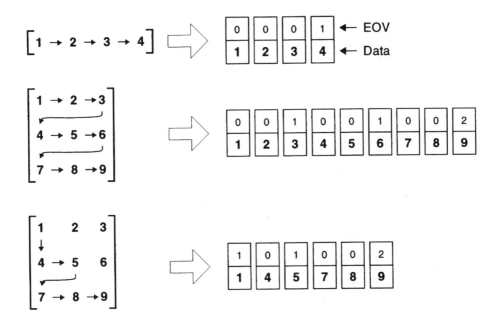

Figure 17.10.   Data stream examples for accessing vectors and matrices

higher dimensions can also be created by allowing the EOV flag to take on more than three values. This was not deemed necessary for any of the Pleiades processors that were considered. EOV flags are inserted into data tokens by either an address generator producing the address sequence that is used to access the required data structure or by the control processor. Memory units simply copy the EOV flag of an incoming address token into the corresponding data token being read from memory. This is illustrated in Fig.11. How the EOV flags are used by a functional unit depends entirely on the instruction being executed by that functional unit, and the instruction being executed by a functional unit must specify the type of and the manner in which an incoming data stream that is to be processed. Some examples of how data streams can be consumed by a satellite processor are shown in Figure 17.11 for the case of the MAC satellite processor. For any given instruction of a satellite processor, if the dimensionality of all input streams is increased by one, then the dimensionality of the output data stream is automatically increased by one, without the need to specify a new instruction. For instance, if the input data streams of the vector dot product instruction of the MAC processor are two-dimensional vectors instead of the one-dimensional vectors shown in Figure 17.11, then the output will automatically be a vector, with the proper EOV delimiters, instead of a scalar.

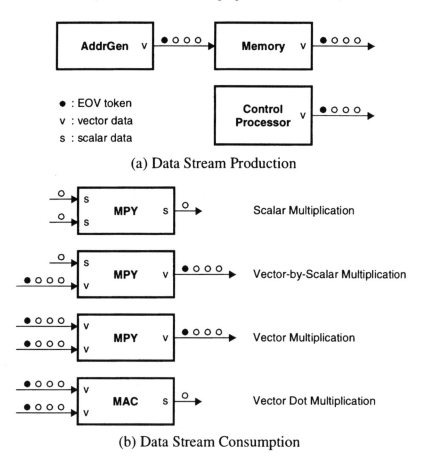

Figure 17.11.   Examples of data stream production and consumption

17.7.2 Summary

With its distributed data-driven control mechanism, the Pleiades architecture avoids the energy and performance overheads of communicating instructions and control signals globally across large distances, while providing modular and scalable support for highly concurrent implementations of signal processing algorithms. The control mechanism used in the Pleiades architecture provides support for handling common signal processing data structures such as vectors and matrices efficiently.

17.8 The Pleiades Design Methodology

The Pleiades approach is not only a hardware architecture for domain-specific processors, but it also involves an associated design methodology that is used to create domain-specific processor instances based on the Pleiades architecture template.

The Pleiades design methodology has two separate, but related, aspects that address different design tasks. One aspect of the methodology addresses the problem of designing a domain-specific processor for a given algorithm domain. The other aspect of the methodology addresses the problem of mapping a given algorithm onto an existing domain-specific processor instance. Both of these tasks involve analyzing algorithms and mapping them onto hardware resources. The chief difference between these two tasks is that in one of them, i.e. the problem of creating a domain-specific processor instance, architectural parameters (i.e. types and numbers of satellite processors and the detailed structure of the communication network) are not fixed and are to be determined by the algorithm analysis and mapping process.

The design flow begins with a description of a given algorithm in C or C++. The baseline implementation is to map the entire algorithm onto the control processor. The power and performance of this baseline implementation are then evaluated and used as a reference during subsequent optimizations, during which the objective will be to minimize energy consumption while meeting the real-time performance requirements of the given algorithm. The key task at this point is to identify the dominant kernels that are causing energy and performance bottlenecks. This is accomplished by dynamic profiling of the algorithm. Dynamic profiling establishes the function call graph of the algorithm and tabulates the amount of time and energy taken by each function and each basic block of the program. With this information, the dominant kernels of the algorithm can then be identified. The energy consumption of the baseline implementation is estimated using a modeling approach in which each instruction of the control processor has an associated base energy cost, and the total energy of a given program is obtained by adding the base costs of all executed instructions [27]. More accuracy can be obtained by taking account of inter-instruction energy consumption effects into the base costs of the instructions. A basic optimization step at this point, before going further into the rest of the design flow, is to improve the algorithm by applying architecture-independent optimizations and rewriting the initial description.

Once dominant kernels are identified, they are ranked in order of importance and addressed one at a time until satisfactory results are obtained. One important step at this point is to rewrite the initial algorithm description, so that kernels that are candidates for being mapped onto satellite processors are distinct function calls. The next step is to implement a candidate kernel on an appropriate set of satellite processors. This is done by directly mapping the

```
1       int dot_product(int x[], int y[], int n)
2       {
3           int i;
4           int s;
5
6           s = 0;
7           for (i = 0; i < n; i++) s += x[i]*y[i];
8           return s;
9       }
```

Figure 17.12.   C++ description of vector dot product

dataflow graph of the kernel onto a set of satellite processors. With this approach, each node or cluster of nodes in the dataflow graph corresponds to a satellite processor. Arcs of the dataflow graph correspond to links in the communication network, connecting the satellite processors. Mapped kernels are represented using an intermediate form as C++ functions that replace the original functions. The advantage of this approach is that mapped kernels can be simulated and evaluated with the rest of the program within the same environment that was used to simulate and evaluate the original program. In the intermediate form representation, satellite processors and communication channels are modeled as C++ objects. Each object has a set of methods that captures the functionality of the object during configuration and execution. This can be illustrated by an example. Figure 17.12 shows a C++ function implementing the vector dot product kernel. Figure 17.13 shows a mapping of the vector dot product kernel onto a set of satellite processors. Note that in this particular implementation of the vector dot product, both input vectors are stored in the same memory, and are communicated to the MAC satellite through the same communication channel in a time-multiplexed fashion. The MAC satellite is configured to accept both input vectors from the same input port (the other input port is unused). Figure 17.14 shows the intermediate form representation of the same function. The intermediate form representation is functionally identical to the original function but captures details of the actual implementation of the

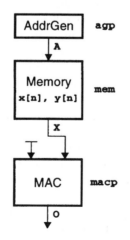

Figure 17.13.   Mapping of vector dot product

```
1      int dot_product(int x[], int y[], int n)
2      {
3          Memory mem;
4          AGP agp;
5          MACP macp;
6          Queue A;    // output of agp, address input of mem
7          Queue X;    // data output of mem, X input of macp
8          Queue O;    // output of macp
9          Queue unused;   // dummy Queue for unused ports
10         int x_base;   // base address of x[] in mem
11         int y_base;   // base address of y[] in mem
12         int i;
13         int rval;
14
15         // create memory map for x[] and y[] and initialize mem
16         x_base = 0;
17         y_base = x_base + n;
18         for (i = 0; i < n; i++) {
19             mem.write(x_base + i, x[i]);
20             mem.write(y_base + i, y[i]);
21         }
22
23         // configure agp and macp
24         agp.load_program("dot_product.pgm");
25         agp.config(x_base, 1, 0, y_base, 1, 0, n, 0, 0);
26         macp.config(MAC, 1, 1);
27
28         // create connections between satellites
29         agp.connect(A);
30         mem.connect(A, unused, X);   // data input is unused
31         macp.connect(X, unused, O);  // Y input is unused
32
33         // run
34         for (i = 0; i < n; i++) {
35             agp.exec(); mem.exec(); macp.exec();
36             agp.exec(); mem.exec(); macp.exec();
37         }
38         // macp has written its results to O
39         rval = O.read().data();
40         agp.exec();   // last execution cycle of agp
41         return rval;
42     }
```

Figure 17.14.   Intermediate form representation of vector dot product

original function on satellite processors. In the intermediate form representation, first the required satellite processors and communication channels are instantiated. The satellite processors are then interconnected by configuring the communication channels. Finding the most efficient way to connect the required set of satellite processors through the communication network is a routing problem that is an important part of the overall design methodology [28]. The satellite processors are configured next. Configuration of the satellite processors and the communication network switches is performed by code running on the control processor. Automatic generation of this configuration code is an important part of the Pleiades design methodology [29]. The overhead of the configuration code must be minimized by scheduling the configuration code such that the amount of overlap between execution of the current kernel and configuration for the next kernel is maximized. The kernel is then executed. Notice that the execution of the kernel in this particular example is scheduled statically, but this is not a requirement, and by employing a thread library, the kernel can be executed as a set of concurrent processes, representing the concurrent hardware components. The energy and performance of the mapped kernels can then be estimated during simulation with macromodels that are captured into the C++ objects representing the satellite processors and the communication network. Further details of the Pleiades design methodology can be found in Refs. [30,31].

17.9 The P1 Prototype

The P1 prototype was designed and built to evaluate and verify the validity of the architectural approach taken in the Pleiades architecture. Lessons learned from the P1 design were used to refine the Pleiades architecture template. The P1 design was also used as an initial driver for the Pleiades design methodology.

The block diagram of P1 is shown in Figure 17.15. The satellite processors employed in P1 include a MAC unit, two memory units, two address generators, two input ports, and one output port. All data and address tokens are 16-bit quantities and are handled by 16-bit datapaths in the satellite processors and 16-bit data buses in the communication network. P1 was not designed for a particular domain of algorithms, but its design was influenced by the properties of CELP based speech coding algorithms. The chip can be used to implement three kernels: vector multiply, vector dot product, and Finite Impulse Response (FIR) filter.

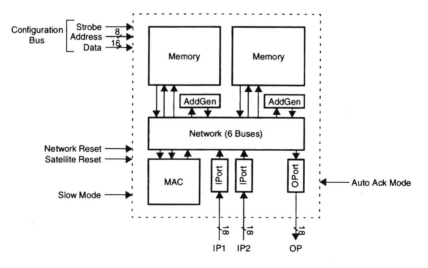

Figure 17.15. Block diagram of P1

Table 17.1 Energy results for dot product kernel

Circuit module	Energy (pJ/cycle)		$E^{measured}/E^{simulated}$
	Simulated	Measured	
Address generator satellite	5.0	4.4	0.88
SRAM satellite (PHA read)	27.8	25.4	0.91
MAC satellite (multiply-accumulate)	107.1	90.5	0.85
Network channel (A input of MAC)	9.1	7.5	0.82
Total (chip core)	207.1	179.1	0.86

P1 was fabricated in a 0.6-μm, 3.3-V CMOS technology. The chip was designed to operate at a minimum cycle time of 50 ns with a 1.5-V supply voltage. The choice of the supply voltage was motivated by the desire to minimize power dissipation while maintaining acceptable performance. The chosen supply voltage results in an energy-delay product that is near the minimum for the CMOS technology used for P1.

Energy and cycle-time measurement results for the P1 chip are shown and compared to simulation results in Tables 17.1 and 17.2. Simulated energy results were obtained by running PowerMill [32] simulations on the extracted layout of the chip, using the exact same vectors that were used for the measurements. Simulated cycle-time results were obtained using the HSpice circuit simulator [33]. All measurements were done at room temperature, using a 1.50-V supply voltage.

Energy measurement results for the dot product kernel are shown and compared to simulation results in Table 17.3. The simulated cycle time for the dot product kernel was 71.4 ns.

Table 17.2 Energy measurement and simulation results

Circuit module		Energy (pJ/cycle)		$E^{measured}/E^{simulated}$
		Simulated	Measured	
Address generator satellite (random mode)		8.1	7.3	0.90
MAC satellite (multiply)	Zero input	11.9	10.5	0.88
	Random input	92.2	72.4	0.79
MAC satellite (multiply-accumulate)	Zero input	14.1	11.6	0.82
	Random input	116.5	95.1	0.82
SRAM satellite (read)	Random data	33.7	32.4	0.96
SRAM satellite (PHA read)	Random data	27.9	25.7	0.92
SRAM satellite (write)	Random data	25.8	23.5	0.91
Network channel	Random data	8.3	6.8	0.82

Table 17.3   Cycle-time measurement and simulation results

Circuit module	Cycle time (ns)		$T^{\text{measured}}/T^{\text{simulated}}$
	Simulated	Measured	
Ring oscillator			
Inverter (51-stage)	35.2	40.0	1.14
Delay cell (15-stage)	39.1	49.8	1.27
Address generator satellite	40.0	47.3	1.18
SRAM satellite	35.6	42.4	1.19
MAC satellite	70.8	87.4	1.23

The measured cycle time of the kernel, based on the period of the input acknowledge signal of the MAC satellite processor, was 88.3 ns, i.e. 1.24 times the simulated value.

17.9.1. P1 Benchmark Study

In this section, results of the P1 benchmark study [36], comparing the Pleiades architecture to a variety of programmable architectures that are commonly used to implement signal processing algorithms, will be presented. This study was based on the results obtained from the P1 prototype. The kernels that were used as benchmarks represent three of the most commonly used DSP algorithms: the FIR filter (FIR), the Infinite Impulse Response (IIR) filter, and the Fast Fourier Transform (FFT). The reference architectures that were considered in this study included the StrongARM microprocessor [34], the Texas Instruments TMS320C2xx and TMS320LC54x programmable signal processors [35], and the Xilinx XC4003A FPGA [8]. Since the architectures being compared are implemented in different fabrication technologies and have different operating supply voltages, the energy and delay metrics of all architectures should be normalized to a common reference, so that a meaningful comparison can be made. We chose to normalize all figures of merit to the 0.6-μm, 3.3-V CMOS process that was used to implement P1. All energy and delay values are calculated for a 1.5-V operating supply voltage. Details of the methodology used in the P1 benchmark study can be found in Refs. [36,2].

Comparison results for the FIR benchmark are shown in Figure 17.16. Results for the IIR benchmark are shown in Figure 17.17. As expected, the StrongARM microprocessor has the worst performance among the architectures considered in this study, as it requires many instructions and execution cycles to execute a given kernel in a highly sequential manner. The lack of a single-cycle multiplier exacerbates this problem. Furthermore, the execution of each instruction is burdened by a great deal of energy overhead. All other architectures have more internal parallelism which allows them to have much better performance than the StrongARM processor. Pleiades and the TI processors can execute an FIR tap in a single cycle. Pleiades performs much better on the energy scale than the TI processors because the TI processors have a general-purpose design, incurring a great deal of energy overhead to each instruction. Pleiades, on the other hand, has the ability to create a hardware structures optimized for a given kernel and can execute operations with a relatively small energy overhead. Features such as zero-overhead looping reduce the instruction fetch overhead

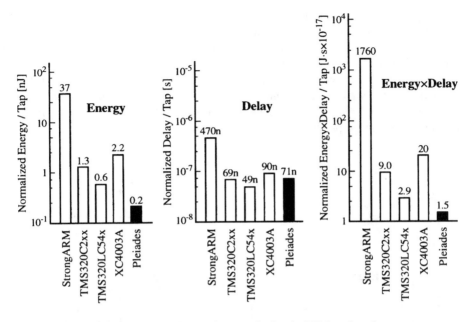

Figure 17.16.   Comparison results for the FIR benchmark

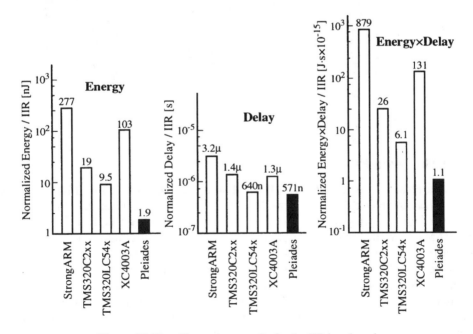

Figure 17.17.   Comparison results for the IIR benchmark

for the TI processors, but they still fall short of the performance achieved by Pleiades. The XC4003A implementation of the FIR benchmark executes five taps in a single cycle. The XC4003A is not very energy efficient, but it has the ability to use optimized shift-and-add multipliers, instead of the full multipliers used in the other architectures.

Comparison results for the FFT benchmark are shown in Figure 17.18. Compared to the FIR and IIR benchmarks, the FFT benchmark is more complex. Pleiades outperforms the other processors by a large margin, owing to its ability to exploit higher levels of parallelism by creating an optimized parallel structure with minimal energy overhead.

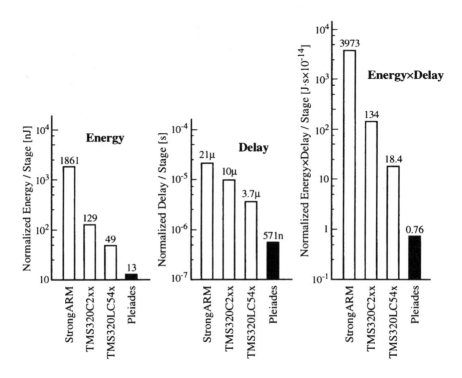

Figure 17.18.   Comparison results for the FFT benchmark

17.10 The Maia Processor

Maia is a Pleiades processor for CELP based speech coding applications [37,38]. Figure 17.19 shows the block diagram of the Maia processor. The computational core of Maia consists of the following satellite processors: eight address generators, four 512-word 16-bit SRAMs, four 1024-word 16-bit SRAMs, two MAC units, two arithmetic/logic units, a low-energy embedded FPGA unit, two input ports, and two output ports. To support CELP based speech coding efficiently, 16-bit datapaths were used in the satellite processors and the communication network. The communication network uses a two-level hierarchical mesh structure. To reduce communication energy, low-swing driver and receiver circuits are used in the communication network. Satellite processors communicate through the communication network using the two-phase asynchronous handshaking protocol. Each link through the

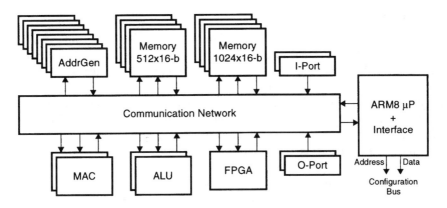

Figure 17.19. Block diagram of the Maia processor

communication network consists of a 16-bit data field, a 2-bit EOV field, and a request/acknowledge pair of signals for data-driven control and asynchronous handshaking. The EOV field can have one of three values: 0, 1, 2. As a result, the control mechanism used in Maia can support scalar, vector, and matrix data types. The I-Port and O-Port satellites are used for off-chip data I/O functions.

17.10.1 Control Processor

The control processor in Maia is a custom implementation of the ARM8 microprocessor core that has been optimized for low-power operation [39]. The control processor communicates with the satellite processors through an interface module. The interface module enables the control processor to send and receive data tokens through the communication network, configure the satellite processors and the communication network, reset the satellite processors and their handshake control circuits, and initiate the execution of kernels and detect their completion.

17.10.2 Address Generator Processor

The Address Generator Processor (AGP) is responsible for generating the address sequences that are needed to access data structures from the memory units while executing a kernel. The architecture of the AGP is based on a programmable datapath with a small instruction memory. The AGP has a simple but flexible instruction set that allows the programmer to scan the elements of a vector or a matrix in complex but structured patterns.

Execution of a kernel is initiated by the control processor by sending a request signal to the relevant AGP. The request signal triggers the execution of a pre-loaded program in the AGP. AGP programs are typically very short (just a few instructions at the most). Multiple programs can be stored in the instruction memory, which can store up to 16 instructions in the Maia implementation. The request signal that initiates the execution of an AGP program is accompanied by a data token that specifies which one of the pre-loaded AGP programs is to be executed. When the last address token has been generated and sent, the AGP returns an acknowledge signal that can be used to interrupt the control processor.

17.10.3 Memory Units

The functionality of the memory unit is quite simple. It has three ports: address (A), data in (DI), and data out (DO). An input address token on A includes a memory address, a read/write flag, and an EOV flag. The address input is typically generated by an address generator. If the address token specifies a read operation, then the memory location specified by the address is read and sent to the DO output. The EOV flag of the address token is copied onto the EOV flag of the output data token. Thus, a memory unit preserves the type of data structure specified on the input address stream. If an address token specifies a write operation, then the data token on the DI input is written to the memory location specified by the address token. Two memory sizes were chosen for Maia: 512-word and 1024-word. Both sizes can be used for all vector and matrix calculations. The 1024-word memories were selected for storing and manipulating the codebook structures that are commonly used in the CELP based speech coding algorithms. The smaller memories consume less power and are favored for most kernels that do not need the larger memories.

17.10.4 MAC Unit

The core of the MAC satellite processor consists of a multiplier, followed by an accumulator. The MAC unit has two inputs, A and B, and one output, Q. The MAC unit performs one of two basic tasks: multiply and multiply-accumulate. The MAC unit has two pipeline stages in the Maia implementation. The MAC unit can perform one of four possible functions on the A and B streams: scalar multiplication, scalar-by-vector multiplication, scalar-by-matrix multiplication, and vector dot multiplication.

The dimensionality of the output data stream is derived from the EOV flags of the input data streams. The MAC unit automatically delimits its output data stream with the proper EOV flags. The MAC unit also has the ability to shift, round, and saturate the output result, as specified by the configuration state of the MAC unit.

17.10.5 Arithmetic/Logic Unit

The Arithmetic/Logic Unit (ALU) processor performs a variety of arithmetic, logic, and shift operations. It has two inputs, A and B, and one output Q. It has three basic types of instructions:

- Single-input scalar operations: absolute value and logical not.
- Two-input scalar operations: add, subtract, shift, min, max, compare, logical and, logical or, logical xor.
- Two-input vector-to-scalar operations: accumulate, vector max, and vector min.

If the dimensionality of the input data streams of the ALU is increased, then the executed functions will automatically become vector or matrix operations.

17.10.6 Embedded FPGA

The FPGA unit consists of a 4 × 9 array of five-input, three-output logic blocks. The design of the FPGA unit has been highly optimized for energy-efficient operation [40,41]. The FPGA can have up to two input ports and an output port. The port behavior of the FPGA units is completely programmable and can be set by four of the 36 logic blocks. The FPGA unit has two important functions that give the Maia architecture a great deal of flexibility:

- In addition to being able to implement the functions performed by the ALU processor (albeit at a higher cost), the FPGA can implement irregular bit-manipulation and arithmetic operations that cannot be supported by the MAC and ALU processors. The FPGA can also implement finite-state machines.
- The FPGA can be used to implement irregular address generation patterns that are not supported by the AGP instruction set. This can be done either in stand-alone fashion, or in conjunction with an AGP, in which case the FPGA performs a transformation function on the stream produced by the AGP. A good example of the latter is the bit-reversed addressing mode needed for performing FFT functions.

17.10.7 Maia Results

The Maia processor was fabricated in a 0.25-μm CMOS technology. The chip contains 1.2 million transistors and measures 5.2×6.7 mm^2. It was packaged in a 210-pin PGA package. A Die photo of Maia is shown in Figure 17.20. With a 1.0-V supply voltage, average throughput for kernels running on the satellite processors is 40 MHz. The ARM8 core runs at 40 MHz. The average power dissipation of the chip is 1.5–2.0 mW. Table 17.4 shows performance parameters of the various hardware components of the Maia processor.

Table 17.5 shows the energy profile of the VSELP speech coding algorithm, running on Maia. Six kernels were mapped onto the satellite processors. The rest of the algorithm is executed on the ARM8 control processor. The control processor is also responsible for configuring the satellite processors and the communication network. The energy overhead of this configuration code running on the control processor is included in the energy consumption values of the kernels. In other words, the energy values listed in Table 17.5 for the kernels include contributions from the satellite processors as well as the control processor executing configuration code. The power dissipation of Maia when running VSELP is 1.8 mW. The lowest power dissipation reported in the literature to date is

Table 17.4 Performance data for hardware components of Maia

Component	Cycle time (ns)	Energy per cycle (pJ)	Area (mm^2)
MAC	24	21	0.25
ALU	20	8	0.09
SRAM (1K × 16)	14	8	0.32
SRAM (512 × 16)	11	7	0.16
Address generator	20	6	0.12
FPGA	25	18[a]	2.76
Interconnect network	10	1[b]	N/A

[a] This value is the average energy for various arithmetic functions.
[b] This value is the average energy per connection.

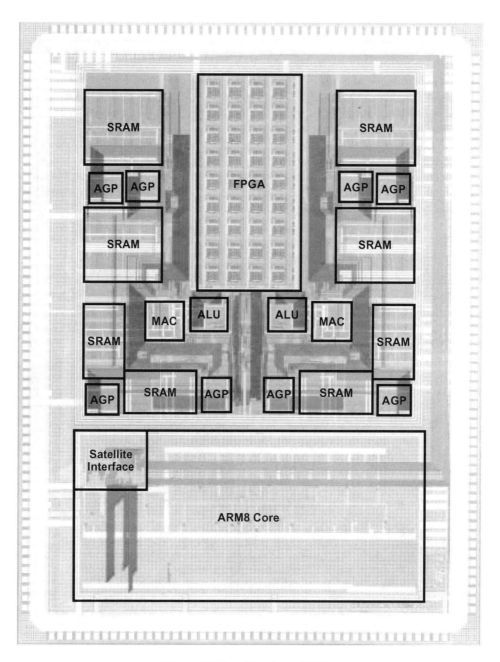

Figure 17.20.   Die photo of Maia

17 mW for a programmable signal processor executing the Texas Instruments TMS320LC54x instruction set, implemented in a 0.25-μm CMOS process, running at 63 MHz with a 1.0-V supply voltage [42]. The energy efficiency of this reference processor is

Table 17.5   Energy profile for the VSELP algorithm running on Maia

Function		Power (mW)
Kernels running on satellite processors	Vector dot product	0.738
	FIR filter	0.131
	IIR filter	0.021
	Vector sum with scalar multiply	0.042
	Code-vector computation	0.011
	Covariance matrix computation	0.006
Program running on control processor		0.838
Total		1.787

270 μW/MHz, whereas the energy efficiency of Maia is 45 μW/MHz, which corresponds to an improvement by a factor of six.

17.11 Summary

The Pleiades architecture was presented in this chapter. The architecture has been designed for energy-efficient implementation of domain-specific programmable processors for signal processing applications. The key features of the Pleiades architecture template are:

- A highly concurrent, scalable multiprocessor architecture with a heterogeneous array of optimized satellite processors that can execute the dominant kernels of a given domain of algorithms with a minimum of energy overhead. The architecture supports dynamic scaling of the supply voltage.
- Reconfiguration of hardware resources is used to achieve flexibility while minimizing the overhead of instructions.
- A reconfigurable communication network that can support the interconnection patterns needed to implement the dominant kernels of a given domain of algorithms efficiently. The communication network uses a hierarchical structure and low-swing circuits to minimize energy consumption.
- A data-driven distributed control mechanism that provides the architecture with the ability to exploit locality of reference to minimize energy consumption. The control mechanism provides special support to handle the data structures commonly used in signal processing algorithms efficiently.

Prototype designs have proven that the approach indeed leads to substantial energy savings. The idea of using reconfigurability to realize low-energy programmable implementations of signal processing functions has since become mainstream, and has been adopted by quite a number of start-up and semiconductor companies.

References

[1] Abnous, A. and Rabaey, J., 'Ultra-Low-Power Domain-Specific Multimedia Processors', *Proceedings of the 1996 IEEE Workshop on VLSI Signal Processing*, 1996, pp. 461–470.

[2] Abnous, A., Low-Power Domain-specific Architectures for Digital Signal Processing, Ph.D. Dissertation, University of California, Berkeley, CA, 2001.

[3] Schroder, M.R. and Atal, B.S., 'Code-Excited Linear Prediction (CELP): High Quality Speech at Very Low Bit Rates', *Proceedings of the International Conference on Acoustics, Speech, and Signal Processing*, 1985, pp. 937–940.

[4] Spanias, A.S., 'Speech Coding: A Tutorial Review', *Proceedings of the IEEE*, October 1994, , pp. 1541–1582.

[5] Hoare, C.A.R., 'Communicating Sequential Processes', *Communications of the ACM*, Vol. 21, No. 8, August 1978.

[6] Laplante, P.A., *Real-Time System Design and Analysis: An Engineer's Handbook*, Second Edition, IEEE Computer Society Press, New York, 1997.

[7] Hwang, K. and Briggs, F.A., *Computer Architecture and Parallel Processing*, McGraw Hill, New York, 1984.

[8] *The Programmable Logic Data Book*, Xilinx, Inc., 1994.

[9] Zhang, H., Wan, M., George, V., and Rabaey, J., 'Interconnect Architecture Exploration for Low-Energy Reconfigurable Single-Chip DSPs', *Proceedings of the IEEE Computer Society Workshop on VLSI '99*, 1999, pp. 2–8.

[10] Yeung, A.K., A Data-Driven Multiprocessor Architecture for High Throughput Digital Signal Processing, Ph.D. Dissertation, University of California, Berkeley, CA, 1995.

[11] Tsu, W., Macy, K., Joshi, A., Huang, R, Walker, N., Tung, T., Rowhani, O, George, V., Wawrzynek, J. and DeHon, A., 'HSRA: High-Speed Hierarchical Synchronous Reconfigurable Array', *Proceedings of the 1999 ACM/SIGDA Seventh International Symposium on Field Programmable Gate Arrays*, 1999, pp. 125–134.

[12] Zhang, H. and Rabaey, J., 'Low-Swing Interconnect Interface Circuits', *Proceedings of the 1998 IEEE Symposium on Low-Power Electronics and Design*, 1998, pp. 161–166.

[13] Zhang, H., George, V., and Rabaey, J., 'Low-Swing on-Chip Signaling Techniques: Effectiveness and Robustness', *IEEE Transactions on VLSI Systems*, June 2000, pp. 264–272.

[14] *XC4000E and XC4000X Series Field Programmable Gate Arrays*, Xilinx, Inc., 1999.

[15] Segars, S., 'The ARM9 Family: High Performance Microprocessors for Embedded Applications', *Proceedings of the International Conference on Computer Design*, 1998, pp. 230–235.

[16] DeHon, A., 'DPGA-Coupled Microprocessors: Commodity ICs for the Early 21st Century', *Proceedings of the IEEE Workshop on FPGA Custom Computing Machines*, 1994, pp. 31–39.

[17] Trimberger, S., Carberry, D., Johnson, A., and Wong, J., 'A Time-Multiplexed FPGA', *Proceedings of the IEEE Workshop on FPGA Custom Computing Machines*, 1997, pp. 22–28.

[18] Hauser, J.R. and Wawrzynek, J., 'GARP: A MIPS Processor with a Reconfigurable Coprocessor', *Proceedings of the IEEE Workshop on FPGA Custom Computing Machines*, 1997, pp. 12–21.

[19] Dennis, J.B., *First Version Data Flow Procedure Language, Technical Memo MAC TM61*, MIT Lincoln Laboratory for Computer Science, May 1975.

[20] Fellman, R.D., 'Design Issues and an Architecture for the Monolithic Implementation of a Parallel Digital Signal Processor', *IEEE Transactions on Acoustics, Speech, and Signal Processing*, pp. 839–852, May 1990.

[21] Lee, E.A. and Messerschmitt, D. G., 'Synchronous Data Flow', *IEEE Proceedings*, pp. 1235–1245, September 1987.

[22] Lee, E.A., 'Consistency in Dataflow Graphs', *IEEE Transactions on Parallel and Distributed Systems*, April 1991, pp. 223–235.

[23] Lee, E.A. and Messerschmitt, D.G., 'Static Scheduling of Synchronous Data Flow Programs for Digital Signal Processing', *IEEE Transactions on Computers*, January 1987, pp. 24–35.

[24] Hoang, P. and Rabaey, J., 'A Compiler for Multiprocessor DSP Implementation', *Proceedings of the IEEE International Conference on Acoustics, Speech, and Signal Processing*, Vol. V, 1992, pp. 581–584.

[25] Hoang, P. and Rabaey, J., 'Scheduling of DSP Programs onto Multiprocessors for Maximum Throughput', *IEEE Transactions on Signal Processing*, June 1993, , pp. 2225–2235.

[26] Pino, J.L., Parks, T.M., and Lee, E.A., 'Automatic Code Generation for Heterogeneous Multiprocessors', *Proceedings of the IEEE International Conference on Acoustics, Speech, and Signal Processing*, Vol. II, 1994, pp. 445-448.

[27] Tiwari, V., Malik, S., Wolfe, A., and Lee, M.T., 'Instruction Level Power Analysis and Optimization of Software', *Journal of VLSI Signal Processing*, August/September 1996, pp. 223–238.

[28] Zhang, H., Wan, M., George, V., and Rabaey, J., 'Interconnect Architecture Exploration for Low-Energy Reconfigurable Single-Chip DSPs', *Proceedings of the IEEE Computer Society Workshop on VLSI '99*, 1999, pp. 2–8.

[29] Li, S.-F., Wan, M., and Rabaey, J., 'Configuration Code Generation and Optimizations for Heterogeneous Reconfigurable DSPs', *Proceedings of the 1999 IEEE Workshop on Signal Processing Systems*, October 1999, pp. 169–180.

[30] Wan, M., Zhang, H., Benes, M. and Rabaey, J., 'A Low-Power Reconfigurable Dataflow Driven DSP System', *Proceedings of the 1999 IEEE Workshop on Signal Processing Systems*, October 1999, pp. 191–200.

[31] Wan, M., A Design Methodology for Low-Power Heterogeneous Reconfigurable Digital Signal Processors, Ph.D. Dissertation, University of California, Berkeley, CA, 2001.

[32] http://www.synopsys. com/.

[33] http://www. avanticorp. com/.

[34] *Digital Semiconductor SA-110 Microprocessor Technical Reference Manual*, Digital Equipment Corporation, 1996.

[35] http://www. ti. com/.

[36] Abnous, A., Seno, K., Ichikawa, Y., Wan, M., and Rabaey, J., 'Evaluation of a Low-Power Reconfigurable DSP Architecture', *Proceedings of the Reconfigurable Architectures Workshop*, 1998, pp. 55–60.

[37] Zhang, H., Prabhu, V., George, V., Wan, M., Benes, M., Abnous, A. and Rabaey, J.M., 'A 1-V Heterogeneous Reconfigurable DSP IC for Wireless Baseband Digital Signal Processing', *IEEE Journal of Solid-State Circuits*, November 2000, pp. 1697–1704.

[38] Zhang, H., Prabhu, V., George, V., Wan, M., Benes, M., Abnous, A. and Rabaey, J.M., 'A 1-V Heterogeneous Reconfigurable Processor IC for Baseband Wireless Applications', *International Solid-State Circuits Conference Digest of Technical Papers*, 2000, pp. 68–69.

[39] Burd, T.D., Pering, T.A., Stratakos, A.J., and Brodersen, R.W., 'A Dynamic Voltage Scaled Microprocessor System', *IEEE Journal of Solid-State Circuits*, November 2000, pp. 1571–1580.

[40] George, V., Zhang, H., and Rabaey, J., 'Low-Energy FPGA Design', *Proceedings of the International Symposium on Low-Power Electronics and Design*, 1999, pp. 188–193.

[41] George, V., Low-Energy FPGA Design, Ph.D. Dissertation, University of California, Berkeley, CA, 2000.

[42] Lee, W., Landman, P., Barton, B., Abiko, S., Takahashi, H., Mizuno, H., Muramatsu, S., Tashiro, K., Fusumada, M., Pham, L., Boutaud, F., Ego, E., Gallo, G., Tran, H., Lemonds, C., Shih, A., Nandakumar, M., Eklund, B. and Chen, I.-C., 'A 1-V DSP for Wireless Communications', *International Solid-State Circuits Conference Digest of Technical Papers*, 1997, pp. 92–93.

18

Application Specific Instruction Set Architecture Extensions for DSPs

Jean-Pierre Giacalone

18.1 The Need for Instruction Set Extensibility in a Signal Processor

In the early 1990s, digital signal processing in wireless terminals mainly covered voice compression as well as channel equalization, coding and decoding techniques (also called "modem" function). Corresponding electronic systems built to optimize these applications tried to make the best trade-off between software and hardware in order to minimize system cost and power consumption. Software was mostly used as an efficient means to allow a quick evolution or on-the-fly correction of signal processing functions. Digital signal processor hardware was tailored to minimize power consumption.

A very classical example of such a trade-off was the implementation of convolutional decoders Viterbi trellis on the TMS320C54x processor, in software, by association of specific instructions and very optimized hardware [1]. On previous implementations, dedicated hardware, sitting outside of the processor, took care of the whole trellis execution. Software brought the flexibility of correcting branch metrics computations without the need for expensive communications to send them to this external hardware and for duplication of storage resources.

Since then, more and more applications have been added around the modem, moving it to a commodity function. With third generation wireless networks, the requirements in terms of system flexibility will increase even more due to the increase of data transmission bandwidth from several kilobits to several megabits per second. As a consequence, system design must not only optimize cost and power consumption for a single application but for a wide range of applications (speech recognition, voice memo, video display, video-conferencing, etc.). The modem itself is increasing in complexity and diversity of signal processing operations required to equalize and decode data streams at these rates. The behavior of the network carrying these streams is also becoming more and more flexible with the combination of

Table 18.1   Complexity and flexibility requirements

Application	Complexity (processor MHz)	Flexibility
2G modems + voice	40–50	Tuning, quick multiple standard fanout
3G modem	1000	Channel equalization, surrounding cell measurements, power control
Video-conferencing	100	Multiple audio and video standards (decoding), coding efficiency

configuration cases which are difficult to predict. Table 18.1 summarizes complexity and flexibility requirements on some typical applications.

Although software solutions may seem to be the natural approach to address flexibility requirements, they also require that the Digital Signal Processor (DSP) runs the code in the best way for all possible applications (e.g. complexity requirements are met with best possible power consumption). This, in return, puts heavy constraints on the processor architecture and may compromise its capabilities to be easily programmable with a high level programming language like C, for instance.

The solution to both complexity and flexibility requirements, optimizing system cost and power consumption resides in the capability to open the DSP architecture to external enhancements that can optimize the execution of some applications, while sharing existing resources (memory system, communication buses, control structures) and keeping the same programming model. These Instruction Set Architecture (ISA) extensions remove most of the drawbacks of external co-processors:

- communication overhead to transfer data between the DSP and their internal storage resources;
- duplication of resources like program sequencers and address generation units.

With the TMS320C55x DSP, TI is introducing for the first time in its family of processors, the concept of ISA extensibility which allows the enhancement of the main CPU features with external functions that share internal core resources and that are controlled by the instruction flow of the processor. In the rest of this chapter, we will see how these extensions can be connected to the core and controlled by the software. We will also see what their typical architecture is. Then, we will look at the various domains of applications, through practical examples. Finally we will address the design challenge of such functions and how they can be built to be re-configurable.

18.2 ISA Extension Capability of the TMS320C55x Processor

The TMS320C55x DSP offers a highly optimized architecture for wireless "modem" as well as vocoding applications execution. Particular care has been put into reducing code size and power consumption for these applications. Architecture key features can also benefit from a wider range of applications (speech recognition, voice memo, static image compression and decompression, etc.) with some trade-offs in performance or power consumption compared to modem functions. In order to enhance the capabilities of this architecture to optimize any application execution, it was decided to provide means in the core in order to allow an external hardware addition to be plugged in. These extensions are meant to be seen as a natural part of the processor ISA, hence the name, ISA extension. This concept is also called tightly coupled *HardWare Acceleration* (HWA).

In this concept, the external hardware is seen as being a natural execution unit of the processor, once it has been added. This means that the extension must have access to internal CPU resources just as any other execution unit. Because digital signal processing generally requires intensive arithmetic data processing, it was also decided to focus the new capability towards data computation extensions, but the same approach can be extended to both address and data computation functions. Let us have a look at the resources that need to be shared between the core and the extension.

- Internal data registers (accumulators) of the DSP. This is where intermediate computed values are kept alive. Thus, it is a natural place for the extension to return results. It will also find here the main source point for sending values back to memory.
- Memory access resources, i.e. data buses and addressing units. Intensive data processing requires having access to the maximum data bandwidth with DSP local memory which is the main storage place for data in high performance processing. Hence, data pipes must also reach the extension. Addressing resources are shared by mean of the instruction set.
- Status information (for arithmetic mode definition, for instance).
- Control and sequencing features (also by means of the instruction set).

As one can easily understand, the model of resource sharing described above makes the extension look like a pure datapath which depends on the core for the delivery of its operands and for the description of its controls. Let's keep this simplistic view in mind for the purpose of the description below, but we will see later, while looking at real applications, that much more complex and efficient extensions can still be built with the same concept.

The connection of an extension is performed via the TMS320C55x *ISA extensions interface*. This is the hardware means by which the CPU core allows new functions to be electrically plugged in. This interface consists of following signals

Data flow
Bbus	Data read using B bus port in memory	(16-bit)
Cbus	Data read using C bus port in memory	(16-bit)
Dbus	Data read using D bus port in memory	(16-bit)
Acxr	First accumulator data read	(40-bit)
Acxw	First accumulator data write	(40-bit)
Acyr	Second accumulator data read	(41-bit)
Acyw	Second accumulator data write	(41-bit)

Control flow
Status	Arithmetic status flags	(4-bit)
Inst	Instruction + strobe signal	(9-bit)
Pipe	Pipeline indicators (stall, error, ...)	(4-bit)

The table above shows clearly the emphasis put on delivering data to the extension. We will see in more detail, in the rest of the chapter, how critical this is with respect to performance and power objectives. The new hardware receives detailed operation indicators that keep its execution under the total control of the processor. Combinations of data flows create what are called "dataflow modes'. The various ways of expressing execution control to the hardware are called "control modes'.

18.2.1 Control Modes

The control of an ISA extension must be provided by the instruction set of the processor. Several approaches are possible to build this control. In order to avoid an expensive opcode space expansion and to avoid adding new instruction formats, it was decided to re-use existing opcodes for the definition of new ones. Instructions already controlling data processing operators were chosen because of their obvious properties for manipulating described shared resources. It was decided to use the multi-instruction dispatch capability of the processor in such a way that existing instructions could be re-used.

As already mentioned earlier, a TMS320C55x ISA extension is considered as just another execution unit of the processor. There is one restriction to this definition in the sense that when an accelerator is in use, no other processor instruction can be executed in parallel. This may look like a strong limitation but the reader must remember that the advantage of the accelerator should be a substantial, intrinsic gain in performance (e.g. a floating multiplication accelerator call may bring a net gain of 2 or 3 in the number of cycles versus a pure software execution of the same operation). Thus, benefiting from the fact that once an extension instruction is executed, no arithmetic instruction can be executed in parallel, all these native instructions can be re-used to control the accelerator. In order to distinguish between "internal' and "external' operations, a class of instructions called "copr' was added to the TMS320C55x instruction set.

This new instruction class uses four opcodes to "qualify' existing arithmetic instructions and to create new virtual dataflows that a programmer can use, associated with the corresponding hardware datapath (i.e. the accelerator content). Table 18.2 explains the role of each of these opcodes. The qualification process happens when a normal arithmetic instruction is dispatched in parallel with one of the "copr' opcodes.

Table 18.2   List and functions of the "copr' instructions class

"Copr' opcodes functions	
copr ()	Qualifies instruction
copr(k6)	Qualifies instruction, sends k6 constant to extension control interface
Smem = ACx, copr()	Qualifies instruction, writes accumulator to memory (16-bit)
Lmem = ACy, copr()	Qualifies instruction, writes accumulator to memory (32-bit)

There are two basic operations control modes available:

- simple qualification of an existing arithmetic instruction in order to re-direct its data flow to the extension;

- qualification with parallel writing to memory from accumulators (this provides a way to operate on data stored in memory and return results back to this memory while computing).

As a first result of the control process, the 8-bit instruction word sent to the extension by the CPU is extracted from bits of the "qualified' and "qualifier' opcodes (in the case of the copr(k6) qualifier the "k6' bit field is used with two other bits of the qualified instruction in order to build the instruction sent to the extension). This extraction is performed in the instruction decoder of the processor and the resulting 8-bit instruction is sent through the interface. As a second result, all shared resource controls of the qualified instruction (including data address generation) are activated by this instruction decoder. Thus, addresses are computed, memory data and accumulators contents are fetched or stored as part of the new ISA extension instruction created by the pair of standalone opcodes. Figure 18.1 illustrates the whole control process.

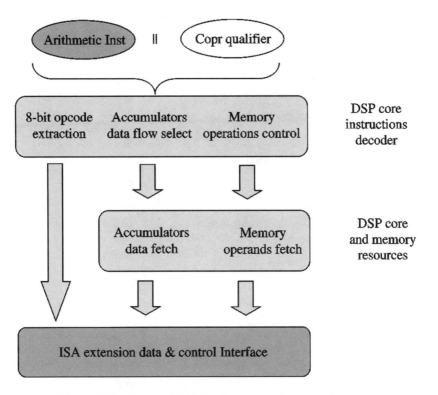

Figure 18.1.   TMS320C55x hardware extension control process

Other control signals exported to the extension are derived from the internal pipeline execution status of the CPU. These signals allow perfect synchronization of executions in the processor core and in the extension.

18.2.2 Dataflow Modes

The ISA extension operations control process described above allows not only the creation of the necessary opcodes for the datapath but, also, the exportation of available data flows to it. These flows are combined together by means of decoded controls from the qualified instruction and from the qualifier opcode (e.g. when the "copr' opcode contains a memory write reference). Resulting combinations of data flows according to the pair of instruction built are called *dataflow modes* and are summarized in Table 18.3. They define how resources are shared between the TMS320C55x DSP core and an extension and how many are available for an instruction. They also add more flexibility to the concept by allowing:

several accelerators to be connected to the interface;
several instructions to run on a single accelerator;
each instruction to exercise a wide variety of dataflow types.

Table 18.3 Dataflow modes available for a TMS320C55x extension and corresponding qualifier

Extension dataflow modes	Qualifier used
ACy = copr(k8,ACx,ACy)	copr(k6)
ACy = copr(k8,ACx,ACy), Smem = Acz	Smem = ACx,copr()
Acy = copr(k8,ACx,ACy), dbl(Lmem) = ACz	Lmem = ACx,copr()
ACx,ACy = copr(k8,ACx,Acy)	copr(k6)
ACx,ACy = copr(k8,ACx,Acy), Smem = Acz	Smem = ACx,copr()
ACx,ACy = copr(k8,ACx,Acy), dbl(Lmem) = ACz	Lmem = ACx,copr()
ACy = copr(k8,ACx,Smem)	copr(k6)
ACy = copr(k8,ACx,Xmem), Ymem = ACz	Smem = ACx,copr()
ACy = copr(k8,ACx,dbl(Lmem))	copr(k6)
ACy = copr(k8,ACx,dbl(Xmem)), dbl(Ymem) = ACz	Lmem = ACx,copr()
ACy = copr(k8,ACx,Xmem,Ymem)	copr(k6)
	copr()
ACx,ACy = copr(k8,ACx,ACy,Xmem,Ymem)	copr(k6)
ACx = copr(k8,Ymem,Coef), mar(Xmem)	copr(k6)
ACx = copr(k8,ACx,Ymem,Coef), mar(Xmem)	copr(k6)
ACx,ACy = copr(k8,Xmem,Ymem,Coef)	copr(k6)
ACx,ACy = copr(k8,ACx,Xmem,Ymem,Coef)	copr(k6)
ACx,ACy = copr(k8,ACy,Xmem,Ymem,Coef)	copr(k6)
ACx,ACy = copr(k8,ACx,ACy,Xmem,Ymem,Coef)	copr(k6)
	copr()

In Table 18.3, "k8' represents the 8-bit opcode sent to the accelerator, "ACx' and "ACy' represent TMS320C55x processor core internal accumulator references and "Xmem', "Ymem' and "Coef' represent the three possible references for reading and writing memory data. From a programmer standpoint, only dataflow modes are visible in the assembler tool of the processor. Hence, he/she never has to worry about how the instruction pairs are built, as they are automatically assembled within this tool (see Section 18.2.4).

A dataflow mode describes the call to the extension datapath from the software. The syntax used in Table 18.3 utilizes the generic keyword "copr()" as a short form of the qualified instruction and qualifier opcode pair. The implicit parallelism syntax (ex: ACy = copr(-k8,ACx,ACy), Smem = ACz) is used for Smem or Lmem writes that are allowed in parallel with the execution in the CPU accelerator.

Usage of dataflow modes of Table 18.3 in the software is the same for any other processor instruction. Dataflow modes can be used in any software control structure, including single- and multi-instruction loops and conditional executions. This provides a very powerful control mechanism for the sequencing of extension datapath operations. Moreover, dataflow modes can be freely mixed with regular DSP core instructions (standalone or pairs). These regular instructions allow, for instance, the preparation of values in accumulators before using them by issuing a dataflow mode. A consequence of this last property is that, unlike co-processors, only the required datapath functions need to be implemented in the extension. As results can be easily shared between the core and the extension, one can imagine the partition in a very fine-grain way of the application kernel between regular and *accelerated software* (i.e. dataflow mode running on the extension datapath).

For the sake of implementation of the hardware connections between the core and an extension datapath, when multiple accelerators are present in an application, the instruction field (8-bit) exported at the interface is divided into two parts:

- bits 7–5 indicate the number of extension datapaths (up to eight can be connected);
- bits 4–0 indicate the instruction code for the selected extension (up to 32 instructions per extension).

18.2.3 Typical C55x Extension Datapath Architecture

In this section, we will see what the key characteristics of an accelerator datapath are and we will understand connection constraints and limitations. In order to go to this level of detail it is important to have a quick overview of the extension interface protocol and timings.

The Interface is synchronous to the CPU clock frequency and execution within the extension datapath is expected to occur in one clock cycle. Hence, operation speed is set to be the same as the operating frequency of the processor core (no wait states are supported through the interface). In order to avoid instruction decoding for internal datapath controls generation negatively impacting the speed, the extension instruction and validation strobe are given one cycle ahead of the cycle at which arithmetic operations and data exchanges occur. This allows the extension to decode and register these controls before issuing them to the datapath. Figure 18.2 shows the whole operating protocol. The cycle after the instruction is sent to the extension; operands, either coming from memory or from internal accumulators according to the dataflow mode selected, and status bits are sent through the interface. All of them almost directly come out of registers in order to minimize timing. At the end of this cycle, two things are expected to occur:

- Internal datapath registers are updated; and
- Values are returned to CPU accumulators, according to the dataflow mode selected.

In this last case, values that are returned to the CPU are registered in the extension, so that set-up timing constraints can be easily met.

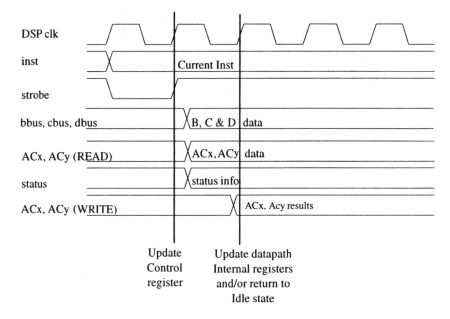

Figure 18.2.   Timing diagram of a typical operation through the interface

For the same reason, operands sent by the CPU are always registered at the entrance of the extension. Figure 18.3 describes the various register levels within a datapath, including the control pipeline.

From Figure 18.3 a few conclusions can be made on the way an extension is used by the core:

- Basic operations require the combination of three steps in a pipelined way: one cycle to load operands, one cycle to execute from previously stored data (results or recent operands) and one cycle to return data to the CPU (one more cycle must be added in the case of memory write).
- An extension datapath containing no local storage would loose cycles to bring in and send out data. This cycle loss can be removed, during steady state operations, by having an intermediate local storage that allows results to be picked up in the extension within a single cycle.
- Similarly, managing pipelining of operands loads and result stored back in the CPU, often requires input and output data buffers in order to take into account differences in data organization (e.g. in memory) versus computation organizations.

Hence, a generic micro-architecture of the extension datapath, can be described as follows (see also Figure 18.4):

1. A register file containing several areas (RF$_i$), for keeping local intermediate results and for operands and I/Os with the core;
2. A set of function units to perform dedicated computations, according to application needs;

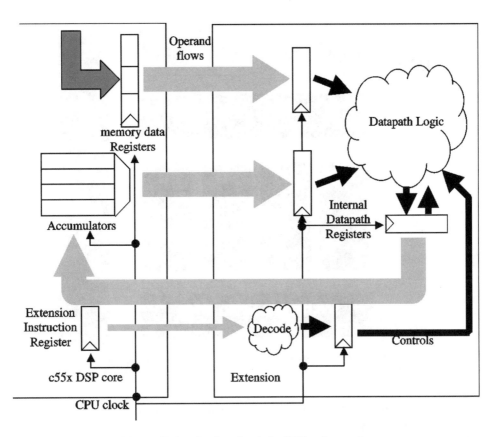

Figure 18.3.   Register levels in CPU and extension

3. A connection network between the register file and the operators, as well as between areas of the register file;
4. An instruction decode and controls generation logic with special features to reduce activity within the datapath (clock domains control).

One consequence of this architecture organization on some of the basic properties of the concept is that, despite the fact that results and operations are built so that they all occur in one CPU clock cycle (this defines the speed constraint), accelerated kernels could be implemented using several cycles by internally pipelining all operations (like in a traditional co-processor). Also, an extension is not only built of pure datapath functions but may contain locally stored elements like status information that can influence processing execution without direct interaction with the core. Hence, virtually any type of application can be addressed with this concept. We will see, in Section 18.3, which parameters will limit the scope of its effective utilization.

By nature, an architecture like the one shown in Figure 18.4 can be built so that it supports a good level of re-configuration. This requires that following properties are introduced:

- Datapath operators must support the various arithmetic operations for each configuration;

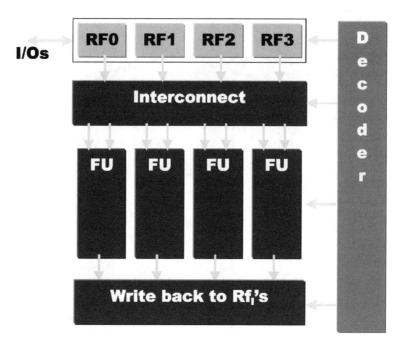

Figure 18.4.   Generic micro-architecture of a TMS320C55x extension

- Register files must be dimensioned so that a wider number of variables are supported, depending on worst case configurations;
- The instruction decoder must provide a wider instruction register in order to accommodate more operators, bigger register files and possibly more complex operand selection networks.

The standard way of controlling operands and generating extension control instructions, via dataflow modes, allows, by definition of an equivalent language that describes how resources are manipulated within the datapath, the minimization of the final amount of distinct internal resources. A study, performed in collaboration with the university of Leuven (Belgium), showed that this was achievable and an example was processed with two tightly coupled accelerators developed for video processing (see Section 18.3). In cases where sharing of resources between configuration is not effective, then re-configuration will lead to significant overheads in area, power and, possibly, performance, compared to separated, optimized implementations.

18.2.4 Integration in Software Development Tools

As the main processor ISA is extended by a new functional unit and new instruction to exercise it; the whole tool chain for software development must naturally be able to integrate it in each of the parts. This implies that assembler, software simulator, debugger and C compiler tools know about the concept and provide corresponding flexibility.

The assembler must understand dataflow mode syntax so that programmers do not have to write a corresponding pair of instructions in order to control the extension. This pair is automatically built for them after correct dataflow syntax is checked. Some code size optimizations can be performed by the assembler when building the pair, if some flexibility is available (by means of the choice of "copr' opcodes).

Generated binary codes must then be simulated. In order to do this, a software simulator is used which must be capable of:

- adding a custom model of an extension (developed with defined templates);
- simulating the added behavior with the core;
- debugging the internal extension resources, in conjunction with those of the core.

These features are usually implemented using shared libraries for the new function that are dynamically linked to the simulator core, in conjunction with utilization of specific data structures, in the custom model, for internal extension registers (RF_is, for instance). These data structures, associated with dump and restore functions, allow the display of internal variables at each cycle (showing the name and value of the variable). By putting software breakpoints on dataflow mode instructions and stepping through the code, users can see the computed results evolve in the new extension. Values can be downloaded for verification at any time.

Moving further down the road of the product life, it is important to re-conduct the same approach on silicon. A user will want to display the same view of internal extension variables on the running part. This requirement is supported by means of the emulation logic that comes with the core. Special extension instructions (i.e. lower 5-bit field of the "k8' constant in the dataflow mode) are always present in order to download or upload data values from the debugger software to the hardware on silicon. These are dedicated emulation instructions that follow the same building process as the regular dataflow modes but that are issued only by the emulator. Any register in the extension is mapped to an address space that can be accessed by both application and emulation logic. Dataflow modes can address a maximum of 32 of these registers. Hence, there cannot be more than 32 functional registers in a TMS320C55x extension. Each extension is uniquely identified by an embedded "ID' code that can be read at address 0 (note: this address does not contribute to the maximum number of locations described previously). This ID indicates the number of valid register addresses available for download and a code to define what the extension is. Figure 18.5 explains how the emulator and extension are coupled.

More and more, higher software development languages, like C, are used to describe DSP applications. Hence, it is natural to give access to this concept of extension to a C compiler. Several levels of support can be introduced. A first step can be to introduce a library of dataflow mode calls that can be filled by code that emulates the extension behavior but that will be replaced, at compile time, by a corresponding pair of instructions with the right resource allocation. Application code would be built by instantiation of these calls and mixing them with regular C expressions. The resulting code can, thus, be easily prototyped on another platform for validation purposes and then compiled on the DSP with the optimum performance provided by the ISA extension. A more advanced approach could be to synthesize dataflow calls from user code. This approach is being actively investigated in various computer research groups. Some of the biggest issues are to find an efficient mapping of the user code structure into the extension datapath organization and to allocate variables to operand resources so that interface protocol constraints are met.

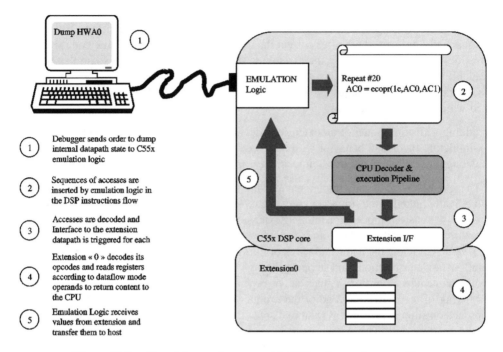

Figure 18.5.   How debugger interacts with CPU and extensions on silicon

18.3 Domains of Applications and Practical Examples

Before going into the detail of some examples, we will study, in this section, what the key parameters are for defining a TMS320C55x tightly coupled accelerator. This will show the application domain limits.

DSP kernels have typical characteristics that make them more or less suitable to acceleration. Most of these kernels have one or more critical loops where we are going to focus in order to develop a new extension. We can summarize characteristics of this loop as:

- Inner operations complexity: multiply-accumulate (FIR, correlations, …), add-compare-select (Viterbi and Turbo decoders), FFT butterfly,... This has an impact on the level of pipelining required to operate at CPU speed.
- Inner operations input operands size and number. This last parameter must be analyzed across iterations. These have an impact on the maximum level of parallelism across iterations that can be extracted (hence, the number of inner operators required).
- Intermediate results usage across iterations (e.g. trellis based functions): this has an impact on the amount of variables that must be stored between iterations. It has also an impact on the level of parallelism that can be achieved. By analyzing how intermediate results contribute to the progression in, let's say, a trellis, significant acceleration can be extracted.
- Dependency of inner operations on the operands content (e.g. variable length decoders): this limits the possible parallelism that can be exploited. Conditional execution of an

extension function is a key support feature for this. Because it may require managing status bits within the extension, this has also an impact on the number of functional registers.

The concept of datapath extension described in Section 18.2 provides parameters to evaluate above criteria. Table 18.4 describes these parameters and their effects. There is one more constraint which is linked to design complexity of extension datapaths. The requirements of matching the speed of the extension datapath to that of the TMS320C55x core leads to a practical gate count limitation. As multiple accelerators can be tightly connected to the CPU, it is reasonable to limit the gate count of an elementary function to a fraction of the core gate count. Fifteen percent is a good target and leads to a gate count limit of 30–40 kgates. This limit will also keep the current consumption budget of the assembly low: DSP core + extension + local memory. The overall current consumption budget (Figure 18.6) for this system is established as follows (average):

- 37% of the total current is spent on memory accesses (data and program);
- 44% is spent on the DSP core; and
- 12% is spent on the extension datapath.

Table 18.4 TMS320C55x interface parameters and their effect on performance of the extension

Parameter	Value for the extension hardware	Effect
Memory operands bandwidth	8-bit reads: 2–6 per cycle – aligned feches	Limits, practically, the amount of parallelism that can be extracted across iterations (max. is 4×8-bit, 2×16-bit and 1×32-bit memory operands per cycle)
	8-bit writes: 2–4 per cycle – aligned	
	16-bit reads: 1–3 per cycle	
	16-bit writes: 1–2 per cycle	
	32-bit reads and writes: 1 per cycle – aligned	
Number of registers	32×32-bit registers	This affects both the number of internal variables that can be re-used, the amount of pipeline registers and the amount of functional status bits that can be defined
Conditional execution	Two-cycle operation	Will have significant cycle impact unless shared by many iterations

In order to illustrate how the TMS320C55x processor can be efficiently enhanced with the right set of tightly coupled functions, the domain of video processing will be considered. As part of the OMAP® platform [2] and in order to address still and moving image compression and de-compression, which is one of the first domains that happens to be a major extension of functionality into wireless terminals, and which is strongly enabled by third generation

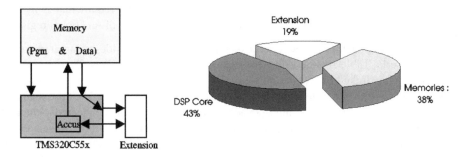

Figure 18.6. Relative distribution of current consumption in a TMS320C55x sub-system with extension

network, a set of TMS320C55x extensions were implemented. We will see how the features and rules described for a generic extension can be applied in real examples.

Video processing covers a wide set of applications ranging from still image display and enhancement to digital still camera and video coding and decoding (video-conferencing). All these follow a set of standard algorithms (mainly JPEG and MPEG4) which are very close in terms of the application sub-tasks they use. This is interesting as it allows the sharing of the main critical routines between applications. Among these routines, motion estimation, Discrete Cosine Transform (DCT) and its inverse function (iDCT) and pixel interpolation are among the most consuming in terms of number of cycles for a pure software implementation using the TMS320C55x processor. Table 18.5 gives their corresponding average contributions, in percent of total cycles executed by the software, for a video Coder-Decoder (CoDec) running on a QCIF color image size (176 × 144 pixels), at 30 frames per second (MPEG4 layer and control tasks cycle count numbers are not taken into account). These critical routines, representing about 80% of the total number of cycles for the CoDec, are, thus, good candidates for being accelerated by tightly coupled acceleration as described in this chapter.

The first step in order to define corresponding extension datapaths is to analyze the most critical part of each routine. This will be supported by dedicated hardware, under software supervision as already described. DCT, iDCT and interpolation are well-isolated pieces of algorithm [3] and hardware software partitioning is easy:

- DCT/iDCT: selection of a fast transform scheme is the first step. Then hardware must fulfill accuracy requirements.

Table 18.5 Most critical routines cost in cycles for an MPEG4 video CoDec

MPEG4 video CoDec, QCIF format @ 30 fps	
DCT	10% of total cycles
iDCT	20% of total cycles
Motion estimation	35% of total cycles (with step-search algorithm – see text)
Pixel interpolation	15% of total cycles

- Half pixel interpolation has a very simple datapath, doing mainly averaged out sums of pixels, with 1/2 and 1/4 coefficients. Rounding of results is normalized for the decoder.

As far as motion estimation is concerned, partitioning is a bit more complicated. This last function consists of a search of the most similar block of 16 × 16 pixels in the previous image corresponding to a block of pixels in the current image. In order to measure how similar the blocks are, a mean absolute error criteria is used. The search area is defined in the range of ± 15 pixels around the block in the current image. This search process is not normalized. Depending on performance trade-offs, either hierarchical full search or step-search approaches can be used. Due to its smaller computation requirements in number of cycles, Step-search algorithm is often chosen, although it generally leads to finding a sub-optimal location of that minimum (Figure 18.7).

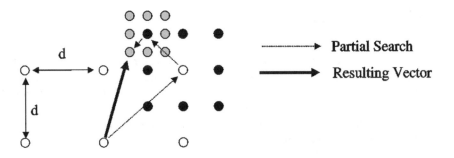

Figure 18.7.   Step-search algorithm by reducing the distance d at each iteration

The search process can be separated into two basic operations, the error computation for a 16 × 16 pixel block and the finding of the minimum error position corresponding to the best block location. This last operation is very well optimized in the DSP so it makes sense to leave it in the software. The most cycle consuming operation is the error computation, as c55x basic ISA has only one ALU to do the absolute distance computation step of only 16-bit data. The accelerator is, thus, concentrating on accelerating this part of the routine by adding more efficient absolute distance computation hardware on 8-bit data by:

- providing more parallelism through inner operator complexity (more absolute distance computation steps per operator);
- minimizing the data storage requirements by manipulating 8-bit data (pixels of searched and current images);
- using the maximum data operand bandwidth possibly available. This combines a smaller data format with maximum bandwidth available for reading.

Pixels of the current image (also called "reference' image) 16 × 16 block are re-used during each step of the search process. More precisely, the error computation is performed row by row. Hence, keeping locally a row of the reference macroblock and re-using it for three macroblocks of the searched image provides the necessary parallelism increase. Finally, the extension datapath consists of:

- three absolute distance computations operators, each processing 2 pixels at a time;
- a register file (RF0) for storing the reference macroblock line;

- a register file (RF1) for storing the search image pixels;
- a register file (RF2) containing three error results.

At the video CoDec application level, three extensions were defined in order to support the required level of speed-up. Table 18.6 summarizes the characteristics of these functions.

Table 18.6   Video extensions characteristics

HWA type	Current consumption (at 1.5 V) (mA/MHz)	Number of gates	Speed-up factor versus software
Motion estimation	0.04	7000	× 5.2
DCT/iDCT	0.06	34000	× 4.1
Pixel interpolation	0.01	7000	× 7.3

The full video CoDec application is accelerated by a factor of 2 versus a pure software implementation. The total gate count for all extensions, including the management of the write path to the interface, represents 51,000 gates of logic, for 200 MHz CPU operation. This gate cost (about 24% of the total core size) must be balanced by the acceleration obtained at the application level, which allows a reduction in the system power consumption by working at lower frequency and supply voltage.

18.4 ISA Extensions Design Flow

TMS320C55x datapath extensions are all built following well-defined constraints and rules. In terms of implementation, each hardware is made up of two major functions: an instruction decoder and a datapath. Hence, hardware design can essentially be built from a template architecture which integrates most of the concept rules. The decoder structure generation can, for instance, be easily automated. Other structures like clock gating control can be also automatically created. Once hardware–software partitioning is known, correct datapath construction and scheduling according to timing constraints can be helped by tools like Behavioral Compiler®, from Synopsys.

Most of the difficulties of developing such an extension are mainly linked to:

- best extension hardware versus TMS320C55x software trade-offs definition; and
- validation of the new hardware.

Optimizing hardware and software requires working with a software simulator including a representation of the TMS320C55x core and the extension. In Section 18.2, we saw that the TMS320C55x simulator allows the importation of the accelerator description and simulates it. In this approach, the user manually defines the architecture and the software to control it. It is a long process which is prone to errors. A lot of research has been conducted in the domain of hardware–software co-design. Some academic developments have led to taking into account estimates of software and hardware performance in order to appropriately schedule the final application while optimizing the resulting cost [4]. These techniques, integrating the rules of connection of an extension to the TMS320C55x core, could really help in making these trade-offs quickly and for comparing the cost of various solutions. Cost is generally measured in terms of gate count or power consumption (Figure 18.8).

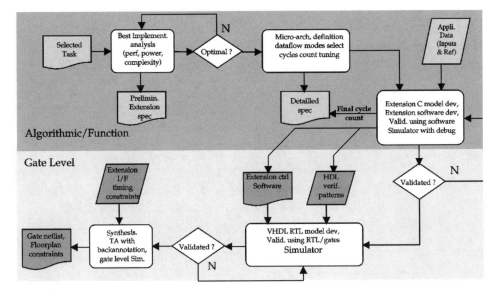

Figure 18.8. TMS320C55x ISA extension typical design flow

On the validation front, it is very important to be able to move test sequences prepared during development and validation of the software, at the simulator level, down to functional and gate level simulation steps. In order to do this, the software simulator must have a mode in order to export patterns, exchanged through the extension interface, in a file that can be further exploited by RTL and gate level simulations. These sequences must be further enlarged by dedicated design sequences to characterize other typical DSP system situations like behavior in the case of memory stalls or in the case of interrupts. Combining application-oriented patterns with standard implementation test cases (that can be re-used from one extension datapath design to another) allow a high level of functional coverage to be kept, for such types of design.

References

[1] *TMS320C54x DSP Processor User's Guide*, Texas Instruments.

[2] Chaoui, J., Cyr, K., Giacalone, J.P., de Gregorio, S., Masse, Y., Muthusamy, Y., Spits, T., Budagavi, M. and Webb, J., *OMAP™: Enabling Multimedia Applications in Third Generation (3G) Wireless Terminals*, Texas Instruments, SWPA001, December 2000.

[3] Bhaskaran, V. and Konstantinides, K., *Image and Video Compression Standards*, Kluwer Academic Publishers, Boston/Dordrecht/London, 1995.

[4] Bianco, L., Auguin, M. and Pegatoquet, A., *A Prototyping Method of Embedded Real Time Systems for Signal Processing Applications*, I3S Laboratory, University of Nice Sophia Antipolis, France and Philips Semiconductors.

19

The Pointing Wireless Device for Delivery of Location Based Applications

Pamela Kerwin, John Ellenby and Jeffrey Jay

19.1 Next Generation Wireless Devices

Wireless devices stand at a new frontier. Reduction in the cost and size of powerful processors allow handsets, PDAs and digital cameras to advance from single purpose devices to complete multimedia platforms. Broader bandwidth networks allow vastly expanded functionality in these devices.

This platform is more than a small television or tiny computer. It is an entirely new medium with its own limitations and advantages. Most importantly it is used in a completely unique way.

Exciting opportunities abound for content providers and users of these new devices.

19.2 The Platform

Phones have daylight viewable color screens and significant processing power. They include technology that allows them to determine exactly where they are. They support Internet connectivity, numerous applications, video and music. However, since they are pocket devices, they remain small and input to these devices continues to be difficult.

The devices put incredible power in a user's hand. The prospect of having immediate access to information and entertainment at any time wherever you may be is truly compelling.

19.3 New Multimedia Applications

Certainly we are seeing new applications and content becoming available for these platforms. Users can download digital music and play interactive games. Enterprise applications, travel programs, video conferencing, movie trailers, etc. are all literally in the hand of the user.

19.4 Location Based Information

The ability to wander about with a position smart and powerful device in one's hands naturally encourages users to want to get information about the world around them. Standard coordinates such as latitude and longitude can index objects in the world. Databases exist which contain information about those objects – everything from cost when last sold through restaurant reviews.

The powerful, position intelligent mobile device with Internet connectivity can summon information to the user – exactly when he/she is interested in it.

For several decades now, we have watched Star Trek crews explore alien universes assisted by their small tricorders. While we are not quite building tricorders yet, we have certainly taken the first step.

19.5 Using Devices to Summon Information

Information exists in various databases and on various sites and the device has the ability to present users with images and audio feedback, still the greatest challenge facing the location industry is making it easy for the user to get the information he/she wants.

Devices have small screens and keypads that have been designed for nothing more complex than number input. Voice command is making progress but an open-ended request for information about unknown objects will not be a reality for many years.

How does the user indicate exactly what object or series of objects interests him/her?

What is required to create a mass market for location based services is a user interface that is intuitive and efficient – like the mouse was to the PC and Mosaic/Netscape was to the Internet. A simple user interface is essential for consumer acceptance of location based services

19.6 Pointing to the Real World

Minor enhancements can be made to devices that allow the device to be pointed at the object of interest to summon information about it. A consumer-friendly interface to location based services can then be offered. Such a method offers a seamless coupling of the real world with relevant information on the Internet or any other existing database. The consumer can simply point his phone at an establishment to say: "Tell me about that!" The device also becomes an intuitive navigational platform.

The device manufacturer adds a small inexpensive heading sensor (a digital compass) to the device. When determining its location, the device also takes a direction reading from the heading sensor.

The wireless network forwards that data and service request to an information server. The server uses the position and pointing direction data to search its database of geo-coded objects, identify the correct object and link to the appropriate information requested about that object.

This interface method was designed to meet the unique information needs of mobile individuals as they inquire naturally about their environment. Pointing technology significantly enhances information about the context of the mobile user by providing not only information about the user's location, but also where his/her attention is focused.

19.7 Pointing Greatly Simplifies the User Interface

Direction is the natural complement to position. The pointing interface operates as a smart filter for location based services. This is an advanced method that considerably simplifies the way users interact with their devices to retrieve information. It leverages the natural human gesture of "pointing," minimizes time-consuming button presses required by current services and allows for targeting of information more effectively than by simple location.

Below are two examples of an individual seeking information about restaurants in San Francisco. He is walking from his office to the Ferry Building and he wants to eat along the way. The example features actual data for restaurants provided by InfoUSA.com. In case one the individual searches for restaurants using the position of his phone alone as a filter. He received 1059 objects as a response. In the second example the user asks for restaurants only in a particular direction – on his way to the Ferry. Now he receives only 56 responses.

The search is simpler and the experience is much more friendly. The phone has become a virtual mouse, allowing the user to inquire about objects in the real world by pointing at them and to gain useful information without having to sort through a great deal of irrelevant data.

Many • Clicks Required to Find Restaurant   **Two or Three Clicks Required**

19.8 Uses of Pointing

People can point their phones to quickly and easily find information about places of interest. Businessmen point PDAs to find restaurants between their meeting place and hotel.

Travelers trust their devices to guide them to points of interest and to automatically annotate the pictures they have just taken with their digital cameras. Young people will leave notes for one another and will play location-enhanced games. Shoppers will point at stores to indicate that they are willing to receive promotional offers.

19.9 Software Architecture

19.9.1 Introduction

The provisioning of pointing-enabled location based services (i.e. services rendered based on position and heading (direction) data of wireless devices) requires, in a very general sense, the integration of a wireless device, a wireless network, and a server capable of retrieving information based on the device's position and heading data.

19.9.2 Assumptions

It is assumed that the fundamental function of the service is to retrieve information about real-life (or imaginary) objects (buildings, businesses, natural landmarks, gaming characters or simply, mere points in space) by pointing at them with a wireless device. This system was specifically designed for pedestrians.

19.9.3 Overview

The generic service environment for a generic service is depicted below.

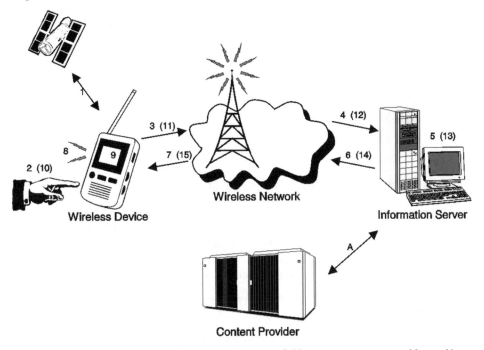

Content Provider

Generic GeoVector service environment. Aspects of this system are patented by and/or are proprietary to GeoVector.

When the wireless device is powered up, it must first determine its position data (1). The user indicates that a pointing service is desired (2) and the device transmits the position data and a service request to the wireless network (3). The wireless network forwards that data and service request to an information server (4). The information server uses the position data to search a database of geo-coded objects surrounding the device and prepares an object list according to the service requested (5). The information server then transmits this object list through the wireless network (6) back to the device (7).

The device gathers its position and heading data (8) and uses that data to select a subset of the object list received from the information server. If more than one object is selected and if it is within the nature of the service offered, the user is presented with the opportunity to select a single object. Information about that selected object is displayed by the device (9).

The user may request greater detail about the object (10) if that detail is available from the information server. That request would take the same path as the initial request for objects (11–15).

It is possible that the information server will need to pull some content from another location to satisfy this (and any subsequent) request. In this case, the information server would communicate with some content provider (A) for that information. In all likelihood, the content provider would be another server accessed via the Internet, though that is not required.

19.9.4 Alternatives

The preceding assumes that the wireless device has the processing and memory capacity to perform step 8. In the case where such capacity does not exist, the information server (or another server) may perform that task. This alternative allows for very "thin" wireless devices to still be used as pointing devices.

On the other end of the "capacity spectrum", if the wireless device has sufficiently large enough memory, a large geo-coded database may be stored on the device and updated only when the user has moved a significant distance. This alternative reduces the number of wireless transmissions needed to provide service (though the transmission of the geo-coded database can be quite lengthy).

19.10 Use of the DSP in the Pointing System

Use of the DSP in pointing designs will vary depending on the actual DSP/MCU combination within the cell-phone and on the scheduling and duration of other DSP tasks.

DSP loading will determine the length of time available for carrying out heading sensing. Available time will vary widely depending on the communications protocol, and other parameters. It is assumed that establishing heading takes place while the transmitter is not operating. In this discussion it is assumed that time available for heading sensing is anticipated to be in the region of 20 ms.

It is assumed that the DSP controls all functions of the circuit. These are:

1. Power on the heading sensing circuit
2. Reset each axis of the heading sensor
3. Take a number of heading readings (number is based on available time)
4. Average the heading readings
5. Power down the circuit

The DSP/MCU can also be used for these functions:

1. To interpret data that monitors user behavior and prepares the device for the next search
2. To compensate for anomalies in position or direction readings
3. To inform the user of reliability of data
4. To monitor user behavior and adapt the device to anticipate user requests

19.11 Pointing Enhanced Location Applications

The natural human gesture of "pointing," minimizes time-consuming button presses required by current services and allows for targeting of information more effectively than by simple location. The simplicity of this information access methodology allows for mobile content delivery to move beyond the most basic service paradigm of "push 360° of surrounding information" to a service structure where individuals have far more control over what information they receive and when they receive it. The result for individuals is a far more useful and friendly experience. For carriers, content developers and location based service providers this higher relevancy leads to higher probabilities of user actions and transactions. This in turn justifies and promotes development of a host of compelling pointing enabled tours, games, community and sponsorship applications that drive value for the wireless user all the way back through service providers to device and component manufacturers.

By pointing at an object of interest this tourist has access to both information and purchasing opportunities. In this example, a book is offered along with payment and delivery options. The point of interest can be at some distance from the tourist.

19.11.1 Pedestrian Guidance

Because this method is not line-of-sight limited, a user can quickly and easily identify locations around the corner, down the street or half way across the city. In addition, a user can refine his search by limiting distance variables or other criteria. A unique capability enabled by the device's knowledge of direction is a highly simplified pedestrian guidance function. This feature can generate a simple arrow and allow the user to follow the arrow to his/her destination. It is not necessary for the user to know his/her way around, to have a sense of direction or to struggle to find information on a tiny map.

19.11.2 Pull Advertising

The pointing interface method allows a user to signify his/her interest and intent.

The user can indicate his/her interest in an establishment by pointing at it. At that time, the ad can be sent. This opt-in ad has a much higher probability of resulting in a transaction.

This interface allows the service to know which establishment is of interest to the user at that moment and can also be used as a means for the user to signal his/her willingness to receive promotional offers. The system offers the highest possible level of targeting sensitivity with respect to location, time and intent. On-demand advertising delivered at the moment of inquiry is far more likely to stimulate a transaction from the user than general-location, push based advertising targeting.

Phones with pointing capability also provide a number of unique marketing opportunities. Advertisers can create virtual storefronts or billboards that contain electronic coupons, or real-time customized specials. This marketing channel can be more tightly focused and accountable than other media. The effectiveness of promotional campaigns can be evaluated immediately when the consumer takes advantage of the offer.

19.11.3 Entertainment

The industry now generally recognizes wireless handsets and PDAs as new media platforms. Location based games add a dynamic new dimension to wireless entertainment. Users can play in the real world with other players in their neighborhoods.

As we have seen in Japan (and on American TV) quality content or programming is necessary to guarantee the business success of the device or service. "The Pointing Access Method™"[1] extends the capability of mobile devices and allows them to support valuable and fun new programs.

Pointing capability allows for unique and compelling implementation of applications such as specialized tours with an emphasis on history, architecture, food, shopping, etc. Pointing also enables engaging entertainment applications: role-playing experiences, adventure

---

[1] A trademark of GeoVector Corporation. Note that the pointing access method is a patented GeoVector technology.

games, treasure hunts and virtual hide and seek. Each structure or location in space is capable of hosting information and can become a message board, a clue, or a character.

A particularly interesting use of pointing is to point to objects and leave graffiti-like messages on them for access by public or private groups. This capability offers a meaningful extension of the Internet's messaging features and buddy groups into a mobile environment.

19.12 Benefits of Pointing

Some of the applications discussed above can certainly be created without implementing a pointing feature but interfacing with them would be cumbersome and pointing brings a number of extremely valuable enhancements. These are outlined below.

19.12.1 Wireless Yellow Pages

Pointing provides an automatic, intuitive filter that reduces button presses and makes searching for information more efficient.

A pointing system allows users to inquire about remote or unknown objects. Location alone will not all allow users to inquire about objects that are far away. The increase in data required to respond to a request involving a long range would overwhelm the user. Further, if the user does not know what category the object is in (hospital, hotel, university), he/she cannot find it any other way but by requesting information about all the objects within that range.

19.12.2 Internationalization

A pointing interface overcomes language problems. Users do not need to know how to read signs or pronounce foreign words in order to get information.

19.12.3 GIS Applications

Using the directional data provided, the device can guide the user to his/her destination without requiring him/her to read maps.

19.12.4 Entertainment and Gaming

Pointing devices can be used to target people or things and enables directionality in game play. It provides an efficient link between the real world and fantasy.

19.12.5 Visual Aiding and Digital Albums

When implemented in digital cameras, pictures can be labeled automatically.

19.13 Recommended Data Standardization

The following is a recommendation for passing position and direction data that allows for maximum benefit to location applications.

This discussion describes the format and range of the device data necessary to implement various pointing-enabled location based services. In addition, it offers a similar format for non-pointing location based services.

The term "device data" is used to refer to data that describes some aspect of a wireless device. It is important to note that the device itself may not necessarily generate the data. For example, a cellular carrier may decide to use network-based methods for determining the location of a handset. This data will be generated by the network and will likely be passed to a location based service without ever setting foot on a handset.

In fact, there exists the notion of a "mobile location server" that a few infrastructure companies are marketing to cellular carriers now. The mobile location servers will store and make available the location of the handset *even if the handset has an onboard GPS receiver*. Therefore, this section talks about data and its format as it appears to the pointing-enabled location based service, regardless of how, or where, the data is generated.

19.13.1 Consideration of Current Standards Efforts

It is acknowledged that there is ongoing work by standards bodies (e.g. 3GPP and LIF) to standardize the content and format of data to enable location based services. This section refers to non-pointing location based services with that knowledge and offers a format for non-pointing location based services for completeness.

19.13.2 Device Data Types and Tiered Services

The device data is presented such that it supports four types of pointing-enabled location based services, each available in a 2D and 3D environment depending on device hardware capabilities. The four types are *location, pointing, gesturing,* and *pointing with imaging.*

Location is enabled by the single, discreet measurement of a device's location (position). It is expected that for non-pointing location based services, height will not play a major role in service offering, so these types of services will tend to be 2D only.

Simple pointing is enabled by the single, discreet measurement of a device's location and heading (direction in which it is being pointed). Note that *heading* does not imply movement of the device and that the term *pointing* will often be used interchangeably with *heading*. Both

2D and 3D services are enabled depending on the ability of the device to capture 3D data (see Table 19.1).

Table 19. 1

Device data	Description	Tier	Location	Pointing		Gesturing		Imaging
			2D	2D	3D	2D	3D	3D
Latitude	Location y	1	●	●	●	●	●	●
Longitude	Location x	1	●	●	●	●	●	●
Height	Location z	3			●		●	●
Horizontal	Pointing α (compass points)	2		●	●	●	●	●
Vertical	Pointing θ (up or down)	3			●		●	●
Roll	Pointing ϕ	3						●
Path	Sequential list of above	4				●	●	

Gesturing is enabled by multiple, discreet measurements over a small period of time of a device's location and heading. Again, both 2D and 3D services are possible.

Pointing with imaging is enabled by the addition of a roll variable. This allows a pointing-enabled location based imaging device to correctly overlay generated images with real-world images. Imaging services are generally not practical with 2D data, so 3D is always assumed.

Table 19. 1 illustrates the types of pointing-enabled location based services made possible by the different types of device data and their relationship to one another. This structuring of data allows for different devices to access different *tiers* of location based services.

Devices that are only capable of generating tier 1 data are only able to access tier 1 services. For example, a device that uses cell tower triangulation for location determination is only able to access simple 2D non-pointing services.

Devices that are capable of generating tier 2 data are able to access tier 1 and 2 services. For example, a device that uses cell tower triangulation for location determination and a simple 2D digital compass is able to access simple 2D services, both pointing-enabled and non-pointing.

Devices that are capable of generating tier 3 data are able to access tier 1, 2, and 3 services. For example, a device that uses GPS for location determination and a 3D digital attitude orientation sensing subsystem is able to access simple 2D services, 3D services, and services that support imaging.

Devices that are able to sample tier 2 or tier 3 data over time and package that data up in a sequential list (tier 4 data) are able to access 2D or 3D services that support gesturing.

19.13.3 Data Specifications

19.13.3.1 Location Data

The location data includes latitude, longitude, and altitude along with a confidence value (Table 19.2). The confidence value indicates to the service how reliable the values are in pinpointing the location. It may also indicate (possibly) how the position was derived.

Table 19.2

Data	Range	Units	Sample
Latitude	−90.00000–90.00000	Degrees	38.27782
Longitude	−180.00000–180.00000	Degrees	−122.63490
Altitude	−100–10000	Meters	41
Confidence	0–100		10

Latitude: measures position relative to the equator with positive values indicating north and negative values indicating south. The range is from − 90 to + 90°. Precision is to 1/100,000°, which is roughly equivalent to 1.11-m accuracy at the equator. Note that if the precision is reduced by a factor of 10 (four numbers to the right of the decimal instead of five), position accuracy can never be better than 11.1 m.

Longitude: measures position relative to the meridian with positive values indicating east and negative values indicating west. The range is from − 180 to + 180°. Precision is same as latitude.

Altitude: measures height relative to sea level. While not necessarily practical, some below level values should be allowed; − 100 m is mentioned here, though that value is arbitrary. Similarly, maximum altitude is somewhat arbitrary; 10,000 m is roughly equivalent to the cruising altitude of major airlines. Certainly, some consideration must be made for readings taken at the top of buildings, on the side of mountains, and so forth.

Confidence: indicates the confidence level of the accuracy of the position information where a higher number indicates a greater level of confidence. For example, a GPS reading might include a confidence value of "95" while a positioning method is using only the cell ID might include a value of "5". The values need to be determined.

19.13.3.2 Pointing Data

The pointing data includes a horizontal measurement, a vertical measurement, and a roll measurement along with a confidence value. For the truly technical, the horizontal measurement is called yaw and is the measurement of the rotation around the z-axis. The vertical measurement is called pitch and is the measurement of the rotation around the y-axis. The roll measurement is called roll and is the measurement of the rotation around the x-axis.

The horizontal and vertical measurements are sufficient to pinpoint a point in 3D space. The roll measurement is necessary to correctly orient the view of that point (Table 19.3).

Table 19.3

Data	Range	Units	Sample
Horizontal	0–359	Degrees	15
Pitch	−90–90	Degrees	45
Roll	0–359	Degrees	0
Confidence	0–100		95

Horizontal: measures direction such that 0 refers to magnetic north and 180 refers to magnetic south, with 90 indicating east and 270 indicating west. Precision beyond whole degree units is not critical. Note that these values are to indicate "magnetic" directions not "true" directions. It is sufficient for the pointing-enabled location based service to obtain these values along with the position and the date (the server will supply this) to calculate "true" directions.

Vertical: measures angle of direction relative to the plane that is tangential to the surface of the earth at the position of measurement. -90 indicates straight down; 90 indicates directly up.

Roll: measures angle of rotation of the device relative to "up"; 0 indicates that the device is level while 180 indicates that the device is upside-down.

Confidence: indicates the confidence level that the pointing values are not interfered with by non-natural fluctuations in the Earth's magnetic field (as caused by a large iron structure). For example, if the compass reading is wildly fluctuating at the time this data is passed on to a pointing-enabled location based service, this value might be "5". If the compass reading is extremely stable, this value might be "100". How the value is to be determined is to be, well, determined.

19.13.3.3 Gesturing Data

Gesturing data is in addition to location and pointing data. It is the indication of movement of the device over a short period of time.

The horizontal and vertical measurements are sufficient to pinpoint a point in 3D space. The roll measurement is necessary to correctly orient the view of that point (Table 19.4).

Samples: measures the number of location and pointing data samples taken.

Rate: measures the rate at which the samples are taken. The number of samples multiplied by the rate indicates the total elapsed time of the gesture.

Table 19. 4

Data	Range	Units	Sample
Samples	> 1		2
Rate	> 0	Milliseconds	100

19.13.4 Data Format

19.13.4.1 Format Preferences

The preferred format for all data is a single Unicode string. While binary values may require less storage space, there is a greater chance of encountering certain problems related to the differences in processors. Two common problem areas are byte ordering differences and data type size differences. Byte ordering problems occur when, in binary formats, one processor expects the high-order bytes first and the other expects them last. Size problems occur when one processor expects, for example, a signed integer to use 4 bytes and the other expects a signed integer to use 2 bytes. Unicode strings mitigate these problems since they do not differ from one processor to another and their lengths are inherently variable.

Specifically, the string should contain all of the previously mentioned data, separated by commas.

19.13.4.2 Tier 1

The fields in this string are (in order) the format type of "L", latitude, longitude, and location confidence. The format type indicates that the data represents a single, discreet measurement. For example:

"L,38.27782,-122.63490,95" indicates

a single location sample;
a latitude of 38.27782° north;
a longitude of 122.63490° west;
a location confidence value of 95.

This single Unicode string format is uniform for all tier 1 services.

19.13.4.3 Tiers 2 and 3

The fields in this string are (in order) the format type of "P", latitude, longitude, height, horizontal, vertical, roll, location confidence, and pointing confidence. If certain data are unavailable, that field should be left empty. The format type indicates that the data represents a single, discreet measurement. For example:

"P,38.27782,-122.63490,41,47,,,95,80" indicates

a single pointing sample;
a latitude of 38.27782° north;
a longitude of 122.63490° west;
a height of 41 m;
a horizontal direction of 47°;
no vertical measurement;
no roll measurement; and
a location confidence value of 95;
a pointing confidence value of 80.

This single Unicode string format is uniform for all tier 1, 2 and 3 services. Tier 1 services can ignore tier 2 and tier 3 data while tier 2 services can step down to a tier 1 service if tier 2 data is missing (depending on the requirements of the services) and so on.

19.13.4.4 Tier 4

The fields in this string are (in order) the format type of "G", number of samples, rate, the location and pointing data repeated the appropriate number of times, location confidence, and pointing confidence. If certain data is unavailable, that field should be left empty. The format type indicates that the data represents a series of discreet measurements. For example:

"G,2,50,38.27782,-122.63490,41,47,,,38.27782,-122.63490,41,76,,,95,80" indicates a gesture, two samples, 50 ms between samples first sample...

a latitude of 38.27782° north;
a longitude of 122.63490° west;
a height of 41 m;
a horizontal direction of 47°;
no vertical measurement;
no roll measurement; and

second sample...

a latitude of 38.27782° north;
a longitude of 122.63490° west;
a height of 41 m;
a horizontal direction of 76°;
no vertical measurement;
no roll measurement; and
a location confidence value of 95;
a pointing confidence value of 80.

This gesture indicates a device that remained stationary while sweeping from 47 to 76° in a 50-ms time frame.

Note that this is somewhat inefficient since the location of the device did not change but was included twice. We can improve this by introducing the "r" character to indicate the re-use of a particular field. The "r" character cannot exist in the first data sample. So the previous string becomes:

"G,2,50,38.27782, − 122.63490,41,47,,,r,r,r,76,,,95,80"

and a shorter string is passed to the tier 3 service.

19.13.5 Is it sufficient?

Assuming the pointing-enabled location based service is aware of the location of objects in the real world, this data is sufficient (and necessary) to identify one or more objects that a device (or user) is expressing interest in. Confidence levels allow the service to modify its search algorithms to compensate for inaccuracies inherent in some methods of data collection (such as TOA for positioning).

19.14 Conclusion

Pointing is the most natural way of selecting objects of interest. When users want information about something they can simply point their phones in its direction. This is the most efficient and user-friendly way to inquire. Just "point and click on the real world™".[2]

---

[2] A trademark of GeoVector Corporation. All other marks belong to their respective owners.

Index

AAC, 279
accelerate, acceleration, 44, 48
access, 41–49, 54
acquisition, 46, 49, 266, 278, 280
adaptive array, 58
adaptive differential pulse code modulation
    (ADPCM), 141, 147, 153
algorithm, 41–42, 47–51, 54
algorithm domains, 327–328
allocation, 42, 46, 48–49
ALU, 20
AM, 253, 254, 279, 280
analysis-by-synthesis, 139, 140, 143–145,
    151
antenna, 42
    array, 85
    fixed beam, 58
AR|T Designer tool, 294
architecture, 41, 46, 48–54
ARM, 131, 132
ASIC, 44, 46, 48, 54

backplane, 43
bandwidth, 44, 50, 52, 255, 260, 262, 263,
    273, 280, 281
baseband, 42, 43, 53
basestation, 41–55
biometric, 217
    ability to acquire, 229
    accuracy, 229
    system reference architecture, 233
    testing, 231
    user identification, 217
Bit Error Rate, 254, 256, 270
BLAST (Bell labs LAyered Space Time), 87
blind adaptation, 65, 71
blind adaptive detection, 89
Bluetooth, 5, 6

capacity, 41–42, 52
Capacity Unit (CU), 269
CardVM, 132
cellular, 41
centralized control, 340
centralized modem architecture, 32
channel, 41–55
channel capacity, 86, 87
channel coding, 138, 146–148, 152, 155
chip, 44, 45, 47, 54
chip-rate, CR, 44, 46–49, 53, 54
Cholesky factor, 68, 70
code division multiple access (CDMA),
    149–153, 156, 157
Code Division Multiple Access, CDMA,
    41–44, 46, 48, 52, 54
code excited linear prediction (CELP), 143–
    145, 147, 148, 150–153, 156, 157
    algebraic CELP (ACELP), 145, 147, 148,
        150–153, 156
codeword, 49, 52
coherence bandwidth, 62, 63

coherent and non-coherent accumulation, 45, 47, 52, 53
command, 54
communication vs. computation tradeoff, 313
complex baseband, 59
configuration state, 331, 333, 338
constant modulus (CM), 90
Constant Modulus Algorithm (CMA), 72, 73
constrained minimum output energy (CMOE), 89
control of architecture, 43, 45–53
convolutional code, 259, 269
coprocessor, 33, 41, 42, 45–52, 55
correlation, 41–42, 44–45, 47–48, 52–54
Correlator coprocessor (CCP), 48–49, 52–54
crossbar network, 334
CVM, 132
cyclic prefix, 257, 266
cyclostationarity, 75

DAB, 253, 254, 255, 280
data aggregation algorithm, 319–320
data auto-correlation matrix, 67
data-driven execution, 341
dataflow graph, 331
decoder, 41–42, 47–52, 54
decoding, 261, 266, 269, 270, 274, 276, 277, 278, 281, 285
decoupled Capacity, 79
Delay Locked Loop (DLL), 45, 46, 53–54
demodulation, 264, 266, 267, 268, 269, 274, 275, 277, 280, 281, 285
despreader, 42, 44–49
Digital Baseband Platform, 16
Direct Memory Access (DMA, EDMA), 41–44, 46, 48, 50–52, 54
discontinuous transmission (DTX), 146, 148, 150, 152
Discrete cosine transform, 20
distributed arithmetic, 323
distributed control, 330, 339
distributed modem architecture, 33
DoCoMo, 43
domain-specific architectures, 327–329
downlink, 42, 44–46, 55

DQPSK, 256, 258, 263, 264, 270, 275, 280
Dual Mode, 1, 23, 28, 31
DSP, 269, 274, 275, 277, 278, 279, 281, 282
dynamic voltage scaling, 304–306, 315–316, 329, 332

EFR, 15
encode, 47, 50–51
encoding, 256, 259, 261, 271, 285
energy
    measurements, 45, 47, 53–54
    dissipation, 303–324
    efficiency, 327
    energy efficient system partitioning, 313–317
    energy scalable architectures, 320–324
    energy-agile algorithm, 320
    energy-quality scalability, 317–319
equalization, 256, 269, 278, 284
error correction, 54
estimation, 45–49, 53–54
Eureka 147, 254–267, 269, 270, 273, 275–280, 285
external to the DSP, 44, 46, 48

fading, 253, 254, 273
false accept rate, 229
false reject rate, 229
Fast Fourier Transform (FFT), 256, 267, 268, 269, 272, 273, 275, 276, 306–7, 314–316
Fast Information Channel (FIC), 263–266, 270, 274, 276
FCP, 41–42, 46–48, 54
feedback, 53
Field Programmable Gate Arrays, 328, 335
finger, 42, 45–46, 48–49, 53–54
fingerprint
    measurements, 45, 47, 53–54
    dissipation, 303–324
    efficiency, 327
    energy efficient system partitioning, 313–317
    energy scalable architectures, 320–324
    energy-agile algorithm, 320
    energy-quality scalability, 317–319

FIR filtering, 318–319
fixed point arithmetic, 290
FM, 253, 254, 279, 280, 281, 283
Forward Error Correction (FEC), 279, 280, 281, 285

generalized mesh network, 336
Givens rotation, 70
GPS, 8
gradient descent, 66
guard interval, 257, 258, 267, 272

Hadamard Transform, 45
handshaking, 341
hardware reconfiguration, 338
Harvard Architecture, 20
Henry classification system, 242
heterogeneous processor array, 327–329, 332
High Speed Circuit Switched Data, 12
Householder reflection, 70
HR, 15
HW/SW partition, 29

In-Band-On-Channel (IBOC), 254, 279, 280, 281, 282, 283, 284
instruction profiling, 303–304
Inter-Carrier Interference (ICI), 256, 272, 273, 275
interference, 42, 55, 253, 254, 255, 256, 258, 259, 263, 272, 275, 279, 280, 284
interleaving, 42, 46–47, 51–53, 259, 263, 269, 273, 276, 279, 280
internet, 42
Inter-Symbol Interference (ISI), 255, 257, 258, 264, 269
iteration, 51, 52
IXI Mobile, 6

Java
    CDC Java specification, 131
    CLDC Java specification, 131, 134
    CLDC, 205
    KVM (K Virtual Machine), 132, 134, 205
    J2ME (Java 2 Micro Edition), 132, 205

RNG, 211, 212, 213
    PKCS, 211
    HMAC, 212
    Distributed Java Virtual Machine, 132
    Java Execution Environment, 119
    Java opcodes, 120, 126
    Jworks, 132
    Scratchy Modular Java Virtual Machine, 119, 127
    SDE (Scratchy Development Environment), 130, 133

leakage control techniques, 308–311
leakage current, subthreshold leakage current, 306–311
Least Mean Square (LMS), 66
Least Squares (LS), 65
Line of Bearing estimation, 314–316
linear estimation, 87
linear MMSE detection, 88
linear predictive coding (LPC), 141–147, 149, 151, 154, 156
Linux, 131
locality of reference, 329, 335–336
lock, 45

Main Service Channel (MSC), 263, 264, 265, 269, 270, 276
maximal ratio combining, MRC, 42, 45–49, 53, 54
maximum likelihood detection, 90
maximum obtainable SINR, 65
mean square error (MSE), 63, 65
memory, 44, 47, 49–50
MF, 280, 281, 282, 283
microsensors, 299–302
MIMO channels, 83
modem, 41, 43, 45, 47, 49, 51, 53, 55
Modified Graham Schmidt Orthogonalization (MGSO), 68
Modularity, 127, 129
modulation, 255, 256, 260, 261, 263, 269, 280, 285
most significant transform, 318
MP3, 253, 262

MPEG, 260, 261, 262, 266, 270, 271, 274, 276, 277, 285

MPEG4, 114, 116

Multi-mode, 23

Multipath, 45, 84, 253, 254, 255, 256, 258, 269, 270, 273, 275, 284, 285

Multiple Threshold CMOS (MTCMOS), 308–309

multiplex, 45–46

multistage detector, 92

MUSICAM, 260, 261, 262

mutual information, 79

narrow-band antenna, 60, 61

network, 41–45
    protocols, 311–313

non-linear estimation, 90

Normalized Mean Square Error (NMSE), 65

null symbol, 258, 264, 267

Ocapi tool, 294

OFDM/COFDM, 255, 256, 258–260, 264–267, 269, 272–280, 283, 284

OMAP™, 38, 107, 111, 112, 113, 114, 116, 117, 120, 133, 134

optimum detection, 90

orthogonal/orthogonality, 256, 257, 263, 266, 272, 273, 275

OS, operating system, 111, 113, 114, 207, 209, 210

Oversampling, 85

parallel interference cancellation, 92

Perron Root, 80

Perron Vector, 80

Personal Digital Assistant (PDA), 3, 4, 6, 114

Phase Locked Loop (PLL), 213, 214

Phase Reference Symbol (PRS), 267, 268, 269, 272

pilot, 45, 54

PKI, 203 and 215

platform, 42, 55

postfiltering, 146

power benchmark, 287, 295

preamble, 45

processor, 43–44, 48, 303
    energy, 303
    programmable, 327
    RISC, 107, 109, 110, 113, 117
    StrongARM SA-1100, 300–307
    TI C54x, 293
    TI C55x, 294
    TI C6x, 294
    TMS320C64, 41, 46, 48, 53–54, 55
    VLIW, 20

profile, 45

protocol, 53

pseudo noise, PN, 45, 47

pulse code modulation (PCM), 141, 147, 149

rake, 44–49, 52–54

rate matching, 46–47

reconfigurable communication network, 330, 333

reconfigurable processing element, 331

RF, 42–43, 253, 255, 264, 265, 266, 273, 274, 276, 278, 281, 285

RMS delay spread, 61

RSCC, 50–52

satellite, 253, 254, 255, 258, 284, 285

Self-Coherence Restoral (SCORE), 77

semi-programmable, 42, 44, 48, 54

sensor networks, 299–302

sequence, 42, 45, 47, 53

Single Frequency Network (SFN), 257, 258, 260

SINR (Signal to Interference Noise Power), 62

Sliding window, 52

SOI (Signal of Interest), 57

speaker verification, 218

linear prediction, 222

cepstral features, 223

Hidden Markov Model (HMM), 223

spreader, 42, 44–49

Standards and generations of phones
    1G, 11
    2G, 25, 41, 43, 50, 111

2.5G, 3, 10, 12, 43, 50, 107, 111, 117
3G, 3, 10, 25, 107, 111, 117, 41–55
AMPS, 12
ARIB, 147
ETSI, 147, 148, 151–154
GPRS, 13
GSM, 25, 149–153, 157
I-Mode, 7
IMT-2000, 41–42
International Telecommunications
    Union, ITU 41–42, 139, 141, 147,
    148, 151, 153, 154, 156
IS-2000, 42, 44–45, 47, 50, 52–54
PDC, 13
Third Generation Partnership Project
    (3GPP), 148, 151–153, 157
TIA, 147, 149–152, 157
UMTS, 7
stochastic gradient descent, 90
stopping criterion, 52
sub-hand, 262, 271, 276, 277
Subspace Constraint, 73
successive interference cancellation, 91
Switched substrate Impedance scheme,
    310–311
symbol-rate, SR, 44, 46–47, 53–54
synchronize, synchronization, 45, 50, 257,
    262, 264, 266–269, 272, 274, 275, 278

temporal correlations, 329, 334
time division multiple access (TDMA), 149,
    151, 152
Time Division Multiplexing (TDM), 284,
    285
time-to-market, 43
tracking, 266, 272, 278
traffic, 44–49
Transmit Beamforming, 77

Turbo decoding, 41–42, 44, 46–48, 50–52, 54
    a priori probability, 51–52
    extrinsic, 51–52
    MAP, 51–52
    parity, 52
    posteriori, 51
    systematic, 50–52
    Turbo coprocessor (TCP), 51–52, 55
    trellis, 49–50

uplink, 42, 44–48, 54

variable computation processing, 305–307
variable length filter, 321–322
variable precision filter, 322–324
Variable Threshold CMOS (VTCMOS),
    309–310
vehicle tracking, 319–320
Viterbi, 259, 266, 267, 268, 269, 270, 274,
    275, 276, 277, 278, 285
Viterbi decoding, 20, 41–42, 47–49, 54
    Add Compare select (ACS), 49, 50
    branch metric, 49
    convolutional, 42, 44, 46–48, 50
    state metric, 49
    traceback, 49–50
    Viterbi Coprocessor (VCP), 49–50, 55
voice, 41–44, 46–47, 50
voice activity detection (VAD), 146, 148,
    150, 152
VSELF, 15
VxWorks, 131

Walsh, 45, 47
weighted MOPs (WMOPs), 154, 155
wideband speech coding, 137, 147, 148,
    152, 153, 155, 157
Windows, 131

zero-forcing detection, 87